Nonlinear Optical Materials and Devices for Applications in Information Technology

NATO ASI Series

Advanced Science Institutes Series

A Series presenting the results of activities sponsored by the NATO Science Committee, which aims at the dissemination of advanced scientific and technological knowledge, with a view to strengthening links between scientific communities.

The Series is published by an international board of publishers in conjunction with the NATO Scientific Affairs Division

A	**Life Sciences**	Plenum Publishing Corporation
B	**Physics**	London and New York
C	**Mathematical and Physical Sciences**	Kluwer Academic Publishers
D	**Behavioural and Social Sciences**	Dordrecht, Boston and London
E	**Applied Sciences**	
F	**Computer and Systems Sciences**	Springer-Verlag
G	**Ecological Sciences**	Berlin, Heidelberg, New York, London,
H	**Cell Biology**	Paris and Tokyo
I	**Global Environmental Change**	

PARTNERSHIP SUB-SERIES

1. Disarmament Technologies	Kluwer Academic Publishers
2. Environment	Springer-Verlag / Kluwer Academic Publishers
3. High Technology	Kluwer Academic Publishers
4. Science and Technology Policy	Kluwer Academic Publishers
5. Computer Networking	Kluwer Academic Publishers

The Partnership Sub-Series incorporates activities undertaken in collaboration with NATO's Cooperation Partners, the countries of the CIS and Central and Eastern Europe, in Priority Areas of concern to those countries.

NATO-PCO-DATA BASE

The electronic index to the NATO ASI Series provides full bibliographical references (with keywords and/or abstracts) to more than 50000 contributions from international scientists published in all sections of the NATO ASI Series.
Access to the NATO-PCO-DATA BASE is possible in two ways:

– via online FILE 128 (NATO-PCO-DATA BASE) hosted by ESRIN, Via Galileo Galilei, I-00044 Frascati, Italy.

– via CD-ROM "NATO-PCO-DATA BASE" with user-friendly retrieval software in English, French and German (© WTV GmbH and DATAWARE Technologies Inc. 1989).

The CD-ROM can be ordered through any member of the Board of Publishers or through NATO-PCO, Overijse, Belgium.

Series E: Applied Sciences - Vol. 289

Nonlinear Optical Materials and Devices for Applications in Information Technology

edited by

A. Miller

Department of Physics and Astronomy,
University of St Andrews, U.K.

K. R. Welford

Defence Research Agency,
Malvern, U.K.

and

B. Daino

Fondazione Ugo Bordoni,
Rome, Italy

Kluwer Academic Publishers

Dordrecht / Boston / London

Published in cooperation with NATO Scientific Affairs Division

Proceedings of the NATO Advanced Study Institute on
Nonlinear Optical Materials and Devices for Applications in Information Technology
Erice, Sicily, Italy
July 13–26, 1993

A C.I.P. Catalogue record for this book is available from the Library of Congress

ISBN 0-7923-3457-4

Published by Kluwer Academic Publishers,
P.O. Box 17, 3300 AA Dordrecht, The Netherlands.

Kluwer Academic Publishers incorporates the publishing programmes of
D. Reidel, Martinus Nijhoff, Dr W. Junk and MTP Press.

Sold and distributed in the U.S.A. and Canada
by Kluwer Academic Publishers,
101 Philip Drive, Norwell, MA 02061, U.S.A.

In all other countries, sold and distributed
by Kluwer Academic Publishers Group,
P.O. Box 322, 3300 AH Dordrecht, The Netherlands.

Printed on acid-free paper

Printed in the Netherlands

Table of Contents

Preface

This volume is the proceedings of the NATO Advanced Studies Institute on *Nonlinear Optical Materials and Devices for Applications in Information Technology*, located in Erice, Sicily from 13th to 26 July 1993. The ASI was the 26th International School of Materials Science and Technology, and was held at the Ettore Majorana Centre for Scientific Culture. Total attendance at the ASI was 82 from 21 countries with a broad mix of post-graduate and post-doctoral researchers from materials science, chemistry, physics and electrical engineering.

Nonlinear optics has become an increasingly diverse and complex topic embracing resonant, nonresonant and hybrid opto-electronic interactions in a large range of inorganic and organic, bulk and low dimensional materials. Much of the recent interest has been driven by the success of optics and lasers in penetrating areas of information technology such as communications, data storage, displays, signal processing and computing. The essentially complementary properties of electrons and photons give optics the potential to expand its role within these technologies via nonlinear optical interactions. This volume reviews and offers comparison of the fundamental mechanisms behind a range of nonlinear optical interactions in different material systems. Devices and systems applications of these phenomena in optical signal routing, communications, signal processing and computing are discussed.

As with the ASI, the intention of this book has been to provide a tutorial coverage at an advanced graduate level in the most important areas in nonlinear optical materials and devices currently under development for applications in information technology. As such, it is hoped that the volume can be used as a text by those entering this research area or those wishing to broaden their knowledge of the field as a whole. The first chapter provides a detailed and up-to-date account of our present understanding of the fundamental nonlinear interactions of laser light with matter. In subsequent chapters, the nonlinear optical properties of liquid crystals, optical fibres and semiconductors, in bulk and low dimensional forms, are emphasised. These materials produce a variety of very large nonlinear optical effects that can be employed in the all-optical control of laser light. The interactions are generally described in terms of $\chi^{(3)}$, the third order nonlinear susceptibility. After establishing the basic laser-material interactions, optical nonlinearities in optics in long-haul fibre optical communications including soliton systems, the potential for all-optical switching in integrated optics, developments in parallel arrays and smart pixels are discussed. Finally, digital optical computing architectures are surveyed from the use of optics within electronic machines to the potential of all-optical systems.

The editors wish to acknowledge Professor Minko Balkanski, Director of the Erice International Schools of Material Science and Technology, for his guidance, support and enthusiasm. We would also like to thank the staff of the Ettore Majorana Centre for Scientific Culture for their assistance in organising this meeting in such beautiful surroundings, with particular thanks to Miss Pinola Saralli for all her help. The school benefited from additional support came from the Italian Ministry of Education, the Italian Ministry of University and Scientific Research, the Italian National Research Council, the Sicilian Regional Government, the Ugo Bordoni Foundation, the European Physical Society, the National Science Foundation (USA) and the International Science Foundation (USA).

A. Miller, St Andrews, UK

K R Welford, Malvern, UK

B Daino, Rome, Italy

FUNDAMENTALS OF NONLINEAR OPTICS

CHRISTOS FLYTZANIS
Laboratoire d'Optique du C.N.R.S.
Ecole Polytechnique
91128 Palaiseau, cedex
France

ABSTRACT. Elementary concepts of nonlinear optics are reviewed with particular emphasis on the nonlinear dipolar polarization and the relevant nonlinear materials for applications in devices.

1. General Introduction

In nonlinear optics we are concerned with the class of phenomena that occur when the response of a material system to an applied electromagnetic field is nonlinear in the amplitude of the field. With response most frequently we understand the induced dipolar polarization and the related nonlinear phenomena can only be observed with intense coherent light sources such as lasers. For this reason it is customary to place the start of nonlinear optics in 1961 when the first observations(1) of the second harmonic and optical rectification were made but nonlinear effects *stricto senso* had been observed and studied long before although in a different context. Such are the electrooptic effects, Pockels or Kerr, and several light scattering effects to name a few ; these now constitute special cases of nonlinear optics.

The essential implication of the nonlinear regime is that the optical properties of the matter become sensitive to the field amplitude and by the same token can be controlled by varying this amplitude. This field amplitude dependence results from the large displacements that the charges in the matter undergo when driven by intense fields ; the impact of such displacements cannot be accounted for within the harmonic approximation which constituted(2,3) the basis of optics with incoherent light sources. Anharmonicity in the interaction between the charges must be included(4) and this leads invariably to a nonlinear behavior in the response ; but this is not the sole mechanism. Indeed, even in the absence of such interactions, the response will be nonlinear because of the intrinsic nonlocality built in the interaction between a charge and the

A. Miller et al. (eds.), Nonlinear Optical Materials and Devices for Applications in Information Technology, 1–74.

electromagnetic field as can be inferred[5] by the expression of the Lorentz force. This is a propagation-retardation effect, an intrinsic feature of Maxwell equations, and because of its cumulative effect it can lead to nonlinearities comparable to the ones due to the anharmonicity. In the last analysis anharmonicity in the charge motion and nonlocality in its interaction with electromagnetic field constitute the two major mechanisms that at the very microscopic level are responsible for the nonlinear response of the matter to intense electromagnetic fields.

The scope of nonlinear optics is to study the nonlinear behavior in the induced polarization and to analyze and control its impact on the propagation[6,7,8] of intense electromagnetic fields in the matter. As a matter of fact the exploitation of the nonlinear optical effects is an essential part of this activity and justifies the spectacular growth that the field witnessed within the span of the three decades of its existence.

The theoretical framework[4] for the description and study of the nonlinear optical phenomena remains the same as for the other optical phenomena in general namely the Maxwell equations

$$\underline{\nabla} \wedge \underline{E} = -\frac{1}{c}\frac{\partial \underline{B}}{\partial t} \qquad \underline{\nabla} . \underline{E} = 4\pi\rho$$
$$\underline{\nabla} \wedge \underline{H} = \frac{1}{c}\frac{\partial \underline{D}}{\partial t} \qquad \underline{\nabla} . \underline{B} = 0 \tag{1}$$

with

$$\underline{D} = \underline{E} + 4\pi \int_{-\infty}^{t} \underline{J}(t')\, dt' \tag{2}$$

where for simplicity we assumed that there are no extraneous charges and currents and $\underline{J}$ is the induced current density which can be related to the fields $\underline{E}$ and $\underline{B}$ through the solution of the equation of motion of the charges in the presence of the electromagnetic field ; in principle this is the Schrödinger equation but classical equations of motions can also be used[4] for a phenomenological description for instance the forced damped anharmonic oscillator equation

$$\ddot{\underline{r}} + \Gamma\dot{\underline{r}} + \omega_{o}^{2}\underline{r} + \beta\underline{r}^{2} + \gamma\underline{r}^{3} = \frac{e}{m}\underline{E}(t) \tag{3}$$

for bound charges in dielectrics or the free charge in a Lorentz force field equation

$$\dot{\underline{v}} + \Gamma\underline{v} = \frac{e}{m}\left(\underline{E}(t) + \underline{v} \wedge \underline{B}(t)\right) \tag{4}$$

for unbound electrons in metals. The two equations (3) and (4) reflect the two basic mechanisms that induce nonlinear behavior namely the anharmonicity and the nonlocality.

The current density $\underline{J}$ in (2) can be expanded in multipole series

$$\underline{J}(t) = \frac{\partial}{\partial t}(\underline{P} - \underline{\nabla}\underline{\underline{Q}}) + c\,(\underline{\nabla} \times \underline{M}) \qquad (5)$$

where $\underline{P}$, $\underline{\underline{Q}}$, and $\underline{M}$, are the electric dipole, electric quadrupole and magnetic dipole polarization densities respectively with $\underline{H} = \underline{B} - 4\pi\underline{M}$. When perturbation theory applies these can be cast in the form of power series expansions in the field amplitudes ; in certain cases analytical expressions in closed form can be derived. In general one may write $\underline{P} = \underline{P}_L + \underline{P}_{NLS}$ and similarly for $\underline{Q}$ and $\underline{M}$ where $\underline{P}_L$ is the part of the dipole polarization linear in the field amplitude $\underline{E}$ and $\underline{P}_{NLS}$ is the nonlinear part. These nonlinear terms act as polarization sources generating new fields with drastically different spatiotemporal characteristics from those expected in the linear regime. Actually the energy transfer involved

$$\frac{dW}{dt} = - < \underline{E} \,.\, \left\{ \frac{\partial}{\partial t} \left\{ \underline{P}_{NLS} - \underline{\nabla} \,.\, \underline{\underline{Q}}_{NLS} \right\} + c\underline{\nabla} \times \underline{M}_{NLS} \right\} > \qquad (6)$$

where the brackets indicate time averaging, can be substantial and the new fields can attain large amplitudes with specific spatiotemporal characteristics.

The nonlinear effects most relevant for applications are those related to the second and third order nonlinear terms in the series expansion of the dipole polarization in powers of the electric field amplitude $\underline{P} = \underline{P}^{(1)} + \underline{P}^{(2)} + \underline{P}^{(3)}$

$$\underline{P} = \underline{\underline{\chi}}^{(1)}\underline{E} + \underline{\underline{\underline{\chi}}}^{(2)}\underline{E}\,\underline{E} + \underline{\underline{\underline{\underline{\chi}}}}^{(3)}\underline{E}\,\underline{E}\,\underline{E}$$

the coefficient $\chi^{(n)}$ is the dipolar susceptibility of order n, a tensor of rank (n+1) with $\underline{P}^{(1)} = \underline{P}_L$; it is customary to introduce a Fourier analysis of these polarisation terms and the fields and write :

$$P_i^{(1)}(\omega) = \chi_{ij}^{(1)}(\omega)\, E_j(\omega) \qquad (7)$$

$$P_i^{(2)}(\omega_1 + \omega_2) = D_2\, \chi_{ijk}^{(2)}(\omega_1, \omega_2)\, E_j(\omega_1)\, E_k(\omega_2) \qquad (8)$$

$$P_i^{(3)}(\omega_1 + \omega_2 + \omega_3) = D_3\, \chi_{ijk\ell}^{(3)}(\omega_1, \omega_2, \omega_3)\, E_j(\omega_1)\, E_k(\omega_2)\, E_\ell(\omega_3) \qquad (9)$$

for the linear, second and third order terms and similarly for the higher order ones. The D_n are degeneracy factors equal to the number of distinct permutations of the applied field ; with this definition the value of $\chi^{(n)}$ for $\omega_i \to 0$ is independent of the chosen path. The main characteristics[9] of the nonlinear polarization terms will be discussed later in section 3 ; here we only mention that a rough order of magnitude can be obtained through

TABLE 1. Nonlinear optical effects

Nonlinear effect	Incident frequency	Created frequencies	Nonlinear	Some applications
Second harmonic	ω,ω	2ω	$\chi^{(2)}(\omega,\omega)E_\omega E_\omega$	Near UV generation
Optical rectification	$\omega,-\omega$	0	$2\chi^{(2)}(\omega,-\omega)E_\omega E_\omega$	Ultrashort electrical pulses
Electrooptic effect (Pockells)	$\omega,0$	ω	$2\chi^{(2)}(\omega,0)E_\omega E_0$	Electrooptic modulation
Frequency summation*	ω_1,ω_2	$\omega_1+\omega_2$	$2\chi^{(2)}(\omega_1,\omega_2)E_{\omega_1}E_{\omega_2}$	Near UV generation Up conversion
Frequency difference*	$\omega_1,-\omega_2$	$\omega_1-\omega_2$	$2\chi^{(2)}(\omega_1,-\omega_2)E_{\omega_1}E_{\omega_2}$	IR radiation, parametric amplification
Third harmonic*	ω,ω,ω	3ω	$\chi^{(3)}(\omega,\omega,\omega)E_\omega E_\omega E_\omega$	Near and far UV generation
Frequency summation*	$\omega_1,\omega_2,\omega_3$	$\omega_1+\omega_2+\omega_3$	$6\chi^{(3)}(\omega_1,\omega_2,\omega_3)E_{\omega_1}E_{\omega_2}E_{\omega_3}$	
Frequency mixing*	$\omega_1,\omega_1,-\omega_2$	$2\omega_1-\omega_2$	$3\chi^{(3)}(\omega_1,\omega_1,-\omega_2)E_{\omega_1}E_{\omega_1}E^*_{\omega_2}$	CARS if $\omega_1-\omega_2\approx\omega_R$ **
Optical Kerr effect	$\omega,-\omega,\omega$	ω	$3\chi^{(3)}(\omega,-\omega,\omega)\lvert E_\omega\rvert^2E_\omega$	Optical modulation, optical phase, optical gratings, two photon absorption if
or Two photon absorption	$\omega_1,-\omega_1,\omega_2$	ω_2	$6\chi^{(3)}(\omega_1,-\omega_1,\omega_2)\lvert E_{\omega_1}\rvert^2E_{\omega_2}$	2ω or $\omega_1+\omega_2=\omega_e$** Stimulated Raman if $\omega_1-\omega_2=\omega_R$**
Static Kerr effect	$\omega,0,0$	ω	$3\chi^{(3)}(\omega,0,0)E_\omega E_0^2$	Electrooptic modulation
Induced second harmonic*	$\omega,\omega,0$	2ω	$3\chi^{(3)}(\omega,\omega,0)E_\omega^2E_0$	
and Optical rectification	$\omega,-\omega,0$	0	$6\chi^{(3)}(\omega,-\omega,0)\lvert E_\omega\rvert^2E_0$	

$$\chi^{(n+1)} \approx 1/E_c^{-n} \qquad (10)$$

where E_c is the effective field that keeps attached the polarizable charges or units to each other. The mechanisms that contribute to the nonlinear polarization are numerous but qualitatively can be introduced through such cohesive fields which also indicate ways to enhance the nonlinearities.

Broadly speaking nonlinear optics(4,10) are used either to shift the optical carrier frequency of the availabe fields or to remodel their spatiotemporal characteristics. Accordingly one can separate the nonlinear effects into two classes. The one encompasses those effects which serve to generate fields at new frequencies in which case the coherence or equivalently the field amplitude matters and one is invariably faced with the problem of phase matching, the classic example here being the second harmonic generation. The other class includes those effects where one modifies the spatiotemporal characteristics of a light beam, in particular its spatiotemporal profile and polarization state, without touching its carrier frequency in which case the phase matching problem is practically absent. Here one has either the light self action effects or photoinduced ones where only the spatiotemporal profile of the beam intensity matters and not the field amplitude *per se* the classic examples being the optical Kerr nonlinearity mediated effects ; or the parametric effects where an external parameter is used to modulate the optical characteristics of the medium and by the same token those of the light beam, the classic examples here being the electrooptic (Pockels) and acoustooptic effects. In Table 1 we summarize the most commonly used second and third order processes.

Nonlinear optics is concerned with all these aspects and in addition with their implementation in devices and other applications. In this respect the enhancement of the efficiency of the nonlinear processes by appropriate choice of nonlinear optical materials and interaction configurations constitute a central concern in all these studies. The future development of nonlinear optics is intimately connected with progress in nonlinear optical materials. Nonlinear spectroscopy is another by-product of these efforts and much ingenuity has been developed to use the nonlinear effects for diagnostic purposes. Below we present the derivation of the basic nonlinear propagation equation and its implications are illustrated with simple cases (section 2). In section 3 we summarize the main phenomenological and microscopic aspects of the nonlinear polarization and review some mechanisms. We give a review of the main classes of nonlinear optical materials in section 4.

2. Nonlinear Propagation

2.1. BASIC NONLINEAR PROPAGATION EQUATION.

As previously stated the nonlinear polarization sources induced in a medium generate new fields and the basic problem[4,6,10] here is to derive the propagation equation for the amplitude of a monochromatic electric field of frequency Ω in a nonlinear medium where a nonlinear polarization has been induced at the same frequency Ω. Restricting to transverse plane waves, or $\underline{\nabla}.\underline{E} = 0$, in a nonmagnetic dielectric we obtain from Maxwell equations

$$\Delta\underline{E} - \frac{4\pi}{c^2}\frac{\partial^2 \underline{D}_L}{\partial t^2} = \frac{4\pi}{c^2}\frac{\partial^2 \underline{P}_{NL}}{\partial t^2} \tag{11}$$

where $\underline{D}_L = \underline{E} + 4\pi\underline{P}_L$ is the linear induction. In reality the intense coherent light sources we are dealing with do not deliver plane waves but pulses namely wave packets whose envelope has a finite spatiotemporal extension ; their temporal extension (pulse duration) can range from a few nanoseconds (1ns = 10^{-9}s) to few picoseconds (1ps = 10^{-12}s) down to few femtoseconds (1fs = 10^{-15}s) ; we remind that in 1ps the light travels 300 μm in vacuum. Accordingly the induced nonlinear polarization may have comparable spatiotemporal extensions and the problem then is to derive the equation that governs the spatiotemporal evolution of the envelope of a coherent light pulse generated by such a pulsed polarization source.

Quite generally the later can be written in the form

$$\underline{P}_{NL}(\underline{r},t) = \mathrm{Re}\left\{\underline{\mathcal{P}}_{NL}(\underline{r},t)\, e^{i\left(\underline{K}_\Sigma.\underline{r}-\Omega t\right)}\right\} \tag{12}$$

where $\underline{\mathcal{P}}_{NL}(\underline{r},t)$ is its envelope that varies slowly over the carrier period $T = 2\pi/\Omega$. The electric field generated in the same frequency Ω can be also written in the form

$$\underline{E}(\underline{r},t) = \mathrm{Re}\left\{\hat{e}\, A(\underline{r},t)\, e^{i\left(\underline{K}.\underline{r}-\Omega t\right)}\right\} \tag{13}$$

where K satisfies the dispersion relation

$$K = \Omega\sqrt{\mu_o\,\varepsilon(\Omega)} \tag{14}$$

and in general $\underline{K} \neq \underline{K}_\Sigma$; $A(\underline{r},t)$ is the envelope of the electric field amplitude and varies slowly over a wavelength $\Lambda = 2\pi/K$ and over a period $T = 2\pi/\Omega$. Actually in (14), although not explicitly indicated $\varepsilon(\Omega) \equiv n^2$ is an appropriately contracted direction dependent form of the dielectric constant $\underline{\varepsilon}(\Omega) = 1 + 4p\, \underline{\underline{\chi}}^{(1)}(\Omega)$ which is a second rank tensor.

Replacing (12) and (13) in (11) introducing the wave variable transformation $\xi = x$, $\eta = y$, $\zeta = z$, $\tau = t\text{-}/v_g$ where the group velocity is defined as $v_g = (\partial K/\partial \Omega)^{-1}$ but is actually a vector relation, and inserting the slow varying envelope approximation by setting

$$\left|K \frac{\partial A}{\partial \zeta}\right| >> \left|\frac{\partial^2 A}{\partial \zeta^2}\right| , \left|\frac{1}{v_g} \frac{\partial^2 A}{\partial \zeta\, \partial \tau}\right| \tag{15}$$

and in addition $\left|\Omega^2 \mathcal{P}_{NL}\right| >> \left|\partial^2 \mathcal{P}_{NL}/\partial t^2\right|$ one obtains

$$\Delta_\perp A + \frac{\partial A}{\partial \zeta} - \frac{K''}{2i} \frac{\partial^2 A}{\partial \tau^2} = \left(\frac{iK}{2\varepsilon(\Omega)} \mathcal{P}_{NL} - \frac{1}{c} \frac{\partial \mathcal{P}_{NL}}{\partial t} \right) e^{i\Delta K \zeta} \tag{16}$$

after projecting over the unit vector ê which we assume is unaffected by the nonlinear interaction. In (16) $\Delta K = K_\Sigma\text{-}K$ is the wave vector mismatch along direction ζ, $\Delta_\perp = \partial^2/\partial x^2 + \partial^2/\partial y^2$ and $K'' = \partial^2 K/\partial \Omega^2 = -(\partial v_g/\partial \Omega)/v_g^2$ is the group velocity dispersion. An alternative but essentially equivalent approach to derive equation (16) is to neglect[11] a backward propagating term in the solution of (11). Equation (16) is the basic nonlinear propagation equation valid for coherent light pulses down to few femtoseconds and finite transverse extension. The term $\Delta A_\perp$ takes care of the diffraction and is important whenever the beam profile varies strongly. Note that the amplitude A is dephased by $\pi/2$ with respect to the polarization term $\mathcal{P}_{NL}$ at $z = 0$.

If $\Delta A_\perp$ is negligeable in (16) one has

$$\frac{\partial A}{\partial \zeta} - \frac{K''}{2i} \frac{\partial^2 A}{\partial \tau^2} = \left(i\frac{2\pi K}{\varepsilon(\Omega)} \mathcal{P}_{NL} - \frac{1}{c} \frac{\partial \mathcal{P}_{NL}}{\partial \tau} \right) e^{i\Delta K \zeta} \tag{17}$$

which for $|\Omega \mathcal{P}_{NL}| >> |\partial \mathcal{P}_{NL}/\partial \tau|$ reduces to

$$\frac{\partial A}{\partial \zeta} - \frac{K''}{2i} \frac{\partial^2 A}{\partial \tau^2} = i\frac{2\pi K}{\varepsilon(\Omega)} \mathcal{P}_{NL}\, e^{i\Delta K \zeta} \tag{18}$$

and for a Kerr nonlinearity medium becomes the nonlinear Schrödinger equation that admits[12] as solutions the temporal solitons[13,14]. For K" ≈ 0 and reverting back to the initial variables one has

$$\frac{\partial A}{\partial z} + \frac{1}{v_g}\frac{\partial A}{\partial t} = i\frac{2\pi K}{\varepsilon(\Omega)} \mathcal{P}_{NL}\ e^{i\Delta K z} \tag{19}$$

which in the stationary regime reduces to

$$\frac{\partial A}{\partial z} = i\frac{2\pi K}{\varepsilon(\Omega)} \mathcal{P}_{NL}\ e^{i\Delta K z} \tag{20}$$

but this equation can actually be used even for pulses in the range of few nanoseconds. If in the stationary regime the diffraction becomes important one obtains

$$-\frac{1}{2iK}\Delta_{\perp} A + \frac{\partial A}{\partial z} = i\frac{2\pi K}{\varepsilon(\Omega)} \mathcal{P}_{NL}\ e^{i\Delta K z} \tag{21}$$

which with appropriate normalization and renaming of variables can be cast in the same form as equ.(18) and indeed equation (21) for a Kerr nonlinearity medium admits as solutions the spatial solitons. For simplicity in the previous equations we tacitly assumed isotropic medium but provisions for anisotropic ones can be easily introduced.

For the solutions of the nonlinear propagation equation (16) and the subsequently derived ones, equs (17-21), one clearly needs the expression of $\mathcal{P}_{NL}$ in terms of a susceptibility tensor and the appropriate product of field amplitudes or a sum of such terms ; each of the field amplitudes involved in the interaction in principle is affected by corresponding nonlinear polarization sources and obeys a similar nonlinear equation coupled to those of all fields involved in the interaction and one must actually solve a coupled amplitude system of equation a rather formidable and more often intractable problem unless some drastic simplifications are made. We have chosen some few key cases to illustrate[10,11] these problems and highlight some major features of nonlinear optical processes.

2.2. SECOND ORDER PROCESSES

The second order processes are strictly coherent processes involving three waves $(\omega_i, \underline{k}_i, \hat{e}_i)$ i = 1,2,3 with $\omega_3 = \omega_1 + \omega_2$ interacting in a medium that lacks inversion symmetry (Fig.1). No absorption loss is associated with such a process and this is true for all even order processes. Without loss of generality then assuming a lossless medium and no distorsion due to dispersion, setting up an equation of the type (21) for each one of the three amplitudes A_i in the presence of a second order polarization term induced by the other two field amplitudes, one obtains a set of three coupled amplitude equations. We shall cursively discuss[4,6,10,15] some aspects and processes related to these equations.

2.2.1. *Manley-Rowe relations*. . Introducing the intensities :

$$S_i = \frac{c}{2\pi} n_i \cos^2\alpha_i A_i A_i^* \qquad (22)$$

where α_i is the direction angle of $\hat{e}_i$ and n_i the refractive index for mode i, and the total number of photons crossing a given plane normal to the z-axis

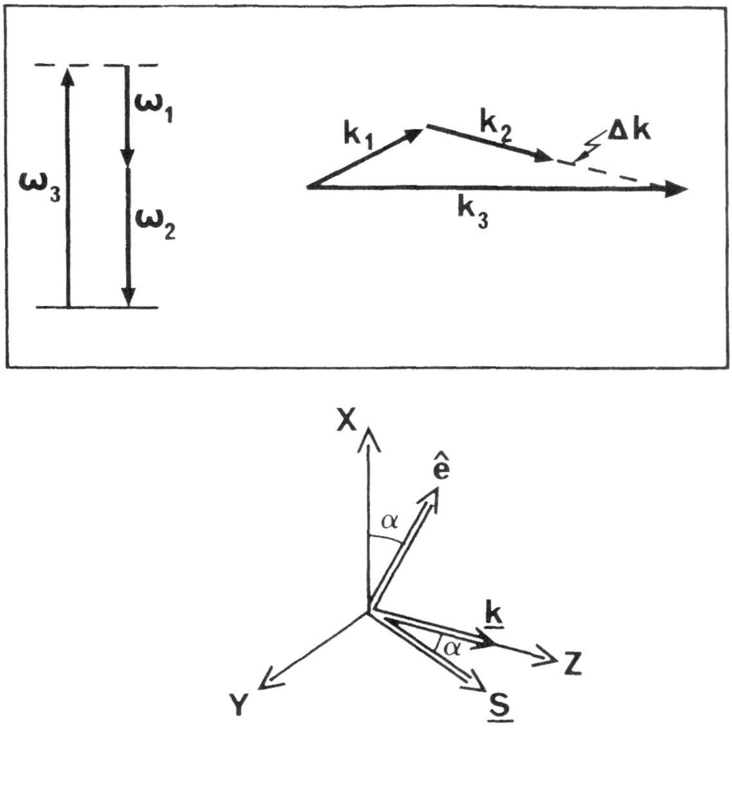

Figure 1

$$N_i(z) = \frac{1}{\hbar\omega_i}\int_{-\infty}^{\infty} dx\, dy\, dt\, S_i(x,y,z,t)$$

one gets after some simple algebraic operations

$$\frac{dN_1}{dz} = \frac{dN_2}{dz} = -\frac{dN_3}{dz} \tag{23a}$$

and

$$\frac{d}{dz}\left(N_1\hbar\omega_1 + N_2\hbar\omega_2 + N_3\hbar\omega_3\right) = 0 \tag{23b}$$

which expresses energy conservation between the waves only ; the medium is taking no part in the energy transfer. One also has

$$\frac{d}{dz}(N_3 + N_2) = \frac{d}{dz}(N_3 + N_1) = \frac{d}{dz}(N_2 - N_1) = 0 \tag{23c}$$

and also for infinite transverse extent

$$\frac{d}{dz}\left(\frac{S_3}{\hbar\omega_3} + \frac{S_2}{\hbar\omega_2}\right) = \left(\frac{S_3}{\hbar\omega_3} + \frac{S_1}{\hbar\omega_1}\right) = \left(\frac{S_2}{\hbar\omega_2} - \frac{S_1}{\hbar\omega_1}\right) = 0 \tag{24}$$

These are the Manley-Rowe[16] relations and simply express the fact that when the wave at ω_3 gains or loses one photon the ones at ω_1 and ω_2 simultaneously lose or gain one a process that symbolically can be represented by $\hbar\omega_3 \rightleftarrows \hbar\omega_1 + \hbar\omega_2$.

2.2.2 *Second harmonic generation..* In this case we have only two frequencies, ω and 2ω, and two coupled amplitude equations

$$\frac{dA_1}{dz} = i\frac{8\pi^2}{\lambda_1 n_1}\chi_e\, A_2A_1^*\, e^{-i\Delta kz} \tag{25a}$$

$$\frac{dA_2}{dz} = i\frac{4\pi^2}{\lambda_2 n_2}\chi_e\, A_1^2\, e^{i\Delta kz} \tag{25b}$$

with $\Delta k = 2k_1 - k_2$ where we simplified by assuming $\cos\alpha_i \approx 1$ and the relation $\hat{e}_1\underline{\underline{\chi}}^{(2)}(2\omega,-\omega)$ $\hat{e}_2\hat{e}_1 = 2\hat{e}_2\underline{\underline{\chi}}(\omega,\omega)\,\hat{e}_1\hat{e}_1 = \chi_e$ was used. This set of equations can be solved[6] quite generally in closed form.

As long as $\Delta k \neq 0$ the solutions will be oscillating in space and will never grow sufficiently to deplete the pump wave at ω_1. If this is the case we may assume $|A_1|^2$ constant and one obtains the solution

$$A_2(L) = i\frac{8\pi^2}{\lambda_1}\frac{1}{n_2}\chi_e\, A_1^2 L\left\{\frac{\sin(\Delta kL/2)}{\Delta kL/2}\right\} e^{\,i\Delta kL/2}$$

If we assume $A_2(0) \approx 0$ and introduce the intensity $I \equiv S$ using equ (22) one gets

$$I_2 = \frac{128\pi^5}{c\lambda_1^2}\frac{\chi_e^2}{n_1^2 n_2} I_1^2 \left\{\frac{\sin(\Delta kL/2)}{\Delta kL/2}\right\}^2 L^2 \tag{26}$$

As anticipated I_2 oscillates in space with period $L_c = \pi/\Delta k = \pi c/|n_1 - n_2|$, the coherence length, and the efficiency of the process decreases as Δk increases as was demonstrated[8] experimentally with the Maker fringes. Actually these provide a very precise method to determine the dispersion of the refractive index.

In the general case where the depletion of the pump at ω_1 cannot be neglected one must solve[6] the set of the two coupled equations (25a,b). Setting $A_j = (2\pi I/n_j c)a_j\exp(i\varphi_j)$ there, with $I = I_1 + I_2$ and a_j real, introducing the interaction length $\ell = \left(n_1^2 n_2 c^3 / 8\pi^3\omega^2 I\, \chi_e^2\right)^{1/2}$ the solution can be obtained in terms of elliptical functions. In the case of phase matching $\Delta k = 0$ one can easily show that with either one of the amplitudes zero at $z = 0$ or $\cos\theta = 0$, one gets[6]

$$\begin{aligned} a_2(z) &= \tan h\zeta \\ a_1(z) &= \sec h\zeta \\ \varphi_2(z) &= 2\varphi_1(z) + \pi/2 \end{aligned} \tag{27}$$

with $a_2^2 + a_1^2 = 1$ which expresses the energy flux conservation. The solution shows that the second harmonic will continuously grow at the expense of the fundamental which after a distance ℓ, the interaction length, will be reduced by about 75% .

Without elaborating further we point out that because of birefringence, which is required to achieve[7] phase matching (see 2.2.3 below), the fundamental and the harmonic will not propagate in the same direction. Hence beams of finite cross sections will eventually cease to overlap after they propagate a certain distance (walk off) and the conversion efficiency will be reduced. The beam profile also plays a certain role. This can be accounted for by including transverse derivatives of A in (25a,b) as indicated in (21) ; this problem has been extensively studied[17] for gaussian beams, as it is crucial for several applications involving nonlinear processes with frequency conversion. A similar effect also occurs in time domain when short light pulses are used for second harmonic generation or other frequency converting nonlinear process ; because the group velocity, say at ω and 2ω, are different, the two pulses eventually will do not overlap in time and the efficiency will be reduced here too. This problem has been studied for gaussian shape pulses.

2.2.3. *Phase matching.* Clearly in order to reach such a situation one must achieve the condition $\Delta k = 0$, namely to phase match[7,8] the two waves. This is a major problem in nonlinear optics and dramatically conditions the efficiency of a given nonlinear process in particular when there is a change of frequency in the process of the interaction .

In the case of three wave interaction with $\omega_3 = \omega_1 + \omega_2$ the phase matching condition is $\underline{k}_3 = \underline{k}_1 + \underline{k}_2$ which for collinear beams becomes

$$n_1\omega_1 + n_2\omega_2 = n_3\omega_3 \tag{28}$$

For an optically isotropic medium this condition cannot be satisfied when all three frequencies ω_1, ω_2 and ω_3 fall within the same transparency region of the medium because of the normal dispersion. This is clearly evident for the special case of second harmonic generation where condition (28) reduces to $n(\omega) = n(2\omega)$ which cannot be satisfied in the normal dispersion region. In principle it can be achieved if one of the frequencies falls in the region of anomalous dispersion which is a rather awkward and unpractical situation.

The most convenient and common method for achieving phase matching is to exploit[7,8] the birefringence that many crystals exhibit, a consequence of their linear optical anisotropy. If such is the case, for a given propagation direction there are two indices for two orthogonal polarizations (Fig.2), namely the ordinary and extraordinary with indices n_o and n_e respectively which can be either $n_e > n_o$ (positive uniaxial) or $n_o > n_e$ (negative uniaxial). By appropriate choice of the polarizations of the beams one can offset the phase mismatch because of the normal dispersion with the birefringence ; one distinguishes type I and type II phase matching as follows

	$n_e>n_o$	$n_e<n_o$
Type I	e,e → O	O,O → e
Type II	O,e → O	e,O → e

Clearly for each set of frequencies different values of the refractive indices are required to achieve phase matching. The change of the refractive indices in the birefringent crystals can be obtained most conveniently either with angle or temperature tuning ; modification of the birefringence by a static electric or magnetic field or pressure can also be used.

We also point out that phase matching can also be achieved[18] in media with circular anisotropy (rotatory power) where the phase velocities for right and left circular polarization states are different ; these can be solids or liquids or even gases.

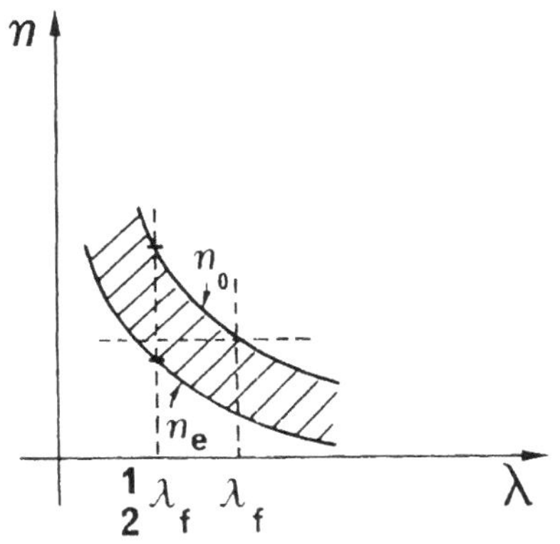

Figure 2

The phase matching actually sets very stringent restrictions on the nonlinear processes and constitutes the most efficient way for selectively enhancing or suppressing a given process with respect to others.

2.2.4.*Second harmonic reflection.* The reflection of second harmonic from an interface or a surface has been given much attention because it constitutes a powerful technique for surface diagnostic. The nonlinear reflection laws can be obtained[20] by the continuity conditions of the fields and inductions on the surface. We only point out here that this leads to

$$n_2 \sin\theta_2 = n_1 \sin\theta_1$$

where θ_1 and θ_2 are the reflection angles for the fundamental and the harmonic ; thus the second harmonic is reflected in a different angle from the fundamental. One can also estimate the reflected harmonic intensity with respect to the transmitted one for normal incidence

$$\frac{I_{2R}}{I_{2T}} = \left(\frac{2k_1-k_2}{2k_1+k_2}\right)^2 = \left(\frac{\Delta n}{2n}\right)^2 \tag{29}$$

which is several orders of magnitude smaller. This was implicitly assumed to be the case in the previous discussion of the second harmonic generation where we assumed that $A_2 = 0$ at the entrance of the crystal which in view of (29) is an approximation albeit a good one.

2.2.5. *Sum, difference frequency generation and parametric amplification.* One can proceed[6] to treat other cases along similar lines ; two such cases are important for device applications namely the sum and difference frequency generation with one beam, the pump, remaining undepleted. The case of sum frequency generation $\omega_1 + \omega_2 = \omega_3$ where one of the two input beams, say ω_1, is strong and remains undepleted while the other ω_2 is weak can be treated analytically in the stationnary regime. Assuming perfect phase matching, $\Delta k = 0$, and $|A_1|$ unaffected by the nonlinear process, setting up the equations and solving for A_2 and A_3 with $A_3(0) = 0$ one obtains

$$A_2(z) = A_2(0)\cos \kappa z \tag{30a}$$

$$A_3(z) = -A_2(0)\frac{\kappa}{\kappa_2}\sin \kappa z \tag{30b}$$

which exhibit an oscillatory behavior with period $L = 2\pi/\kappa$; $\kappa = 4\pi^2\, i\, \chi_e\, A_i/n_i\, \lambda_i$ and $\kappa = |\kappa_2|^2$. The previous discussion is of particular relevance in the case of frequency up-conversion where a weak infrared signal can be amplified and upconverted in frequency, for instance in the visible, by letting it interact with an intense laser beam.

Another important problem is the difference frequency generation $\omega_2 = \omega_3 - \omega_1$, where the beam at ω_3 will be assumed strong and undepleted by the nonlinear process. For the important case of $A_2(0) = 0$ one has

$$A_1(z) = A_1(0)\left\{\cos h\gamma z + \frac{i\Delta k}{2\gamma}\sin h\gamma z\right\} \tag{31a}$$

$$A_2(z) = i\, A_1(0)\, \sin h\gamma z \tag{31b}$$

for the signal and idler respectively $\gamma^2 = \left[\gamma_0^2 - (\Delta k)^2\right]$ and $\gamma_0^2 \sim I_3$. Thus monotonic growth of the signal wave will occur only if γ is a real number that is if the pump intensity I_3 satisfies the condition $\gamma_0^2 > (\Delta k)^2$ or

$$I_3 > I_c = \frac{c^3 (\Delta k)^2}{32 n_1 n_2 n_3 |\chi^{(2)}|\, \omega_1 \omega_2} \tag{32}$$

Thus if phase mismatch is not achieved there will be a threshold for amplification. Above this threshold

$$I_1(z) = I_1(0)\left\{\frac{\gamma_0^2}{\gamma^2}\sinh^2\gamma z + 1\right\}$$

Clearly for perfect phase matching in principle there is no threshold condition.

If in the perfect phase matching case an appropriate feedback is provided, for instance by inserting the system between two mirrors to form an optical resonator, oscillation can occur due to the gain of the parametric amplification process without having beams at frequencies ω_1 and ω_2 initially. The process is initiated by the quantum noise provided by spontaneous break up of pump photons ω_3 into signal and idler photons ω_1 and ω_2 respectively and is called spontaneous parametric scattering ; its treatment needs the quantification of the electromagnetic field modes. Optical parametric oscillators are becoming[22] a very important and successful class of nonlinear optical devides with range of applications as perfectly coherent tunable sources.

2.3 THIRD ORDER PROCESSES

There is a fundamental difference between second and third order processes and for that matter between even and odd order processes which has deep implications in several respects. This difference can be most easily perceived by inspection of the energy transfer $dW^{(n+1)}/dt = \langle \underline{E}.d\underline{P}^{(n)}/dt \rangle$ connected with second and third order processes respectively. As can be seen there the phase of the term related to second order processes cannot be fixed independently of the phases of the involved electric fields. In contrast the phase of the term related to third order processes for certain frequency configurations, namely whenever the condition $\omega_i + \omega_j = 0$ is satisfied for any pair, can be fixed independently of the phases of the electric fields. Otherwise stated if such is the case the imaginary part of $\chi^{(3)}$ can be related to a real energy exchange between the field and the matter through a nonlinear mechanism involving two photons and a material excitation while the real part of $\chi^{(3)}$ is related to an intensity dependent dispersion.

Whenever the condition $\omega_i + \omega_j = 0$ is not satisfied for any frequency pair one cannot make such a separation and one has only coherent processes with energy transfer between the fields without the matter being involved, similar to the second order processes previously discussed although more complex because of the higher order dependence on the fields. Thus third harmonic generation frequency mixing, parametric amplification and oscillation can be discussed along the same lines as outlined in the previous section. The phase matching conditions here can be more easily satisfied than in the corresponding second order processes as they involve the closing of a rectangle and not a triangle. In particular one can achieve perfect phase matching even in the case of isotropic media like gases or liquids[17] by appropriate mixing of two components to introduce adjustable dispersion. These processes can be enhanced with one photon resonances but this to some extent is counterbalanced by the concommittant important linear losses.

Here we shall only consider effects which can occur whenever the condition $\omega_i + \omega_j = 0$ is satisfied for any pair of frequencies ; it is automatically satisfied for the other pair as well. As hinted above in this case the real and imaginary parts of $\chi^{(3)}$ can be related to intensity dependent refractive index and absorption respectively. Indeed from $P(\omega) = P^{(1)}(\omega) + P^{(3)}(\omega) = (\chi^{(1)}(\omega) + 3\chi^{(3)}(\omega,-\omega,\omega)\ |E(\omega)|^2\ E(\omega) \equiv \boldsymbol{\chi}(\omega)\ E(\omega)$ one may define an effective refraction index

$$\tilde{n} = n_o + n_2 I \tag{33}$$

where $n_2 = 12\pi\chi^{(3)}/cn_o^2$ and similarly for the absorption

$$\tilde{\alpha} = \alpha_o + \alpha_2 I \tag{34}$$

These cannot be related to each other in a straight forward manner by a Kramers-Kronig relation as is the case between the linear coefficients n_o and α_o respectively ; note that in α_2 several mechanisms can contribute in addition to two-photon absorption. We shall refer to the cases (33) and (34) as optical Kerr nonlinearities, dispersive and absorptive respectively. In connection with the effective refractive index there is an additional complication compared to the linear case ; in general the nonlinear one does not follow instantly the temporal intensity variations. To a good approximation[10] its temporal evolution obeys a Debye type equation

$$\tau\dot{\tilde{n}}(t) + \tilde{n}(t) = n_2 I(t) \tag{35}$$

or in integral form

$$\tilde{n}(t) = n_o + \frac{n_2}{\tau}\int_{-\infty}^{\infty} I(t)\, e^{-(t-s)/\tau}\, ds$$

where τ is a phenomenological response time for the Kerr nonlinearity.

Another crucial feature of the effective indices (33) and (34) is that these may now become space and time dependent through their dependence on the beam intensity $I(\underline{r},t)$ and one has spatial and temporal lensing or grating effects that can lead to dramatic modifications of the spatiotemporal characteristics of a beam ; these are the light self action or photoinduced effects and include self-focusing and -defocusing, self-phase modulation, spatial or temporal soliton formation, modulational instabilities and other effects. They may lead to coupling of two beams and the appearance of optical gratings which give rise to effects like optical phase conjugation, optical bistability, instabilities and chaotic behavior. We shall briefly describe these effects as they pertain for the dispersive Kerr nonlinearity (33) although the absorptive case can also be treated along similar lines but all these effects cannot occur there because of the inherent energy loss.

2.3.1 Degenerate four wave interactions Optical Gratings. We cursively discuss the optical phase conjugation and optical bistability which are special cases of optically induced gratings.

Nonlinear Phase Conjugation. Phase conjugation in optics is used to correct aberrations and consists in the operation of replacing a wave :

$$\underline{E}_s(\underline{r},t) = \underline{E}_s\, e^{-i\omega t} + c.c$$

by its phase conjugate in space

$$E_c(\underline{r},t) = r\,\underline{E}_s^*\, e^{-i\omega t} + c.c$$

where $\underline{E}_s = \hat{e}_s A_s e^{i\underline{k}_s.\underline{r}}$ or equivalently $E_c(\underline{r},t) = r\,\underline{E}_s(\underline{r},-t)$ and can also be termed time reversal wave front ; thus the polarization state and the amplitude are replaced by their complex conjugate and the wave vector direction is reversed. These operations result in aberration correction as can be easily deduced by the propagation equation or diffraction theory. It should be noted that phase conjugation is also relevant in holography.

The four wave mixing processes(23,24) provide very efficient interaction schemes for real time optical phase conjugation. The classic configuration concerns the degenerate four wave interaction scheme depicted in Fig.3. Two counterpropagating intense beams of frequency ω and wave vectors $\underline{k}_F$ and $\underline{k}_B = -\underline{k}_F$, the forward and backward pumps respectively, interact in a nonlinear medium with a third weak signal beam of same frequency but arbitrary propagation direction $\underline{k}_s$ to form a beam of frequency ω counterpropagating and phase conjugated to the signal beam ; this operation is achieved through the induced third order polarization at frequency ω and amplitude

$$P_{NL} = 6\,\chi^{(3)} A_F A_B A_s^*\, e^{-i\underline{k}_s\underline{r}} \tag{36}$$

since $\underline{k}_F + \underline{k}_B = 0$ and $\underline{k}_s + \underline{k}_c = 0$ so that the phase matching condition is automatically satisfied. Actually the total induced third order polarization contains several other terms but the corresponding fields in general are suppressed because of phase mismatch.

If we assume that the pumps remain undepleted the amplitudes of the signal and its phase conjugated beams in the stationary regime satisfy the coupled equations

$$\frac{dA_s}{dz} = i\kappa A_c^* \tag{37a}$$

$$\frac{dA_c}{dz} = -\,i\kappa A_s^* \tag{37b}$$

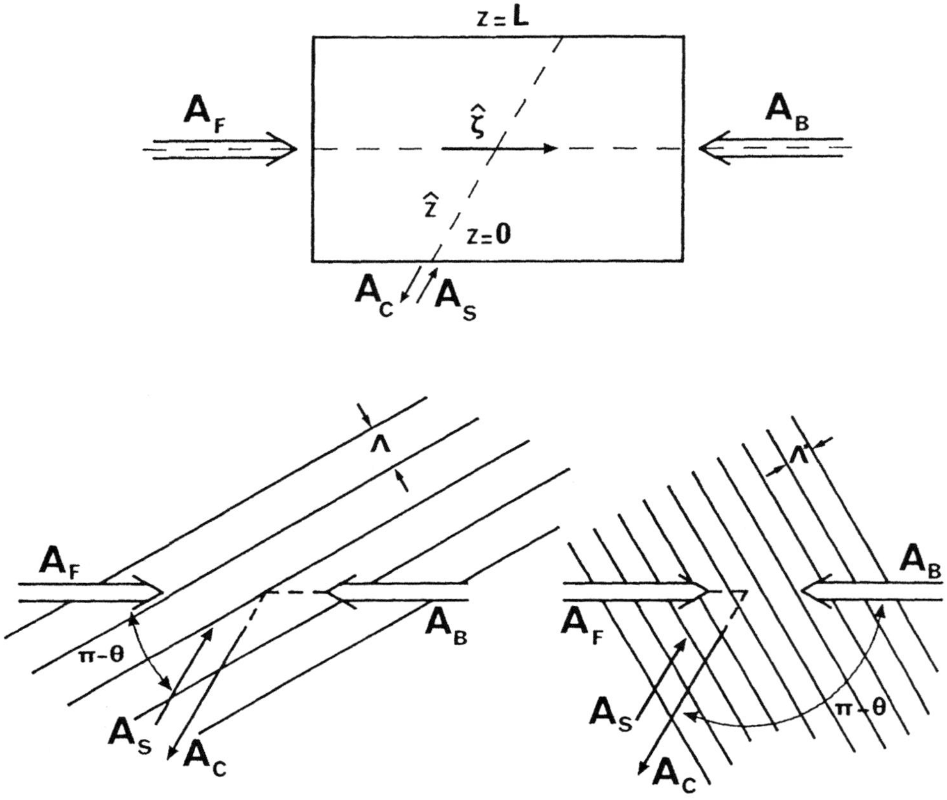

Figure 3

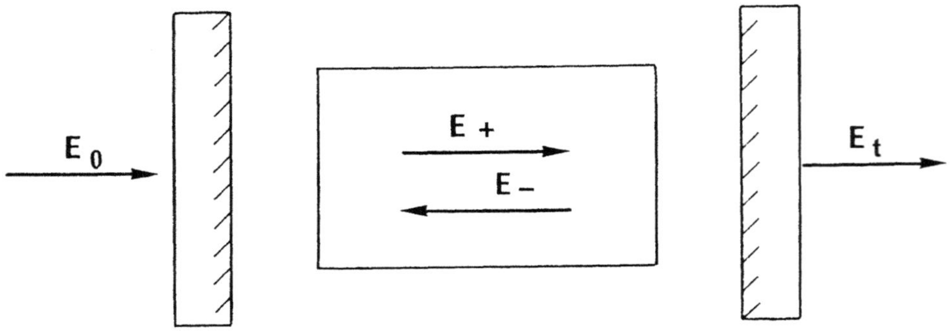

Figure 4

where $\kappa = 12\,\pi\omega\,\chi^{(3)}\,A_F A_B\,/nc$ which can be readily solved and for $A_c\,(L) = 0$ one obtains

$$A_s\,(L) = \frac{A_s\,(0)}{\cos|\kappa|\,L} \quad , \quad A_c\,(0) = \frac{i\kappa^*}{|\kappa|}\,A_s^*\,(0)\,(\tan|\kappa|\,L)$$

Thus the transmitted signal wave is always amplified while its phase conjugated can have a value ranging from zero to infinity depending on the value of $|\kappa|\,L$ and the whole system acts as a phase conjugated mirror with reflectivity $R = |r|^2 = \tan^2|\kappa|\,L$ which can exceed 100 % evidently on the expense of the pumps.

Actually the whole process can be also visualized[23] in terms of optical gratings. Indeed the induced nonlinear polarization can be written quite generally

$$P_{NL}^{(3)} = a\left(\underline{A}_F.\underline{A}_s^*\right)\underline{A}_B + b\left(\underline{A}_B.\underline{A}_s^*\right)\underline{A}_F + c\left(\underline{A}_F.\underline{A}_B\right)\underline{A}_s^* \tag{38}$$

The first two terms correspond to spatial gratings (Fig.3) while the third one to a temporal grating oscillating at frequency 2ω. Note that in analogy with the optical parametric oscillator one can achieve oscillation in any direction by introducing a feedback for the photon noise that results from two twin counter propagationg pump photons being scattered to two twin counterpropagating photons in another direction and the three terms interpreted as the diffraction of each beam in turn by the optical gratings set up by the other two. As previously stated the optical phase conjugation by degenerate four wave mixing can be used to correct beam aberrations in real time in a wide range of applications[23].In particular all operations and effects that were envisaged in holography can be transposed here in real time. Optical phase conjugation by nonlinear processes has been extensively studied[23] for several configurations and in particular in media with periodic refractive index, inside a nonlinear Fabry-Perot cavity where bistable behavior, and different types of instabilities and chaotic behavior can occur.

Optical bistability. Optical bistability or multistability is related to the situation when the output intensity resulting from a given operation on a beam can have two or more walues and accordingly must "decide" on which one will operate. This can be achieved by abruptly rendering the system opaque or transparent by slight changes of some external parameter along different paths respectively. The third order process and in particular the ones related to the optical Kerr nonlinearity, dispersive[25] or absorptive[26], provide a simple scheme to realize such a situation in real time if an appropriate feedback is introduced for instance by inserting the Kerr medium in a Fabry-Perot cavity but other configurations can also be devised for the same purpose.

We cursively discuss the main features and ingredients of the optical bistability in the case of the dispersive nonlinear Fabry-Perot cavity[25] assuming plane monochromatic waves

(Fig.4). A monochromatic wave of frequency ω normally indicent on a Fabry-Perot cavity containing a dispersive Kerr isotropic medium sets up an electric field

$$E(z,t) = \mathrm{Re}\left[\left[A_+ e^{ikz} + A_- e^{-ikz}\right] e^{-i\omega t}\right]$$

inside the cavity whose forward and backward propagating components A_+ and A_- respectively satisfy two coupled amplitude equations. These can be easily solved in the stationary regime and introducing the continuity conditions for the Poynting vector (flux conservation) accross the boundaries give

$$\mathcal{T} \equiv \frac{I_t}{I_o} = \frac{1}{1 + F^2 \sin^2\phi/2} \tag{39a}$$

and $\phi = \phi_o + \phi' I_t$ or

$$\mathcal{T} \equiv \frac{I_t}{I_o} = \frac{\phi - \phi_o}{\phi' I_o} \tag{39b}$$

where $\phi_o = 4\pi n_o L/\lambda$, $\phi' = 4\pi n_2 (1+R) L/(1-R)$, R is the mirror reflectivity, I_o and I_t are the incident and transmitted beam intensities respectively, and $F = 2\sqrt{R}/(1-R)$ is the finesse of the cavity.

The operation point of the system is defined as the intersection of the two transmittivity curves(39a,b) and can be obtained with a graphical procedure. One has bistable operation and hysteresis. As I_o increases one obtains additional hysteresis loops and multistability. The hysteresis is obtained because the intersection of the straight line (39b) and the dashed part of the Airy function (39a) is an unstable operation point and even the smallest deviation will grow monotonically with time. To prove this one must use the non stationary equation for the amplitudes $A_\pm$ and the phase $\phi = 4\pi \tilde{n} L/\lambda$ where $\tilde{n}$ satisfies the Debye equation (35). One finds that there is critical slowing down similar to the one occuring in first order phase transitions. Actually a stability analysis of this system of equation shows that the nonstationary solutions are unstable in a certain region of values of the external parameters and one also has transition to chaos. It is difficult however to characterize this transition for realistic systems where transverse effects become important and lead to insurmontable computational difficulties.

Optical bistability has been extensively studied[27]. There is a whole class of beam interaction configurations that exhibit bistable behavior ; essentially all optical systems that can be switched from opaque to transparent and visa versa exhibit bistable operation if the switching operation can be achieved by a nonlinear process. Thus in the absorptive regime this is achieved by the saturation of absorption[26]. Bistable operation was also observed by exploiting the total reflection effect, in periodic structures, in intracavity optical phase conjugation[23], in hybrid systems, etc... Evidently the interest in the optical bistable operation relies on the possibility to have optical memories in real time by exploiting the hysteresis.

2.3.2. *Light Self-action. Lensing effects.* In contrast to the previous effects that involve the coupling of two fields there is a rich class of nonlinear propagation effects that result through the self-action of a light beam on itself the coupling being provided by the optical Kerr nonlinearity. These self action or lensing effects are of considerable fundamental and technological interest and are relevant in a medium with dispersive Kerr nonlinearity. We shall briefly outline[10] their principles and main features.

Self-focusing,-defocusing. A light beam of finite transverse extension can be focused, defocused or trapped by appropriately reshaping its equiphase surfaces. In conventional optics this is achieved by lens systems or spatial modulation of the optical density of the medium. The optical Kerr nonlinearity induced in the medium by the beam itself can provoke similar effects since the effective refractive index has different values in different parts of the beam profile. This self-action can be readily seen when we consider a gaussian beam

$$\underline{E}(\rho,z,t) = \mathrm{Re}\left\{\hat{e}\, A(\rho)\, e^{ikz - i\omega t}\right\}$$

with $A(\rho) = A_o \exp\{-\rho^2/2a_o^2\}$ where a_o measures the transverse beam extension and ρ is the radial coordinate in a Oxy-plan measured from the propagation axis Oz. From (33) the effective refractive index is now :

$$\tilde{n} = n_o + n_2\, I(\rho)$$

where $I(\rho) = |A_o|^2 \exp\{-\rho^2/a_o^2\}$. This dependance of $\tilde{n}$ on ρ effectively introduces a spatially modulated optical density ; accordingly the optical rays are bent inwards or outwards from the Oz-axis depending on whether $n_2 > 0$ or $n_2 < 0$ leading to self-focusing and defocusing respectively ; both effects have been extensively studied[10].

The self-focusing effect has a threshold because it enters in competition with the diffraction which increases as the beam diameter decreases through the aperture effect. A simple model based on geometric optics gives :

$$I_o > \frac{n_o}{8n_2}\left(\frac{\lambda}{a_o}\right)^2 \tag{40}$$

for self-focusing to occur. The analytical treatment with a gaussian profile beam leads to a similar expression multiplied by a factor 1.22.

Self-trapping. Spatial solitons. If for a certain beam profile the condition $\theta_f = \theta_d$ can be achieved throughout the propagation this particular beam profile can be maintained unchanged ; this is the light self trapping effect. Because of the initially irregular transverse profile this condition may be satisfied simultaneously for different parts of the beam and lead to the beam splitting in several filaments. Although the analytical treatment of the problem is very complex one can gain[28] insight with the following simple model. Assuming $\partial^2 A/\partial y^2 = 0$ and setting $P_{NL} = 3\chi^{(3)} |A|^2 A$ with $\Delta K = 0$ in (21) the equation reads

$$i\frac{\partial A}{\partial z} + \frac{1}{2k}\frac{\partial^2 A}{\partial x^2} + \frac{\omega}{2n_o c}\chi^{(3)}|A|^2 A = 0$$

or

$$i\frac{\partial \psi}{\partial \zeta} = \frac{1}{2}\frac{\partial^2 \psi}{\partial \xi^2} - |\psi|^2 \psi \qquad (41)$$

with $\psi = A\sqrt{\omega\chi^{(3)}/6\pi n_o c} = A\sqrt{\kappa}$, $\zeta = z$ and $\xi = A\sqrt{\kappa}$. Equ (41) is the nonlinear Schrödinger or Landau-Ginzburg equation which admits a class of stable solutions (eigenstates) termed solitons ψ_N with N an integer N = 0,1,2... For N = 1

$$\psi_1 = 1/\mathrm{ch}\xi \equiv \mathrm{sech}\xi \qquad (42a)$$

while the higher order solitons, N > 1, are periodic functions satisfying the boundary condition

$$\psi_N = N/\mathrm{ch}\xi \equiv N\mathrm{sech}\xi \qquad (42b)$$

for $\xi = 0$ and N integer.These spatial solitions play a very important role in guided nonlinear optics and have recently been observed in an increasingly number of nonlinear systems.

White self-transparency. In the previous cases the medium is assumed homogeneous with an index of refraction n_o that has the same value throughout the medium. If n_o fluctuates randomly in space but not in time the spatial fluctuations of n may be compensated with the optical Kerr nonlinearity and the losses due to elastic scattering are then suppressed leading to a new class of self-action effects[30,31] and in particular to self-transparency. To simplify let us assume that the inhomogeneous medium is a composite consisting of small spherical dielectric particles of refractive index n_B and number density N randomly dispersed into a homogeneous dielectric of refractive index n_A, both without absorption losses. A beam propagating in the z direction experiences[32] losses because of elastic scattering which in the Rayleigh and Rayleigh-Debye-Gans regimes, neglecting multiple scattering, is expressed with an elastic diffusion constant $\alpha_s \sim (\Delta n)^2$ where $\Delta n_o = n_A - n_B$. If the incident beam intensity is high then each refractive index is modified by the optical Kerr nonlinearity or $\tilde{n}_A = n_A + n_{2A}I$ and $\tilde{n}_B = n_B + n_{2B}I$. If we assume that the elastic diffusion constant can be written

$$\tilde{\alpha}_s \sim (\Delta\tilde{n})^2 \sim (\Delta n_o + \Delta n_2 I)^2$$

where $\Delta n_2 = n_{2A} - n_{2B}$ one easily sees that for sign $\Delta n_o \Delta n_2 = -1$ there is a light intensity $I_c = |\Delta n_o/\Delta n_2|$ for which $\tilde{\alpha}_s$ vanishes : below this intensity the beam is totally attenuated and above it is trasmitted with constant intensity I_c ; in a certain way the photons in the scattered modes are recycled into the incident one. In particular that part of the spatiotemporal intensity profile $I(\underline{r},t)$ of a beam that has intensity below I_c is attenuated while the one that has higher intensities propagate without further loss once its intensity has been attenuated to I_c ; one can obtain[30,31] square form spatiotemporal profiles. One also can achieve bistable operation with a single mirror and observe several other interesting effects.

Self phase modulation. The temporal counterparts of the previous effects within the class of third order processes lead to quite striking effects of fundamental and technological interest. These occur whenever different parts of a light pulse experience different effective refractive indices because of the time dependence of the refractive index imposed by the light pulse itself through the optical Kerr nonlinearity ; it results[33] to a rearrangement of its spectral density which may have striking repercussions on the pulse form.

From (33) a short light pulse of intensity I(t) experiences an effective refractive index $\tilde{n}(t) = n_o + n_2 I(t)$ where for simplicity we assume that the Kerr nonlinearity is instantaneous and exactly follows the pulse form. After propagation over a distance ℓ the beam experiences a phase modulation $\Delta\phi(t) = \frac{\omega}{c} n_2 I(t)\ell$ and a frequency modulation $\Delta\omega(t) = -\partial(\Delta\phi)/\partial t$ which leads to a spectral broadening with a spectral density

$$S(\omega) = \left| \int_o^\infty I(t)\, e^{i\omega_o t + i\Delta\phi(t)} dt \right|^2$$

If I(t) is symmetric $\Delta\phi$ will also be symmectric while $\Delta\omega(t)$ will be antisymmetric with respect to the carrier frequency ω_o with maximum extension $|\Delta\omega|_{max} = |\partial(\Delta\phi)/\partial t|_{max}$ on either side of ω_o, the Stokes and anti-Stokes regions ; this is the self-phase modulation which plays[10] a very important role in pulse shaping and also in achieving broad band light pulses.

Temporal solitons. Modulational instability. As previously mentionned spatial solitons resulted from the compensation of the self focusing with the diffraction the former being related to the optical Kerr nonlinearity and the latter to the second order transverse spatial derivatives of the amplitude in (21) ; they are special solutions of the nonlinear Schrödinger equation (41). Their temporal analogues, namely the temporal solitons[34,35], exist and result from a compensation of the self phase modulation and the group velocity dispersion the latter being related to a second order partial time derivative of the amplitude. Assuming plane waves, neglecting diffraction, setting $\Delta K = 0$ and inserting $P_{NL} = \chi^{(3)} |A|^2 A$ in (18) one gets

$$i \frac{\partial \psi}{\partial \zeta} = - \frac{1}{2} \frac{\partial^2 \psi}{\partial \tau^2} + |\psi|^2 \psi \tag{43}$$

with $\tau = (t - z/v_g)/\tau_o$, $\zeta = |k''|z/\tau_o^2$ where k'' is the group velocity dispersion and

$$\psi = \tau_o \left(\frac{2\chi^{(3)}}{|k''|} \right)^{1/2} A$$

which by renaming the variables is identical to the nonlinear Schrödinger for the planar spatial solitons. Because of their remarkable stability solitons are being extensively studied(11-14) in connection with short pulse reshaping and shortening and long distance optical signal transmission ; their implications in the latter technology are still difficult to grasp but they may fundamentally revolutionize the field of telecommunications. Higher order solitons may scatter to lower order solitons. One also distinguishes between bright and dark solitons the previous being the bright ones and its complementary in time being the dark one.

The nonlinear Schrödinger equation exhibits a very interesting instability namely the modulational instability(36) where energy is transfered from the primary pump wave to a set of sidebands symmetrically disposed about the fundamental frequency ; it can be used for high rate modulation of a beam and their applications.

2.4. PARAMETRIC PROCESSES

In the previoulsy discussed nonlinear optical effects the frequencies of the interacting fields were tacitly assumed to be comparable : all fall within the same transparency region of the medium so that all fields couple to the same degrees of freedom of the medium. For the transparency region that extends above the vibrational frequencies and below the onset of electronic transitions, for instance, the valence electrons only contribute to the optical non linearities and their response can be assumed practically instantaneous.

If any of the frequencies however is lowered into a different transparency region then the corresponding field couples with additional or even different degrees of freedom. For instance, for frequencies below the vibrational frequencies the ionic motion and its coupling with the electrons sets up addtional mechanisms for nonlinear polarization as also does the molecular reorientation ; these additional mechanisms besides changing the magnitude and even the sign of the nonlinear coefficients it may modify some of their symmetry properties, like the breaking down of the Kleinmann symmetry relations(37), and also introduce different time constants which can be quite crucial when considering very short light pulse interactions.

Similarly when any of the frequencies lies much higher than the electronic transitions and falls in the X-ray region the electrons essentially respond as free and the dipole approximation breaks down and by the same token the symmetry properties as well related to this approximation the most conspicuous being $\chi^{(2n)} \equiv 0$ for media with inversion symmetry. Actually the multipolar expansion must be abandonned and one must proceed from the microscopic Maxwell-Lorentz equations[38] without introducing the space and time averaging that leads to the macroscopic equations (1) ; the relevant quantity now is the microscopic electron density and current distribution.

For such cases where fields of widely different frequencies are interacting it is more convenient to treat differently high and low frequencies fields and regard the latter as external parameters that modify the optical properties of medium felt by the former. This is the case in particular when one of the fields is static or equivalently when its frequency is zero but other situations can be envisaged which involve static magnetic fields or other low frequency disturbances like acoustic waves. The fields of electrooptics, magnetooptics and acoustooptics are related to some very important applications that have great impact on device technology.

2.4.1. *Electrooptics*. This is related to changes of the refractive index of a material brought in by a dc of low-frequency electric field. If this change is linear on the static field one has the linear electrooptic or Pockels effect while when this change is quadratic in the static field one has the quadatric electrooptic effect or static Kerr effect. The former is related to a second order nonlinear polarization

$$P_i(\omega) = 2\,\chi^{(2)}_{ijk}(\omega,0)\,E_j(\omega)\,E_{ok} \tag{44a}$$

and can occur only for materials without inversion symmetry while the latter is related to a third order polarization

$$P_i(\omega) = 3\,\chi^{(3)}_{ijk\ell}(\omega,0,0)\,E_j(\omega)\,E_{ok}\,E_{o\ell} \tag{44b}$$

which can occur in any material. As previously stated one usually proceeds by first deriving the changes of the refractive index induced by the static or low frequency field and then analysing their impact on the propagation properties of an optical field in the linear regime for the latter. One important reason for adopting such a procedure is that the phase matcing is irrelevant for the processes described by (44a,b).

More precisely one actually evaluates the modifications of the optical indicatrix of the material induced by the static electric field. For the case of the Pockels effect one thus writes[39,40]

$$\Delta\left(\frac{1}{n^2}\right)_i = \sum_j r_{ij} E_j$$

where $(1/n^2)_i$ are the optical constants that describe the optical indicatrix in an arbitrary coordinate system and can be expressed in terms of the principal-axis optical constants ; r_{ij} are the electrooptic coefficients and can be related to $\chi^{(2)}_{ijk}(\omega,0)$. An important application of the Pockels effect is the electrooptic modulation(40) of an optical beam where the x and y components of a light beam are forced to experience different phase shifts when propagating through the crystal and its polarization state will be modified in a way that depends on the applied static field along the z-axis. The refractive indices in the x and y directions are $n_x = n_o - n_o^3 r_{63} E_z / 2$ $n_y = n_o + n_o^3 r_{63} E_z / 2$ respectively and the phase difference

$$\Gamma = \left(n_y - n_x\right)\frac{\omega L}{c} = \frac{n_o^3 r_{63} E_z \omega L}{c} = \frac{n_o^3 r_{63} \omega V}{c} \tag{45}$$

where $V = E_z L$ is the voltage applied in the z direction the propagation direction of the optical beam. This effect and its variants have found several applications in optoelectronic devices.

Actually not only the refractive index of a material is modified by a static electric field ; its absorption too can be modified and many applications of the electroabsorption for device operation have been proposed but here one must account with substantial operation losses.

2.4.2. *Magnetooptics*. The oscillating magnetic field associated to a light beam is very weak with respect to the electric field to induce any sizeable effects but a static magnetic fields H_o can be very intense and induce substantial changes in the optical properties as a consequence of the Zeeman splittings that the degenerate magnetic levels undergo and the concomitant changes of the energy spacings and oscillator strengths.

The lowest order magnetooptic effect is related(10) to the polarization

$$P(\omega) = \chi^{(2)}_M(\omega,0)\, E(\omega)\, H_o \tag{46}$$

and gives rise to the Faraday effect(2,38) when the direction of H_o and that of the light beam propagation coincide. As for the Pockels effect previously discussed it is more convenient to first analyse the modifications that the refractive index undergoes in the presence of the static magnetic field. One finds that here the refractive indices for right and left circular polarization become different and these two components experience different phase shifts after propagating a distance L inside the medium ; the polarization state undergoes a rotation by an angle

$$\theta_F = \frac{\omega L}{nc}\left(n_- - n_+\right) \tag{47}$$

the Faraday rotation angle. In terms of susceptibilities we also have

$$n_{\mp}^2 = 1 + 4\pi\left(\chi^{(1)} + \Delta\,\chi_{\mp}^{(1)}\right) = n^2 + 4\pi\,\Delta\,\chi_{\mp}^{(1)}$$

where $\Delta\,\chi_{\mp}^{(1)}$ is the static magnetic field induced change of the linear susceptibility $\chi^{(1)}$ and can be related to $\chi_M^{(2)}(\omega,0)$ in (46).

Very interesting is the photoinduced Faraday rotation effect[41] which is related to the nonlinear polarization

$$P(\omega) = \chi_M^{(3)}\,|E(\omega)|^2\,E(\omega)\,H_o$$

and leads to a photoinduced change of the Faraday rotation angle ; its implications can best be assessed as follows. If the light intensity $I \sim |E|^2$ is high the two refractive indices $n_{\mp}$ in (47) undergo a photoinduced modification which can be written

$$\tilde{n}_{\mp}^2 = 1 + 4\pi\left(\chi^{(1)} + \Delta\,\chi_{\mp}^{(1)} + \chi^{(3)}I + \Delta\,\chi_{\mp}^{(3)}I\right) \tag{48}$$

where $\chi^{(3)}$ is the optical Kerr susceptibility and $\Delta\,\chi_{\mp}^{(3)} \sim \chi_M^{(3)}\,H_o$ is its magnetic field induced modification or equivalently the photoinduced modification of $\Delta\,\chi_{\mp}^{1}$. Accordingly the Faraday rotation angle (47) becomes now

$$\tilde{\theta}_F = \frac{\omega\ell}{nc}\left(\tilde{n}_- - \tilde{n}_+\right) = \frac{\pi\omega\ell}{nc}\left\{\left(\Delta\chi_-^{(1)} - \Delta\chi_+^{(1)}\right) + \left(\Delta\chi_-^{(3)} - \Delta\chi_+^{(3)}\right)I\right\} = \theta_F + \theta_{NL} \tag{49}$$

the second term θ_{NL} is the light induced modification of the linear Faraday rotation angle. The photoinduced Faraday effect can have important applications[41] when fast modulation of the polarization state of a beam is required for instance in nonreciprocal devices.

2.4.3. *Acoustooptics.* A sound wave in a medium is related to density and pressure fluctuations that modulate the refractive index of the material and concomitantly the propagation characteristics of an incident light beam[15,40]. For an anisotropic solid the changes in the refractive index are characterized by

$$\Delta\left(\frac{1}{\varepsilon}\right)_{ij} = \sum_{k\ell} p_{ijk\ell}\,S_{\ell k} \tag{50}$$

where $S_{k\ell}$ is the strain tensor and $p_{ijk\ell}$ the photoelastic tensor ; for an isotropic, medium liquid or gas, one can use a scalar relation. In the case an acoustic wave of wave vector $\underline{q}$ and frequency Ω this effect induces a variation

$$\Delta\varepsilon = \Delta\varepsilon_o \, e^{i(\underline{q}.\underline{r} - \Omega t)} + c.c.$$

in the dielectric constant. An incident beam

$$E_i = \mathrm{Re}\left\{A_i \, e^{i\underline{k}_i.\underline{r} - i\omega_i t}\right\}$$

is scattered off this variation and produces a beam

$$E_s = \mathrm{Re}\left\{A_s \, e^{i\underline{k}_s.\underline{r} - i\omega_s t}\right\}$$

whose characteristics depend on the interaction length. If the interaction length is sufficiently long then phase-matching is important and imposes a single diffracted beam ; this is the Bragg regime. If the interaction length is very short then the phase matching is irrelevant and several diffracted beams are obtained ; this is the Raman-Nath regime. Both cases are connected with important applications[15,40].

3. Nonlinear Polarization

3.1. PHENOMENOLOGICAL DESCRIPTION

Here we summarize[9,10,42] some elementary features of the nonlinear dipolar polarization terms that are not bound to any specific microscopic description but are common to all responses that relate an effect to a cause. These are obtained by invoking, invariance with respect to translation of the time origin, causality, reality, symmetrization and also the symmetry of the material system. When perturbation theory applies, which in practice implies that $E < E_c$, the polarization as previously stated can be expressed in power series as

$$\underline{P}(t) = \underline{P}^{(0)}(t) + \underline{P}^{(1)}(t) + \underline{P}^{(2)}(t) + \underline{P}^{(3)}(t) + \ldots + \underline{P}^{(n)}(t) + \ldots \qquad (51)$$

and each partial polarization term $\underline{P}^{(n)}$ must possess all properties that follow from the above principles. We will in addition assume that the interaction is local or plainly stated the polarization at a given space point is related only to the electric field amplitude at the same point.

3.3.1. *Response functions and susceptibilities.* For the linear term the most general relation between $\underline{P}^{(1)}(t)$ and $\underline{E}(t)$ subject to the above conditions is

$$\underline{P}^{(1)}(t) = \int_{-\infty}^{\infty} d\tau_1 \, \underline{\underline{R}}^{(1)}(\tau_1) \underline{E}(t - \tau_1) \tag{52}$$

where $\underline{\underline{R}}^{(1)}$ is a real second rank tensor that vanishes for τ_1 negative. In the absence of magnetic fields it can be symmetrized and written in diagonal form with three independent components after transforming to the principal axes. The number of independent components can be further reduced by involving the symmetry of the medium ; for an optically isotropic medium a single component only remains.

For the second order term similarly one has

$$\underline{P}^{(2)}(t) = \int_{-\infty}^{\infty} d\tau_1 \int_{-\infty}^{\infty} d\tau_2 \, \underline{\underline{R}}^{(2)}(\tau_1,\tau_2) \, \underline{E}(t - \tau_1) \underline{E}(t - \tau_2) \tag{53}$$

where $\underline{\underline{R}}^{(2)}$ is a real third rank tensor that vanishes when either τ_1 or τ_2 is negative. Furthermore it can always be symmetrized so that $R^{(2)}_{ijk}(\tau_1,\tau_2)$ is invariant under interchange of the pairs (j, τ_1) and (k, τ_2) and this is termed intrinsic permutation symmetry. The number of independent tensor components, at most 27, is however greatly reduced when the symmetry of the material system is invoked : it identically vanishes for a material with inversion symmetry.

Similarly for the higher order terms one has

$$\underline{P}^{(n)}(t) = \int_{-\infty}^{\infty} d\tau_1 \int_{-\infty}^{\infty} d\tau_2 \ldots \int_{-\infty}^{\infty} d\tau_n \, \underline{\underline{R}}^{(n)}(\tau_1,\tau_2,\ldots\tau_n) \, \underline{E}(t - \tau_1) \underline{E}(t - \tau_2) \ldots \underline{E}(t - \tau_n) \tag{54}$$

where $\underline{\underline{R}}^{(n)}$ is a real (n+1)-th rank tensor, that vanishes for negative values of any of its time arguments. It can be symmetrized so that $R^{(n)}_{ij_1\ldots j_n}(\tau_1\ldots\tau_n)$ is invariant under any permutation of the pairs (j_1, τ_1) ... (j_n, τ_n) and this is the intrinsic permutation symmetry.The number of the independent tensor components, there at most 3^{n+1} such components, can be substantially reduced if the material system possess a symmetry ; the even order polarization response functions $R^{(2n)}$ identically vanish for systems with inversion symmetry. The temporal profil of $E(\tau)$ can be arbitrary but in the impulsive regime, with delta-like pulses, it is easily seen that each polarization term coincides with the corresponding polarization function.

By introducing the Fourier representation in frequency space for the polarization and field the previous relation become algebraic. Thus setting

$$\underline{P}^{(n)}(t) = \int_{-\infty}^{\infty} \underline{P}^{(n)}(\omega) \, e^{-i\omega t} \, d\omega \, , \qquad \underline{P}^{(n)}(\omega) = \frac{1}{2\pi} \int_{-\infty}^{\infty} \underline{P}^{(n)}(t) \, e^{i\omega t} \, dt$$

$$E(t) = \int_{-\infty}^{\infty} E(\omega)\, e^{-i\omega t}\, d\omega \,, \qquad \underline{E}(\omega) = \frac{1}{2\pi} \int_{-\infty}^{\infty} \underline{E}(t)\, e^{i\omega t}\, dt$$

one gets

$$\underline{P}^{(n)}(\omega) = \int_{-\infty}^{\infty} d\omega_1 \ldots \int_{-\infty}^{\infty} d\omega_n\, \underline{\underline{\underline{\chi}}}^{(n)}(\omega_1, \omega_2, \ldots \omega_n)\, \underline{E}(\omega_1) \ldots \underline{E}(\omega_n)\, \delta(\omega - \omega_0) \tag{55}$$

with $\omega_0 = \sum_{i=1}^{n} \omega_i$ and

$$\chi^{(n)}(\omega_1, \omega_2, \ldots \omega_4) = \int_{-\infty}^{\infty} d\tau_1 \int_{-\infty}^{\infty} d\tau_2 \ldots \int_{-\infty}^{\infty} d\tau_n\; R^{(n)}(\tau_1, \tau_2, \ldots \tau_n)\, e^{i\sum_{i=1}^{n} \omega_i \tau_i} \tag{56}$$

which defines the n-th order susceptibility $\chi^{(n)}$, a(n+1)-th rank tensor with complex components.

From (56) and the properties of the response functions one shows that because of the causality $\chi^{(n)}$ can be analytically continued in the upper half-plane for all frequencies and

$$\chi^{(n)}(\omega_1, \omega_2 \ldots \omega_n)^* = \chi^{(n)}(-\omega_1^*, -\omega_2^* \ldots -\omega_n^*) \tag{57}$$

because of the reality of $R^{(n)}$; furthermore because of the intrinsic permutation symmetry $R_{ij_1 \ldots j_n}(\omega_1, \omega_2 \ldots \omega_n)$ is invariant under all n! permutations of the pairs (j_i, ω_i) and the symmetry of the material system substantially reduces the number of the independent tensor components. The latter can be obtained by imposing that the tensor components transform to themselves under any transformation of the reference space that leaves the system unchanged (symmetry operation) ; this leads to a number of linear relations between the tensor components from which the independent ones can be fixed. Use of group theoretical techniques leads equivalently to the same result. The tensor properties of $\chi^{(n)}$ are always expressed in orthogonal coordinate systems. In some cases useful insight can be gained by going[43] over to spherical coordinates.

The relations are greatly simplified when the fields are delta-functions in frequency space, namely monochromatic waves ; with $E(t) = \mathrm{Re}\{E_\omega \exp(-i\omega t)\}$ and $P^{(n)}(t) = \mathrm{Re}\{P_\omega^{(n)} \exp(-i\omega t)\}$ one gets

$$P_{\omega_0}{}^{(n)} = D_n\, \chi^{(n)}(\omega_1, \omega_2, \ldots \omega_n)\, \underline{E}_{\omega 1}\, \underline{E}_{\omega 2} \ldots \underline{E}_{\omega n} \tag{58}$$

where D_n is a numerical factor that takes into account the number of distinct permutations of $\omega_1, \ldots \omega_n$. With this convention the limiting value $\chi^{(n)}(0, 0 \ldots 0)$ is independent of the path chosen for the frequencies to reach the values 0.

3.1.2. *Kramers-Kronig Relations. Photoinduced and Parametric effects.* The analyticity of $\chi^{(1)}(\omega)$ in the upper complex plane, that follows from the causality condition through the theorem of Titshmarsh together with Cauchy's residue theorem, leads[3,38] to the Kramers-Kronig relation between the real and imaginary parts of the linear susceptibility along the real axis namely

$$\mathrm{Re}\,\chi^{(1)}(\omega) = \frac{1}{\pi}\mathcal{P}\int_{-\infty}^{\infty} d\omega' \frac{\mathrm{Im}\,\chi^{(1)}(\omega)}{\omega' - \omega} \tag{59}$$

and its reverse relation ; $\mathcal{P}$ stands for the taking of the principal value. These relations which are also related to the dissipation-fluctuation theorem establishes an unambiguous connection between linear absorption and dispersion irrespective of the phase and amplitude of the field or

$$n(\omega) = \frac{c}{\pi}\mathcal{P}\int_{0}^{\infty} d\omega' \frac{\alpha(\omega')}{\omega'^2 - \omega^2} \tag{60}$$

in terms of the absorption and refraction indices.

The causality has similar implications[4,9,44] on the nonlinear susceptibilities but up to present date these have not been cast into any useful and physically transparent form. In some isolated cases extensions of (50) in the nonlinear regime have been proposed[45] and used with scarce attention as far as their applicability is concerned ; this is very limited.

The difficulty in setting up K-K relations for nonlinear susceptibilities and also relating them to physically simple processes is connected[46] with the fact that nonlinear optical processes can both lead to real nonlinear absorption losses to the material system as well as energy transfer from one coherent field (Fourier) component to a large number of other field (Fourier) components through several interaction paths ; as the order of interaction increases these paths become more and more complex since cascading[9] of lower order processes must also be included.

Actually even in the case of even order susceptibilities, the quadratic $\chi^{(2)}$ being the lowest one, which cannot be connected with any nonlinear absorption loss but only to frequency shifting coherent processes involving amplitude transfer between different field (Fourier) components, the attempts to derive K-K relations were not conclusive[44] regarding their usefulness. Thus in the case of quadratic susceptibilities the derived relations mix second harmonic, sum and difference generation as well as the electrooptic and rectification effects.

For the reasons previously evoked the situation is worse in the case of odd order susceptibilities, the cubic $\chi^{(3)}$ being the lowest one. Here besides all the possible coherent frequency shifting processes one can identify nonlinear absorption losses related to real multiphoton transitions, these are two-photon absorption, saturation, and light scattering processes (Raman, Brillouin etc) ; in addition cascading of second order processes introduce additional contributions to all these processes.

It is customary to assume that the K-K relation (56) or (60) can be set up for parametric processes when the linear susceptibility is modulated by an external "static" field ζ_0, for instance pressure or static electric or magnetic fields : one simply substitutes there the modulated susceptibility $\chi^{(3)}(\omega, \zeta_0)$, an analytic function of ζ_0. Such a procedure has been employed without any preambule[45] to the case of the photoinduced effects namely the ones that qualitatively can be described by an intensity dependent effective susceptibility $\tilde{\chi}^{(1)}(\omega, I)$ disregarding the fact that the intensity I may me related to a field component of frequency $\omega' \neq \omega$. The assumed K-K relation then writes

$$\Delta n(\omega, I) = \frac{c}{\pi} \mathcal{P} \int_0^{\infty} d\omega' \frac{\Delta\alpha(\omega, I)}{\omega'^2 - \omega^2} \tag{61}$$

There is no rigorous proof for the validity of this relation but certainly stationarity[46] must apply since expressions (61) is a direct extension of the linear K-K relation which is derived under equilibrium conditions. Since in a nonlinear regime this is hardly the case this relations is of very limited validity. In addition even this approximate relation implies that the optical Kerr coefficient is essentially related to several nonlinear absorption processes, two-photon, light scattering processes (Raman, Brillouin, Rayleigh etc) and not to the first one only as has been customary[45] to assume ; this is rather formidable task and effort. An indication of how such a relation can be handled could be obtained by working out a specific quantum or classical model say the two levels model or the anharmonic oscillator.

The problem of nonlinear K-K relations is also related[47] to the efforts to set up a thermodynamic approach and formally derive the susceptibility from a postulated "free energy". This too encounters difficulties however when the physical interpretation is attempted since the stationarity and equilibrium condition must be properly formulated.

3.1.3. *Nonlocality. Retardation and cascading*. The previous discussion concerned the dependence of the macroscopic nonlinear optical properties on the temporal variations of the field ; the spatial variations were disregarded. This is consequence of the assumption[9] that the interaction is local when setting up the expression of the polarization ; in the microscopic level this approximation can be justified whenever the characteristic charge displacement is much smaller than the optical wavelength λ and all terms higher than zeroth order are neglected in a power series development in (a/λ).

This however leads to the suppression of an essential characteristic of the Maxwell equations and the Lorentz force, namely the nonlocality which gives rise to a spatial dispersion of the optical properties ; it also disregards such fundamental properties as the optical rotatory power. The spatial dispersion and optical rotatory power are closely interrelated manifestations of the nonlocality. At the microscopic level[9] the simplest way to incorporate them is to include terms higher than zeroth in a (a/λ)-power series development of the charge-field interaction. At the macroscopic level this corresponds to also taking into account the variation of the electromagnetic field in the vicinity of a point when setting up the constitutive relation between $\underline{D}$ and $\underline{E}$ at this point ; otherwise stated in the development of $\underline{D}$ one must also include spatial derivatives of $\underline{E}$. The quadru-polarization and the magnetic polarization in (55) result from such considerations. Their inclusion in the Maxwell equations opens up a whole class of phenomena which besides the natural spatial dispersion and optical activity (gyrotropy) also includes artificial gyrotropy (Faraday rotation), Cotton-Mouton effects but also the photoinduced ones and others. Because of their different nature with respect to the dipolar ones we shall not dwell further upon these effects here except to point out that in the near future a growing attention and studies will be focused on them with important implications in a number of applications.

On the other hand we wish to expand on an other manifestation of nonlocality which results from the very essence of the propagation equation namely the retardation effect[9]. In nonlinear optics the retardation effects have several implications the most conspicuous one being the phase matching problem. Another implication is that the total nonlinear polarization of a given order and given frequency, besides the direct contribution will contain indirect contributions arising from many-step processes involving, at the intermediate stage, nonlinear interactions with fields generated by lower order polarization terms. Since the latter fields are strongly affected by the phase matching conditions, at the intermediate stages the effective susceptibility for the global polarization, totalling both the direct and all indirect processes, becomes wave vector dependent. These indirect processes are also termed cascading processes and must be included in an ad hoc way unless one uses from the beginning the true eigenstates of the matter-field system which in the case of crystals are the polariton states.

To illustrate the problem we focus[9,48] our attention on the case of the third order polarization in a crystal without inversion symmetry. In such a crystal besides the direct contribution given by (9) there will be additional contributions to the total third order polarization at frequency $\omega_1 + \omega_2$, $\omega_2 + \omega_3$, $\omega_3 + \omega_1$, generated by two-wave mixing. These fields may beat with the fields at frequencies ω_3, ω_1, ω_2 respectively to generate polarizations at $\omega_1 + \omega_2 + \omega_3$; furthermore these polarizations are created with the same total wavevector as the direct one namely $\underline{k}_1 + \underline{k}_2 + \underline{k}_3$. One can define an effective third order susceptibility for this total third order polarization which however depends on the wave vectors of the primary and intermediate field in contrast to the genuine third order susceptibility related to the direct third order polarization contribution which does not.

Cascading processes play an important role in third and higher order effects and draw a growing attention in applications[49].

3.2. MICROSCOPIC DESCRIPTION

We wish now to give a microscopic description of the induced polarization and relate it to the induced dipoles in the material. This microscopic description clearly must be compatible with the phenomenological one we outlined in the previous section ; it is based on the equation of charge motion in the presence of intense coherent electromagnetic fields and the interactions with the other charges. Because of the complexity of the problem the steps[9] that lead from this equation to the expressions of the experimentally accessible macroscopic nonlinear susceptibilities previously introduced are several and all involve certain simplifications and approximations in order to cope with the many-body interactions and obtain meaningful and useful informations.

3.2.1. *Dipole Approximation. Induced dipole and polarizabilities.* Let us assume that the material system in the absence of external fields is described by a hamiltonian $H_o(p_i,\underline{r}_i)$ with known eigenvalue and eigenstate spectrum, where i runs over all the charges. Under the influence of an intense electromagnetic field represented by a vector potential $\underline{A}(\underline{r},t)$ the state of the charges making up the medium will be perturbed and oscillating currents will be induced in the system. The hamiltonian describing the system is now $H(t) = H_o(\underline{\pi}_i,\underline{r}_i)$ where $\underline{\pi}_i = \underline{p}_i - e\underline{A}_i/c$ with $\underline{A}_i = \underline{A}(\underline{r}_i,t)$ and assuming that H_o is quadratic in the momenta $\underline{p}$ one has

$$H(t) = H_o(\underline{p}_i,\underline{r}_i) - \sum_i \frac{e_i}{2m_i c}\left\{\underline{A}_i\underline{p}_i + \underline{p}_i\underline{A}_i\right\} + \sum_i \frac{e_i^2}{2m_i c^2} A_i^2 \tag{62}$$

A Coulomb gauge with zero scalar potential is used and spin effects were neglected. Beyond this stage one must resort to simplifying considerations in order to obtain tractable and useful expressions for the response of the system and take into account the actual constitution of the medium and in particular the diversity of the charges and polarizable units involved as well as their mutual interactions. At the outset we have electrons, ions and polarizable units of molecular size the latter in general being non-spherical. Because of their vastly different masses their motion is characterized by eigenfrequencies that fall into well separated spectral regions.

If we for simplicity consider monochromatic waves characterized by a wavelength λ and period T the electromagnetic field described by $\underline{A}(\underline{r},t)$ actually samples[9] the space and time over extensions less than λ and T and averages out fluctuations in the charge motion that vary faster in space and time. Otherwise stated the response of the charges to the electromagnetic field is insensitive to all details of their motion that occur within λ and T. This qualitative argument can also be cast into a rigourous form and sets a scaling procedure.

For fast oscillating electromagnetic fields with frequencies much higher than all eigenfrequencies the charges behave as free and the response is directly related to the electronic charge and current density distribution as is the case in X-ray scattering where the wavelength is much less than the extension of electronic wave functions.

As the frequencies of the electromagnetic field are lowered into the optical spectrum and the wavelength becomes much longer than the charge extension in a dielectric the dipole approximation can be introduced ; the dipolar response to the electromagnetic field is described in terms of the electronic polarizabilities of polarizable units. As a consequence of the adiabatic Born-Oppenheimer approximation these coefficients are functions of the ion coordinates and the orientation of the polarizable unit ; accordingly they may be modulated by the corresponding vibrational and orientational motion. In addition the polarizable units may change their relative positions and this motion introduces fluctuations in the delectric constant. The low frequency fluctuations due to intra- and intermolecular motions result in the different light scattering effects but they can also be directly or indirectly coupled to the electromagnetic fields and strongly modulate[9] the induced polarization.

It is evident from the above qualitative discussion that for most purposes in nonlinear optics one may start by setting up the expression of the induced polarization in the dipole approximation. We will restrict ourselves in the most commonly recurring case where the material is composed of identical repeat polarizable entities, atoms, molecules or clusters of dimensions much smaller than the optical wavelength ; these will be either randomly but homogeneously distributed, like for instance in a gas, a liquid or an amorphous solid or set in a periodic array, for instance a crystalline solid.To the extent that the charges are more or less localized in each such entity the total induced polarization density $\underline{P}$ can be set as

$$\underline{P} = \Sigma \underline{p} = N\underline{p} \tag{63}$$

namely as the sum of the effective dipoles $\underline{p}$ induced on each entity in a volume unit by the electromagnetic field that actually acts on each entity.

The expression of the dipole moment $\underline{p}$ in power series of this field will be obtained using perturbation theory in the dipole approximation namely

$$\underline{p} = \underline{\alpha}^{(1)}\underline{E} + \underline{\underline{\alpha}}^{(2)}\underline{E}\,\underline{E} + \underline{\underline{\underline{\alpha}}}^{(3)}\underline{E}\,\underline{E}\,\underline{E} \tag{64}$$

and the coefficients $\alpha^{(n)}$ of the different powers are the microscopic polarizabilities. Their relation to the previously defined macroscopic polarizabilities and susceptibilities requires the knowledge of the relation between the effective field locally acting in the charges within each entity and the macroscopic field that enters the Maxwell equations. The later in the optical frequency domain is actually the average of the microscopic field over volumes of dimensions much less than the optical wavelength but still containing several polarizable entities. The two fields differ because of the mutual interactions that arise between the induced dipoles and other induced interactions. Because of the complexity of these interactions some simple models have been devised to account for their effect.

We shall now discuss in some detail the above procedure and give the expressions of the microscopic and macroscopic polarizabilities and susceptibilities. The outlined procedure actually pertains to dielectrics with charges more or less well localized within each entity, for instance gases, liquids, ionic or amorphous dielectrics. For semiconductors or metals where electrons are very delocalized one must use a different procedure based[9] on band theory and in addition it is more appropriate to calculate there the induced current density $\underline{J}(t)$ which is related to the polarization density P(t) through

$$\underline{J} = -\frac{\partial \underline{P}}{\partial t}$$

We shall cursively discuss aspects related to the nonlinearities of such systems as well.

From (62) one can show that the hamiltonian of each entity in the presence of an electric field in the dipole approximation can be written as

$$H = H_o + H_c + H_R$$

where H_o represents the hamiltonian of the entity in the absence of the field, H_c is the interaction term with the electric field in the dipole approximation

$$H_c = -\underline{\mu}.\underline{E}_e \tag{65}$$

where $\underline{E}_e$ is the effective field in the entity and $\underline{\mu}$ is the total dipole moment operator of all charges of the entity. H_R stands for all random perturbations of the charges in the entity provoked by the environment which is represented as a bath with an infinite number of degrees of freedom : They are usually assumed to be stationnary markoffian processes uncorrelated to the coherent interaction H_C ; their correlation time τ_c is assumed to be very short, much shorter than any relevant time of the problem. These random perturbations provoke relaxation processes which under certain simplifying approximations can be incorporated into the density matrix operator equation which then reads

$$\frac{d\rho}{dt} = \frac{1}{i\hbar}[H_o + H_c,\rho] + \left.\frac{\partial\rho}{\partial t}\right|_R \tag{66}$$

with

$$\left.\frac{\partial \rho_{mn}}{\partial t}\right|_R = -\Gamma_{mn}\rho_{mn} = -\frac{1}{T_{mn}}\rho_{mn} \tag{67a}$$

$$\left.\frac{\partial \rho_{mm}}{\partial t}\right|_R = \sum_{n\neq m}\left[W_{mn}\rho_{nn} - W_{nm}\rho_{mm}\right] \tag{67b}$$

where Γ_{mn} and W_{mn} are positive quantities that are related to the coherence and population relaxation rates or equivalently to the transverse and longitudinal relaxation times T_2 and T_1 respectively. The induced dipole moment is obtained from $\underline{p} = \mathrm{Tr}\,\rho\,\underline{\mu}$. Assuming H_c a perturbation and expanding in powers of its strength one can set $\rho = \rho^{(0)} + \rho^{(1)} + \rho^{(2)} + ...$in (66). Making a Fourier analysis of the field and the induced dipole and solving the resulting hierarchy of equations by an iterative procedure one finds $\underline{p} = \underline{p}^{(0)} + \underline{p}^{(1)} + \underline{p}^{(2)} + ...$ where

$$\underline{p}^{(n)} = \mathrm{Tr}\,\rho^{(n)}\underline{\mu} = \underline{\alpha}^{(n)}\underline{E}\,\underline{E}...\,\underline{E} \tag{68}$$

which defines the n-th order microscopic polarizability. The relation (68) is tensorial and requires explicit reference to the n Fourier components of the field that are involved and the n-th order dipole they induce together. For the three lowest order terms, the linear second and third order they read

$$P_i^{(1)}(\omega) = \alpha_{ij}(\omega)\,E_j(\omega) \tag{69a}$$

$$P_i^{(2)}(\omega_1+\omega_2) = D_2\beta_{ijk}(\omega_1,\omega_2)\,E_j(\omega_1)\,E_k(\omega_2) \tag{69b}$$

$$P_i^{(3)}(\omega_1+\omega_2+\omega_3) = D_3\gamma_{ijk\ell}(\omega_1,\omega_2,\omega_3)\,E_j(\omega_1)\,E_k(\omega_2)\,E_\ell(\omega_3) \tag{69c}$$

where

$$\alpha^{(1)}(\omega) \equiv \alpha(\omega) = \frac{1}{h}\sum_{g,r} f_g\left\{\frac{\mu_{gr}\,\mu_{rg}}{(\omega_{rg}-\omega-i\Gamma_{rg})} + \frac{\mu_{gr}\,\mu_{rg}}{(\omega_{rg}+\omega+i\Gamma_{rg})}\right\} \tag{70a}$$

$$\alpha^{(2)}(\omega_1,\omega_2) \equiv \beta(\omega_1,\omega_2) = \frac{1}{\hbar^2}\sum_{g,r,s} f_g\left\{\frac{\mu_{gr}\,\mu_{rs}\mu_{st}}{(\omega_{rg}-\omega_1-\omega_2-i\Gamma_{rg})(\omega_{sg}-\omega_2-i\Gamma_{sg})} + 5\text{ terms}\right\} \tag{70b}$$

$$\alpha^{(3)}(\omega_1,\omega_2,\omega_3) \equiv \gamma(\omega_1,\omega_2,\omega_3) = \frac{1}{\hbar^3}\sum_{g,r,s,t} f_g$$

$$\left\{ \frac{\mu_{gr}\mu_{rs}\mu_{st}\mu_{tg}}{(\omega_{rg} - i\Gamma_{rg} - \omega_1 - \omega_2 - \omega_3)(\omega_{sg} - i\Gamma_{sg} - \omega_2 - \omega_3)(\omega_{tg} - i\Gamma_{tg} - \omega_3)} + 47 \text{ terms} \right\} \tag{70c}$$

are the linear, second and third order microscopic polarizabilities and we also introduced their more conventional notations α, β and γ instead of $\alpha^{(1)}$, $\alpha^{(2)}$ and $\alpha^{(3)}$ respectively ; these coefficients and the higher ones are tensors of second, third, fourth and higher order respectively. It is clear that in contrast to the linear polarizabilities the nonlinear ones possess a very rich and complicated multiresonant behavior. Furthermore in the third order polarizability, and this is actually the case for all odd order polarizabilities, a new type of intermediate resonance appears, the one where $\omega_1 + \omega_2 = \omega_e$, which is quite distinct from the primary ones that occur whenever any of the incident and created frequencies are close to a dipole allowed transition of the medium. The behavior of the nonlinear polarizability close to this intermediate resonance can be related to the stimulated counterpart of the different spontaneous light scattering effects which will be discussed in section 3.3.

3.2.2. *Susceptibilities and local field corrections.* As previously stated the connection between macroscopic susceptibilities and the microscopic polarizabilites requires[9] the relation between the macroscopic and effective fields. This is a central problem in the whole theory of the dielectrics and touches to its very fundamental aspects. To gain an insight into the problem let us assume that the charge clouds of the different entities do not overlap and we subdivide the space into identical boxes each containing one polarizable entity. The field E_e that acts on a given entity is the one that results from all sources outside its box ; it is the microscopic field E_{mic} which is the same in every box with the exclusion of the field from the entity[50] itself E_s or $\underline{E}_e = \underline{E}_{mic} - \underline{E}_s$. The macroscopic field that enters Maxwell equations is the space average of E_{mic} over many boxes which in our case reduces to its average within any box or $\underline{E} = \langle E_{mic} \rangle_c$. The field E_e is actually not constant throughout its box and the effective field will be defined as a weighted average of E_e over the box the weight being roughly the charge density distribution and similarly for E_s so that

$$E_f \equiv \langle E_e \rangle = \langle E_{mic} \rangle - \langle E_s \rangle$$

Without going into lengthier analysis it is quite evident from the above considerations that the effective field corrections is a measure of the inhomogeneity of the charge density distribution within a box ; the smoother this distribution is the less important are the local field corrections. In the extreme case of a metal where the charge density is uniform throughout each box the local field corrections disappear while in the extreme case of point like induced dipoles the Lorentz model may apply and one finds an isotropic dielectric

$$\underline{E}_f = \underline{E} + \frac{4\pi}{3}\underline{P} \tag{71}$$

For the cases intermediate between these two extremes the calculation of the local field corrections become exceedingly difficult because of the overlap of charge distributions in different boxes. Let us assume[9] for simplicity that

$$\underline{E}_f = \underline{E} + \underline{\underline{L}}\,\underline{P} \tag{72}$$

where L is a frequency independent second rank tensor whose components vanish for a metal and reduce to (71) in a dielectric with point like non overlaping induced dipoles. Thus the effective field contains a contribution from the induced polarization whose dynamics in particular are related to those of equations (67a,b). A more rigorous theory in crystals relates[9] the local field corrections to the Fourier series expansion of the microscopic field in the inverse lattice space.

Assuming isotropic medium and using expression (72) with L a scalar, inserting in (64) and substituting in (63) one obtains

$$\underline{P} = N\alpha\,(\underline{E}+L\underline{P}) + N\beta\,(\underline{E}+L\underline{P})(\underline{E}+L\underline{P}) + N\gamma\,(\underline{E}+L\underline{P})(\underline{E}+L\underline{P})(\underline{E}+L\underline{P}) \tag{73}$$

where for simplicity no explicit reference to the frequencies is made. Setting

$$f(\omega) = 1/(1-N\alpha(\omega)\,L) \tag{74}$$

rearranging and keeping track of terms of the same order in the electric field one can cast this expression into the macroscopic form

$$\underline{P} = N\underline{\underline{\tilde{\alpha}}}\,\underline{E} + N\underline{\underline{\underline{\tilde{\beta}}}}\,\underline{E}\,\underline{E} + N\underline{\underline{\underline{\underline{\tilde{\gamma}}}}}\,\underline{E}\,\underline{E}\,\underline{E} \equiv \underline{\underline{\chi}}^{(1)}\underline{E} + \underline{\underline{\underline{\chi}}}^{(2)}\underline{E}\,\underline{E} + \underline{\underline{\underline{\underline{\chi}}}}^{(3)}\underline{E}\,\underline{E}\,\underline{E} \tag{75}$$

where the macroscopic polarizabilities $\tilde{\alpha}$, $\tilde{\beta}$ ... are

$$\underline{\underline{\tilde{\alpha}}}(\omega) = f(\omega)\,\underline{\underline{\alpha}} \tag{76a}$$

$$\underline{\underline{\underline{\tilde{\beta}}}}(\omega_1,\omega_2) = f(\omega_1+\omega_2)\,f(\omega_1)\,f(\omega_2)\,\underline{\underline{\underline{\beta}}}(\omega_1,\omega_2) \tag{76b}$$

The expression of $\tilde{\gamma}$ is somewhat involved[9] and will not be reproduced ; it suffices to state that it contains additional terms to γ which correspond to microscopic cascading processes. For optically anisotropic media f is a second rank tensor, the inverse of $(1 - N\,\underline{\underline{L}}\,\underline{\underline{\alpha}})$, and the previous relations must be accordingly transcribed. Local field corrections may be substantial[9] and must be included ; in the following we shall assume that most often lead to a renormalization of the oscillator strengths and will be lumped together with them in the expressions of the susceptibilities. These oscillators strengths in the following will be understood to be the renormalized ones.

At this level we wish to remind that for third or higher polarizations additional contributions must be included[9] which result from the so-called cascading processes. These processes result from the interaction of the primary beams with intermediate ones that are generated by lower order induced nonlinear polarizations compatible with the symmetry of the medium. At the macroscopic level these cascading processes can be lumped together with the direct contribution since they have the same total wave vector, frequency and electric field dependence ; the resulting effective susceptibility has the same symmetry properties as the direct one but also depend on the wave vector on the phase matching at the intermediate interactions of the charge density that determine β and γ and by the same token $\chi^{(2)}$ and $\chi^{(3)}$ with appropriate provisions for local field corrections.

3.2.3. *Resonant regime. Stimulated nonlinear processes.* The resonant behavior of the nonlinear susceptibilities is related to several very important nonlinear optical phenomena. As can be inferred from their expressions, the nonlinear susceptibilities can be resonantly enhanced if one or more of the complex frequency /energy denominators have their real part close to or equal to zero. A simple inspection shows that these resonances fall into two categories. The first one concerns the "direct" or "primary" resonances where one or more frequencies of the fields involved in the interaction are close to electric dipole allowed transition frequencies of the medium ; these resonances either lead to trivial resonant enhancement of the nonlinearity which however is counterbalanced with the concomitant linear absorption losses at the same frequency or, in the case of very intense beams they lead to a class of nonlinear phenomena which can be accounted for only by a non perturbative analysis of an equivalent two-level system resonantly interacting with a monochromatic field. The other category concerns the "intermediate" resonances where an algebraic sum of primary frequencies is close to an appropriately allowed transition frequency of the medium ; under certain conditions then energy can be transferred between the field modes through the intermediary of the material mode. These indirect or secondary resonances and related energy exchange processes between the field and material modes are distinct features of the odd order nonlinearities ; as a matter of fact one can show[4,10] that the related processes can even grow from photon noise and give rise to stimulated effects.

In this section we shall give a brief account of these intermediate resonances in the third order nonlinearities and the effects they give rise to ; these occur whenever

$$\omega_i + \omega_j = \omega_e \qquad (77)$$

and the relevant susceptibility is $\chi^{(3)}(\omega_1,\omega_2,-\omega_1)$ where ω_e is any two-photon allowed excitation of the medium. Since condition $(\omega_i + \omega_j = 0)$ is satisfied, one can actually define a nonlinear energy transfer between the two electromagnetic modes ω_1 and ω_2 the difference in energy being stored in the material excitation. As was previously stated these effects can grow even from photon noise and give rise to stimulated effects : the stimulated two photon stimulated emission, Raman, Brillouin, Rayleigh and other light scattering effects.We first derive the expression of the two photon transition matrix element and relate it to the light scattering cross section.

From (70c) one can easily show that for $\omega_i + \omega_j = \omega_{fg} = \omega_e$ and keeping only resonant terms

$$\alpha_R^{(3)}(\omega_1,\omega_2,-\omega_1) = \frac{\alpha\,(\omega_1,\omega_2)\;\;\alpha^*(\omega_1,\omega_2)}{\hbar(\omega_e - (\omega_1-\omega_2 - i\Gamma_{fg})} \tag{78}$$

where

$$\alpha_{ij}(\omega_1,\omega_2) = \sum_n \left\{ \frac{<g|\mu_i|n>\;<n|\mu_j|f>}{\hbar(\omega_{ng} - \omega_1)} + \frac{<g|\mu_j|n>\;<n|\mu_i|f>}{\hbar(\omega_{ng} - \omega_2)} \right\} \tag{79}$$

is a generalized polarizability which for $f \equiv g$ and $\omega_1 \equiv \omega_2$ reduces to the linear polarizability (70a) ; it can also be regarded as the matrix element between states f and g of an effective two-photon operator $H^{(2)}$ whose diagonal elements for $\omega_1 \equiv \omega_2$ are the linear polarizabilities. The two photon resonant susceptibility (78) is complex because of the combined effect of the energy denominator and the generalized polarizabilities (79). If the product $\alpha_{ij}(\omega_1,\omega_2)\;\alpha_{ij}^*(\omega_1,\omega_2)$ is a real number then the separation of (78) into real and imaginary parts is related only to that of the energy denominator. The associated energy transfer

$$\frac{dW_i^{(4)}}{dt} = \left\langle \underline{E}_i \cdot \frac{d\underline{P}_i^{(3)}}{dt} \right\rangle \tag{80}$$

is proportional to the energy density flows $I_i \sim \underline{E}_i\underline{E}_i^*$, i = 1,2, only and unaffected by the phases of the electric fields. Hence one can unambiguously define a nonlinear energy gain or loss in the involved modes 1 or 2 of the electromagnetic field the difference being taken or given up by the material ω_e mode so that energy conservation is satisfied.

The energy density flow can be expressed as photon density flow or

$$I_i = N_i \hbar\omega_i \frac{c}{n_i}$$

where N_i is the photon number density in mode i and n_i is the refractive index at frequency ω_i ; if one introduces the compound light-matter states $|N_1N_2a\rangle$ of energy $(N_1\hbar\omega_1 + N_2\hbar\omega_2 + E_a)$ then one can interpret the whole energy transfer process as transitions between such isoenergetic states. As can be inferred from the previous comments one can describe and analyse these effects within an unified frame. This however is quite cumbersome to set up and we only present a brief analysis(4,10,15).

The relation between the spontaneous and stimulated light scattering processes can be easily established starting(4,10,15) with the expression

$$\frac{dP}{dt} = BN_1 (N_2 + 1) \tag{81a}$$

for the probability per unit time for mode i losing one photon, mode j gaining one and the material system transiting from state g to state f such that $\omega_f - \omega_g \equiv \omega_e = \omega_1 - \omega_2$ or equivalently the transition from state $|N_1N_2g\rangle$ to state $|N_1 - 1, N_2 + 1, f\rangle$; the coefficient B can be derived from Fermi's golden rule using the appropriate two photon transition operator $H^{(2)}$. If the propagation direction of mode i coincides with the z-axis and we assume stationary regime then the above expression can be cast into the form

$$\frac{dN_2}{dz} = B'N_1 (N_2 + 1) \tag{81b}$$

For $N_2 << 1$ the spontaneous scattering is dominant and $N_2(L) \sim B'N_1L$ where L is the interaction length while for $N_j >> 1$ the stimulated effect becomes dominant and grows as

$$N_2(L) = N_2(0)e^{GL} \tag{82}$$

where $G = B'N_1$. This growth can be initiated by injected photons at frequency ω_2 or by the photon noise provided by the spontaneous process.

Before proceeding to analyse an example of a stimulated process we wish to make here more explicit their connection with the light scattering processes(51). The latter are due to the temporal and spatial inhomogeneities of the dielectric constant because of the intrinsic fluctuations of the different degrees of freedom of the polarizable units. The fluctuations introduce additional terms in the induced polarization that oscillate at frequencies shifted below (Stokes side) and above (anti-Stokes side) the incident frequency by amounts equal to the eigenfrequencies of the degrees of freedom. Because the dielectric properites of a medium at a given frequency ω responding to a wavelength λ are essentially obtained by averaging the charge motion over space and time domains of extensions less then λ and $T = 2\pi/\omega$ respectively only fluctuations with eigenfrequencies much less than ω are relevant in light scattering. With this in mind we may write

$$\varepsilon = 1 + 4\pi N\alpha \tag{83}$$

for the dielectric constant of a dielectric medium formed by identical polarizable units of effective polarizability α and number density N ; both α and N are fluctuating quantities because they are modulated by internal and external degrees of freedom respectively. The internal degrees of freedom, for instance vibration or rotation, being confined in space, the one available to a single polarizable unit, are local and essentially possess a discrete spectrum ; in contrast the external ones like translation of the polarizable unit, involve extended motion of large number of units and can be related to thermodynamic fluctuations like density, pressure and temperature whose evolution is described by the hydrodynamic equations of continuity, momentum transfer and heat transport.

The incident light then is scattered off the fluctuations of either the polarizability $\Delta\alpha$ or the density ΔN to produce Stokes and ansi-Stokes components in addition to the incident one. For an intense incident light these additional components can become intense because each added photon in the scattered light modes enhances the probability for getting an additional one in the same mode provoking an avalanche of photons in a single mode which leads to the build up of a large amplitude coherent electric field. The generation of this additional field on the Stokes side is accompanied by the build up of a large amplitude fluctuation at the relevant material eigenfrequency which because of its resonant coupling enhances the energy transfer. This coupling can be best understood by reminding that the energy stored in a polarizable unit interaction with an electric field is

$$W = -\frac{1}{2}\alpha\, E^2(z,t) \tag{84}$$

where $\underline{E} = \underline{E}_i + \underline{E}_s$, $\underline{E}_i$ and $\underline{E}_s$ being the electric fields in the incident and scattered mode. Since this energy is a function of the internal and external degrees of freedom of the polarizable unit it can be regarded as potential energy and one can define a force conjugated to each degree of freedom q by

$$F_q = \nabla_q \left(\frac{1}{2}\alpha\, E^2\right) \tag{85}$$

or

$$F_q = \frac{1}{2}\frac{\partial\alpha}{\partial q} E^2 \tag{86a}$$

$$F_E = \frac{1}{2}\alpha\, \underline{\nabla}\, E^2 \tag{86b}$$

for an internal and external (position) degrees of freedom respectively. This force may drive the corresponding conjugated degree of freedom to a large amplitude. This is the origin of the stimulated process.

We illustrate this effective force approach(52) with the stimulated Raman Scattering. We consider the case of a dielectric in the presence of an electromagnetic field

$$E(z,t) = \hat{e}_i A_i e^{ik_i z - i\omega_i t} + \hat{e} A_j e^{ik_j z - i\omega_j t} + \text{c.c.} \tag{87}$$

We assume for simplicity that the dielectric is isotropic, possesses inversion symmetry and its scalar dielectric constant is given by (41) while the induced dipole is $p = \alpha E$ and the polarization density

$$P = N\varepsilon_o \alpha E \tag{88}$$

Let us assume now that the polarizability α can be modulated by a single internal degree of freedom of coordinate q and can be written $\alpha = \alpha_o + q\, \partial\alpha/\partial q$ to first order in this coordinate ; this can be a vibration or rotation coordinate. The force (86a) drives its conjugated coordinate according to equation

$$\ddot{q} + \Gamma_q \dot{q} + \omega_q^2 q = F_q/m_q = \frac{1}{2m_q} \frac{\partial\alpha}{\partial q} E^2 \tag{89}$$

where we also introduce a phenomenological damping for the material mode coordinate q and ω_q and m_q are the eigenfrequency and effective mass associated to q.

From (87) E^2 contains components at several frequencies and in particular at frequency $\omega_1 - \omega_2$; if $\omega_1 - \omega_2 \approx \omega_q$ only this component is relevant for driving the coordinate q to appreciable amplitude which can be written as

$$q(z,t) = \text{Re}\left\{q\, e^{i(k_i - k_j) z - i(\omega_i - \omega_j) t}\right\}$$

with

$$q = \frac{1}{2m_q} \frac{\partial\alpha}{\partial q} \left\{ \frac{A_i A_j^*}{\omega_q^2 - (\omega_i - \omega_j)^2 - i(\omega_i - \omega_j)\Gamma_q} \right\}$$

This modulation of the polarization (88) leads to a nonlinear term

$$P_{NL}^{(3)} = N \frac{\partial\alpha}{\partial q} q E \tag{90}$$

cubic in the field intensities. With a collinear geometry for the wavevectors k_i and k_j the Stokes component in this polarization,

$$P_j^{(3)} = \mathrm{Re}\left\{\chi^{(3)} A_i A_i^* A_j\, e^{ik_j z - i\omega_j t}\right\}$$

with

$$\chi_R^{(3)}(\omega_2) = -\frac{i\,N}{2m_q\omega_q\Gamma_q}\left(\frac{\partial\alpha}{\partial q}\right)^2$$

for the exact resonant case $\omega_1 - \omega_2 = \omega_q$, is phase matched ; note that $\chi^{(3)}$ is purely imaginary and indeed can be related to a nonlinear absorption loss that contributes to α_2 in (34). If we disregard the other components in (90) then from (20) one has

$$\frac{dA_2}{dz} = g\, I_1 A_2$$

with

$$g = \frac{2\pi^2\omega}{n_1 n_2 c^2}\,\frac{N}{m_q\omega_q\Gamma_q}\left(\frac{\partial\alpha}{\partial q}\right)^2 \qquad (91)$$

for the spatial evolution of the field in the stationary regime. Assuming undepleted pump one gets

$$A_2(z) = A_2(0)\, e^{gz}$$

which implies exponential growth the same as the one derived by (82) if we notice that G = 2g. It is more convenient to related[4,10] the gain coefficient to the Raman cross section related to mode q; one has

$$G = \frac{4\pi^3 N c^2}{\hbar\omega_1^2\omega_2 n_1^2}\left(\frac{\partial^2\sigma}{\partial\omega\partial\Omega}\right) \qquad (92)$$

where $\partial^2\sigma/\partial\omega\partial\Omega$ is the differential spectral cross section.

The above approach is valid whenever the coordinate q is related to a motion with well defined frequency ω_q, vibrational or rotational, of the polarizable unit ; in quantum mechanics this implies discrete level spacing and their dynamics in the presence of the driving fields can still be described by equation (89). If the thermal energy kT however becomes comparable to or larger than the level spacing as is commonly the case for rotation where $kT \geq \hbar\omega_q$, then the motion becomes diffusive and appropriate provisions must be made in the equation of motion (89). In the case of rotation this leads to the classical reorientational motion of the polarizable units if these are non-spherical which can be described by the Debye model. One can simularly proceed with the Stimulated Brillouin and Rayleigh Scattering, which result from driven fluctuations of the dielectric constant as the result of fluctuations in the external degrees of freedom of the polarizable units namely changes of their position which can be related to density fluctuations $\Delta\varepsilon = (\partial\varepsilon/\partial\rho)\,\Delta\rho$; since from thermodynamics $\Delta p = (\partial\rho/\partial p)_s\,\Delta p + (\partial\rho/\partial s)_p\,\Delta s$ and s and p are independent, to a good approximation one can treat independently the two effects.

3.2.4. *Non resonant regime and nonlinear polarization mechanisms* Of particular interest in most applications are the nonlinear optical properties of materials in their transparency region that extends above the vibrational frequencies and below the onset of electronic transitions. These are due to the polarization of the valence electrons and the relevant nonlinear susceptibilities can be related to some important structural characteristics of their density distribution. Besides its intrinsic fundamental interest this relationship plays a key role in the ongoing nonlinear optical material research. Several computational techniques have been devised[9] to account for the magnitude and behavior of these susceptibilities but of equal value are also some simple models that have been worked out to illustrate the nonlinear polarization mechanism in particular the one underlying the quadratic polarization.

a.*Quadratic polarizabilities.* These mainly result from heteropolar bonds whose polarizabilities, first and second order, can be assumed to be additive and transferable. We may then focus our attention to such a bond which for simplicity will be assumed unidimensional.

Unsöld approximation With $\omega_i = 0$ in (70a,b) we set[53] all energy denominators $E_{rg} = \hbar\omega_{rg}$ equal to an average energy E_a that is determined by using the same approximation in the Thomas-Kuhn rule

$$\sum_r |<r|x|g>|^2 E_{rg} = \hbar^2/2m$$

per electron. One obtains $E_a = \hbar^2 / 2m <\Delta x^2>$ and

$$\alpha_{xx} = 4<\Delta x^2>^2 / a_o \tag{93a}$$

$$\beta_{xxx} = 6<\Delta x^3> <\Delta x^2>^2 / e\, a_o^2 \tag{93b}$$

where $\Delta x = x - <x>$; Expression (94) clearly shows that the bonds must be very polarizable and possess a large octupole moment in order to have large second order polarizability ; the energy E_a is also a measure of the electron delocalization along the bond. This simple approximation reproduces quite well the essential trends. Note that only the ground electronic distribution matters.

Charge transfer model. To gain more insight we represent[54] the ground electronic distribution of an heteropolar bond AB of length R with two point like charges $(1-\varepsilon)$ e and $(1+\varepsilon)$e at its two ends. When an electric field E is applied in the direction from A to B a charge amount $e\Delta\varepsilon$ flows from B to A such that $\alpha E = e\Delta\varepsilon R$ where $\alpha = (1-\varepsilon^2)^2 R^4/4a_0$ is the bond polarizability (93) in the Unsöld approximation with the assumed charge distribution.The intrabond charge flow results in a modification of this polarizability by replacing ε with ε-$\Delta\varepsilon$ which now becomes electric field dependent. Developing into powers of the electric field one can identify

$$\beta = (1 - \varepsilon)^3 \varepsilon R^7 / 4ea_0^2 \qquad (94)$$

which can also obtained directly from (93b) with the assumed charge distribution. This expression clearly shows that β vanishes for $\varepsilon = 0$ (homopolar bond) and $\varepsilon = 1$ (ionic bond) the ions being spherical. It acquires a maximum value for $\varepsilon = 1\ \sqrt{7}$ a result which can used as a guide for the optimal choice of heteropolar bonds ; ε can be related to the bond polarity and electronegativity difference of atoms A and B. One also sees that β increases with the polarizability and the bond length.

In organic[55] systems because of the strong overlap of the π-electrons the relevant units where additivity and transferability can apply are larger than a bond ; this is the case for instance in monosubstituted benzenes X- or charge transfer complexes.

Equivalent field model. In the first case one can relate[56] the magnitude of β to the distortion of the π-electrons of the symmetric benzene ring caused by the substituent radical which is measured by the mesomeric dipole moment μ_M. This is done by assuming that this distortion is induced by an equivalent electric field E_R such that

$$\mu_M = \alpha_R\ E_R$$

where α_R is the linear polarizability of the π-electrons of the symmetric ring. In the presence of a field E then, the induced dipole moment results from the compound effect of $E + E_R$ on the symmetric ring and one obtains

$$\beta = 3\left(\frac{\gamma_r}{\alpha_r}\right)\mu_M$$

where γ_r is the third order polarizability.

Two level model. For the charge complexes an estimation of β can be obtained[57] by simulating the system as a quantum two level system the ground (g) and the excited (e) with wave functions $\psi_g = a\psi_0 + b\psi_1$ and $\psi_e = a^*\psi_1 - b^*\psi_0$, the Mulliken states, where ψ_0 and ψ_1 are the "no bond" and "dative" states respectively. For a two level system one has from (24) with $\omega_i = 0$.

$$\beta = 3\ \mu_{ge}^2\ \Delta\ \mu_{eg} / E_{eg}^2 \qquad (95)$$

μ_{eg} is the transition dipole moment, $\Delta\mu_{eg} = \mu_e - \mu_g$, the dipole difference in the ground and excited states, and E_{eg} the energy difference between them. The important point to notice in (95) is that for such a system β can be large although μ_g might be small or even vanish as long as μ_e is large. This remark is essential(55) in designing molecular crystals.

b. *Third susceptibilities*. This case is more complex to analyse with microscopic models both because of the increased computational complexity but also because of more fundamental grounds. If we assume for simplicity centrosymmetric media it evident by inspection of the expression of γ with $\omega_i = 0$ (see (70c)) that large values can be obtained if the electrons are highly delocalized not only within a molecular unit but across several such identical units . This delocalization over several identical repeat units in a periodic array is a sine qua non perquisite for large cubic susceptibilities. Aditivity and transferability then does not apply. This is in particular the case in semiconductors and conjugated polymers.

An extreme case then is to consider the electrons as moving along a chain of N identical bonds. Treating this model either as a free electron in a box(58) or in the Hückel approximation(59) one gets

$$\alpha \sim N^3 \tag{96a}$$

and

$$\gamma \sim N^5 \tag{96b}$$

which stikingly shows the breakdown of additivity which would lead to $\alpha \sim N$ and $\gamma \sim N$.

If the bonds are not identical but alternately so, with resonance energies β_1 and β_2 respectively for the "long" and "short" bonds, and one assumes an infinite chain one gets(60)

$$\chi^{(1)} \sim N_d^2 \tag{97a}$$

and

$$\chi^{(3)} \sim N_d^6 \qquad \text{(97b}$$

where $N_d = (\beta_1 + \beta_2) / (\beta_1 - \beta_2) \equiv E_F / E_g$ where $E_F = (\beta_1 + \beta_2)$ and $E_g = (\beta_1 - \beta_2)$ are the Fermi and band gap energies for the one-dimensional semiconducteur. The delocalization parameter N_d defines a length N_d a where a is a unit all cell consisting of a bond pair and indicates that expressions (97a,b) are valid even for a finite bond alternated chain as long as its length is larger than N_d a ; if its length is reduced below N_da one recovers the free-electron-in a box behavior (96a,b). This is actually a manifestation of quantum confinement.

We shall not expand more on these simple models except to stress again that for large second order susceptibilities is the charge asymmetry within a unit cell that matters and additivity applies while for large third order susceptibilities is the charge delocalization over several identical units in a periodic array that matters and additivity then does not apply. Below we review the performances of several classes of nonlinear optical materials and analyse them in the light of the previous models.

4. Nonlinear Optical Materials

4.1. GENERAL ASPECTS

Despite the remarquable progress that nonlinear optics has witnessed over the three decades of its existence the implementation of related devices in large scale technology is still not satisfactory. This can be traced to several causes, both structural and conceptual, but the most severe ones can be traced to the present day performances(61) of nonlinear optical materials. Nature has not been generous with the optical nonlinearities of bulk optical materials ; in the case of integrated nonlinear optics the situation is aggravated by additional requirements(62) on the materials as regards their processability, adaptability and interfacing with other materials. These additional requirements are intrinsicly related to the fabrication of nonlinear integrated devices, which besides efficiently performing the expected nonlinear operation, must be miniaturized, compact, reliable and with precisely reproducible characteristics in large scale production and long term operation. We succinctly present the main representatives of different classes of nonlinear optical materials that hold promise in this field and discuss the relation between optical nonlinearities and structural and dynamic characteristics of valence electron distribution.

In order to introduce some systematics in the presentation we remind that the nonlinear optical effects irrespective of their order can be divided into two major classes.

<u>the frequency preserving</u> effects like the optical Kerr effect that can be used for bistable operation, phase conjugation, soliton formation and other operations on the spatiotemporal profile of coherent light pulses ; here one can also include the hybrid effects like electrooptic, magnetooptic and acoustooptic effects that can be used for parametric devices,

<u>the frequency shifting</u> effects like the sum or difference frequency generation processes, which allow one to up/down convert the frequencies of the existing light sources.

There is a major reason for this division. Indeed the efficiency(10,15) of the latter effects is conditioned by the ability to match optical characteristics in the material, like the phase and group velocities, at widely different frequencies. This is a formidable task and drastically restricts the possible choices and the situation is further complicated by the fact that the material characteristics can not be optimized over the wide frequency range often required in frequency up and down conversion processes.

In contrast the efficiency of the frequency preserving effects is not affected by such problems since only frequencies within a very narrow spectral range are involved and the matching conditions can be automatically satisfied ; as a matter of fact to a large extent the interactions here can be assumed local and are in general much easier to analyze.

As a consequence the nonlinear frequency preserving effects are the ones that are most seriously considered(62) in integrated optical devices. They can be either all optical or hybrid (parametric) effects. The all optical nonlinearities(9) essentially involve valence electron motion and are in general weaker than the hybrid ones where the ionic motion, vibrational, orientational or translational can set up very large nonlinearities ; the situation however is reversed as regards the speed of establishing and erasing these nonlinearities the electronic one being much faster than the ionic ones. The magnitude and speed of the nonlinearities are essential characteristics in any assessment of the materials for nonlinear optical devices and must be properly introduced(63) in any figure of merit.

The magnitude and speed of the nonlinearity however are strongly dependent on how close to a material resonance is the operating frequency, or a multiple thereof, as this introduces resonant enhancement of the nonlinearity and absorption losses ; furthermore the speed is limited by the relevant relaxation processes close to a resonance. One may have linear (one photon) or nonlinear (multiphoton) absorption losses which can be related to imaginary parts of odd order susceptibilities ; we stress the fact that there are no absorption losses related to even order nonlinear processes. Thus the amount of linear and/or nonlinear absorption losses is another important criterium for assessing nonlinear optical materials along with the magnitude and speed of the nonlinearity at the frequency of interest.

Clearly the above criteria are relevant to the extent that the material is not irreversibly modified by the optical radiation. Such modifications can result from photochemical or thermal bond breaking, state selective photochemistry, optical breakdown and many others directly related to the interaction of light and matter. In addition material growth, doping and aging considerations, processability, interfacing and packaging and many other aspects will heavily weight on the final choice since for a wide class of materials the nonlinear optical figures of merit are of comparable magnitude.

On the basis of the types of cohesive forces that bind the charges and polarizable units together, the nonlinear optical materials can be assigned to one of the following classes :

- ionic crystals, essentially oxygen-polyedra based solids,
- covalent crystals, essentially semiconductors,
- molecular crystals, in particular organic and polymeric,
- disordered and amorphous solids, in particular glasses and polymers,
- composites and inhomogeneous artificial solids.

We shall succinctly review the main features of some materials representative of these classes. At this stage however it is appropriate to make certain remarks concerning and the evaluation criteria for the corresponding nonlinearities.

To the extent that in device applications one essentially exploits(63,64) the phase shifts provoked by the nonlinear terms, the relevant parameters in the assessment of the nonlinear materials are not the bare nonlinear susceptibilities but appropriately renormalized expressions thereof that measure the induced relative phase shift ; these are called figures of merit and are also directly related to the efficiency of the nonlinear interaction. Their precise definition is actually closely related to the operation or the device one has in mind.

Thus if we concentrate(64) our attention on all optical guided wave switching devices like nonlinear directional coupler, nonlinear Bragg reflector, nonlinear Mach-Zehnder interferometer, nonlinear mode sorter or nonlinear X-junction the relevant figure of merit can be derived by the requirement that the nonlinear phase shift per unit absorption length is larger than 2π. The total phase shift over a length L of the nonlinear material being

$$\Delta\phi = (n_0 + n_2 I)\, k_0\, L = \Delta\phi_0 + \Delta\phi_{NL} \qquad (98)$$

where n_0 and k_0 are the linear refractive index and wave vector while n_2 is the optical Kerr effect coefficient (33). The absorption coefficient being $\alpha_0 + \alpha_2 I$
where α_0 is the one photon (linear) loss and α_2 the "two photon" (nonlinear) loss the above criterium implies that

$$\Delta\phi_{NL}/2\pi = n_2 I/\lambda_0\, \alpha_0 \geq 1 \qquad (99a)$$

if one photon absorption is dominant and

$$n_2/\lambda_0 \alpha_2 > 1 \qquad (99b)$$

if two photon absorption is dominant. Thus possible figures of merit are $F_0 = n_2/\alpha_0$ and $F_2 = n_2/\alpha_2$ respectively ; these considerations pertain to steady-state operation regime. In actual devices using high repetition pulse rate regime one rather uses(65) the figure of merit

$$F_d = \frac{n_2\, I}{\alpha_0\, c\, \tau} \qquad (100)$$

where τ is the recovery time of the nonlinearity which in practice is the population relaxation time T_1. One can similarly introduce figures of merit for quadratic nonlinearities. One such figure of merit can be derived from (45) as

$$C = d_{eff}^2 / n^3 \qquad (101)$$

where $2\,d_{eff} \equiv \chi^{(2)}$ and n is the refractive index. Arguments based on figures of merit should be made with some caution as they may lead to controversies.

4.2. SECOND ORDER NONLINEARITIES

There is only one second order frequency preserving effect namely the electrooptic or Pockels effect [39] which was discussed in section 2. It is related to the nonlinear second order polarization

$$\underline{P}_{\omega}^{(2)} = 2\underline{\underline{\chi}}_{EO}^{(2)}(\omega,0)\,\underline{E}_{\omega}\,\underline{E}_{0}$$

where ω is an optical frequency within the transparency region of the crystal and "0" is a frequency well below any vibrational or orientational frequency. The second order susceptibility $\chi_{EO}^{(2)}(\omega,0)$ contains[9] two contributions : one purely electronic, the same as for the optical second order harmonic generation $\chi_{E}^{(2)}(0,0)$ and another which arises from a rearrangement of the valence electron distribution consecutive to nuclear displacements induced by the static field. The first is directly related to the characteristics of the valence electronic charge distribution while the second is related to the modulation of the latter by the phonons and can be evaluated from infrared and Raman spectroscopic data. Except for the ferroelectrics[39], where the ionic displacements related to the driven soft mode can be substantial, the latter contribution is in general smaller although comparable to the purely electronic one and of either relative sign ; the overall signs of $\chi_{EO}^{(2)}$ and $\chi_{E}^{(2)}$ (or $\chi_{Q}^{(2)}$) are of certain relevance[9] and must be explicitly stated together with the adopted axis orientation. In the following we will concern ourselves only with the electronic contribution $\chi_{E}^{(2)}(0,0)$ to the electrooptic coefficient which is also relevant for second harmonic generation.

In the early days of nonlinear optics it was suggested[66] that an estimate of $\chi_{ijk}^{(2)}$ could be obtained through the conjecture that the coefficient

$$\delta_{ijk}(\omega_1+\omega_2) = \chi_{ijk}^{(2)}(\omega_1,\omega_2)\,/\,\chi_{ii}^{(1)}(\omega_1+\omega_2)\,\chi_{jj}^{(1)}(\omega_1)\,\chi_{kk}^{(1)}(\omega_2) \tag{102}$$

is a constant for all crystals. More careful examination[53] however of the nonlinear polarization mechanisms and values of $\chi^{(2)}$ for different crystal classes showed that this conjecture does not apply and δ may differ by orders of magnitude among the different crystal classes and does not constitute a useful figure of merit for identifying crystals with large nonlinearity. However the fact that the spread of δ_{ijk}-values is narrower than that of $\chi_{ijk}^{(2)}$ indicates that among the noncentrosymmetric crystals those with large $\chi^{(1)}$, which implies crystals with covalent bonding, would also have large $\chi^{(2)}$ in general. The question is then to what extent the bond configuration can be optimized for a crystal to show a large second order nonlinearity ; this is the key problem in noncentrosymmetric material research.

One may also assume[9] that the second order polarizability coefficients of these complexes, after certain provisions regarding local field corrections are made, satisfy the additivity and transferability property as was found to be the case for the linear polarizabilities. This implies that the second order susceptibility is entirely determined by the asymmetric charge distribution within the constituent repeat polarizable unit and not by charge delocalization across such units. This greatly simplifies the search of materials with large second order nonlinearities on one hand and, on the other, points the road for fabricating new ones starting with molecules that possess large second order polarizabilities. These remarks have been essential in improving the second order nonlinear coefficients of existing materials by judicious modification of the constituent unit and in conceiving new materials by using appropriate molecular engineering techniques.

4.2.1. *Covalent crystals. Heteropolar semiconductors.* This is the simplest class of noncentrosymmetric crystals[67]. The building block unit here is one heteropolar bond that serves to form the whole crystal through tetrahedral connection ; the bonding is sp^3-type and the structure is cubic zinc-blende (ex GaAs) or hexagonal wurzite (ex. CdS) the former being optically isotropic while the latter is uniaxial and differs from the former by compression or expansion of the bonds along the 111-direction which then becomes the c-axis. Their valence electronic structure has been extensively studied by different approaches. Using the bond additivity conjecture and making provisions for local field corrections one may express[9,53] their linear and second order optical coefficients in terms of those of the heteropolar bonds namely the linear and second order polarizabilities α and β respectively ; one actually may assume the bonds unidimensional. The values of $\chi^{(2)}$ can be thus predicted to high accuracy and related to the charge asymmetry along the heteropolar bond. Using (94) One can show[9,54] that for an isoelectronic series of IV-IV, III-V, II-VI and I-VII compounds the III-V compound has the optimal second order nonlinearity ; furthermore the d-hybridization of the sp^3-orbitals, (ex.CuCl), leads to a reversal of sign of $\chi_E^{(2)}$.

As a general rule $\chi^{(2)}$ is large in these materials but so is also $\varepsilon_\alpha = n^2 = 1 + 4\pi\chi^{(1)}$ so that their figures of merit are not impressively larger than those of other materials we will discuss below. Their most severe drawbacks however are their optical isotropy, which prevents phase matching, and their narrow transparency region that hardly extends beyond 2 eV for any of them. Although the first one can now be compensated by fabricating periodic stacked structures with the polar axis direction periodically reversed from layer to layer, the second one remains a major drawback because real transitions and photocarrier generation can easily take place in these materials which greatly reduce their electrooptic modulation efficiency. With new fabrication techniques[68] that have been introduced in semiconductor technology however these materials have surfaced again in the competition for the development of nonlinear integrated optical devices and compounds like InP, which possess a large electrooptic coefficient and a wide energy gap, are now seriously considered there.

4.2.2. *Ionic crystals. Crystals with oxygen polyedra.* This is the class of materials that presently sets the standards in designing prototype integrated nonlinear devices. These can be viewed as ionic crystals[39,67] where one of the ions is replaced by an oxygen polyedron AO_n, in a ionized state, and stabilized in a noncentrosymmetric configuration by the surrounding ions (usually of metal elements). The second order optical nonlinearity here results from the unbalanced system of heteropolar AO- bonds which has a rather complex configuration around the element A and in contrast to the heteropolar semiconductors the bonds here cannot be assumed unidimensional. Indeed in all models, in order to account for the magnitude and sign of $\chi^{(2)}$, one must introduce[39] for each bond a substantial transverse contribution $\beta_{\perp}$ for the second order polarizability ; this is because the AO bonds are usually counterdirected in pairs and their longitudinal components $\beta_{\parallel}$ cancel to a large extend. Actually the additivity of bond polarizabilities has never been convincingly established for these compounds. Rather one should treat the whole AO_n polyedron as a single unit but this makes the calculation and prediction of the nonlinearities quite cumbersome and subtle since the modelisation of the electronic distribution must be compatible with the one expected from the phase transitions that almost all these materials undergo, e.g. para-to ferroelectric.

As a general rule the values of $\chi_E^{(2)}$ for these materials are lower than those of the heteropolar semiconductors but most of their other bulk properties are superior to those of the semiconductors. In particular they are optically anisotropic allowing for phase matching of nonlinear interactions, their transparency region can extend up to the near ultraviolet, they are robust against photochemical degradation or optical breakdown and can be produced in high optical quality. Because of their complex chemical structure, a superposition of ionic and covalent ones, however they cannot be processed by the same sophisticated fabrication techniques as the semiconductors ; furthermore their surfaces, because of their high polarity and the establishment of charged layers are easily attacked by moisture and other degradation and must be properly treated, encapsuled or packaged.

By weigthing all these factors at presently only $LiNbO_3$ seems to gain the favor of the potential manufacturers and actually constitutes the reference for comparing and evaluating all other potentially useful materials. The progress however in growing other oxygen polyedron based crystals with larger second harmonic efficiency than $LiNbO_3$ has been spectacular[69] over the last years and has led to many applications in ultra short laser pulse technology[22] and may well lead to similar improvements in nonlinear integrated optical devices that exploit the electrooptic effect. Certainly here the introduction of new crystal growth and fabrication techniques will be needed in order to reach the standards required for integrated optical devices.

4.2.3. *Molecular crystals. Organics*. Much expectations were laid[70-73] on noncentrosymmetric molecular crystals, in particular the organic ones, for second order nonlinear effects and a considerable effort has been concentrated the last two decades on identifying such crystals or growing new ones with second order nonlinearities higher than those of $LiNbO_3$. The results have been mitigated ; such molecular crystals with large values of $\chi^{(2)}$ formed by organic molecules have now been grown but their other performances fall short of all expectations and their potential implementation in nonlinear integrated optical devices seems problematic and still open to debate. The major limitation is their photochemical instability and aging against long term exposure to intense light pulses at high repetition rate as required in most applications.

All advantages but also disadvantages of the molecular organic crystals with respect to the covalent and ionic ones for applications in nonlinear integrated optics can be traced to the difference between inter- and intramolecular forces that prevail in these systems. As a consequence the constituent molecules preserve their main characteristics even when assembled to form the molecular crystal. Indeed, in contrast to the covalent and ionic crystals where inter- and intracell forces are similar and of equal strength, in the molecular ones the intramolecular forces are covalent directive and already saturate with the formation of the molecule while the intermolecular ones are essentially unsaturated van der Waals or dipole-dipole interactions which are more than one order of magnitude weaker than the intramolecular ones and do not significantly alter the intramolecular electron distribution. In several cases other forces like the hydrogen bonding[70], intermediate in strength between the previous two, are operative.

Thus the process of search and growth of noncentrosymmetric molecular crystals with large second order nonlinearities at first sight is greatly facilitated by these facts since it is reduced to that of the identification and synthesis of stable asymmetric molecules with large second order polarizabilities β and wide transparency range between the highest vibrational and lowest electronic transitions. The latter requirement is very restrictive regarding organic molecules and excludes the majority of them for further consideration in nonlinear optics. Among the remaining ones a large proportion is also excluded because they cannot form stable noncentrosymmetric crystals. This is mainly because asymmetric molecules most frequently carry a dipole moment in their ground electronic state and in order to reduce the dipole-dipole interaction, which is dominant over the van der Waals in the lattice, a head- to-tail antiparallel configuration will be favored when forming the crystal most frequently resulting in centrosymmetric crystalline structures and consequently vanishing $\chi_E^{(2)}$ unless certain precautions[70] are taken to prevent this to happen. There are several features that can be exploited in this respect.

* polling in an externally applied field,

* synthesis of chiral molecules,
* synthesis of asymmetric molecules with vanishing dipole moment in the ground electronic state,
* hydrogen bonding.

The first one is not very favorable for crystalline structures but has been successfully applied and exploited in polymers(74-76) as will be discussed below. The other three, on the other hand, have been successfully used for growth of good optical quality asymmetric crystals with large second order susceptibilities. We cursively present the most representative crystal cases namely MAP (methyl-(2,4-dinitrophenyl)-aminopropanoate), POM (3- methyl-4-nitropyridine-1-oxide) NPP (N-(4-nitrophenyl)-(L)-prolinol) and urea ($CO(NH_2)_2$).

Because of the wide difference between inter- and intramolecular forces in molecular crystals one may safely assume that the molecular polarizability coefficients α and β satisfy the additivity property. For most organic molecules the dominant contribution to β comes from an asymmetric part of the electronic distribution which can be modelled with a quantum two-level system ; the expression of β was derived in (95). The coefficient β is a third rank tensor and clearly its maximum value is obtained for directions where $\Delta\mu$ and μ_{eg} are maximal but not necessarily collinear. For most organic molecules the most favorable case occurs when a charge transfer complex is formed(70) and this can be modelled with a two level system involving a nonbonding π-donor orbital and a vacant π-antibonding acceptor. The important point to notice in (95) is that β only depends on the difference $\Delta\mu$ and not on the individual dipole moments μ_{gg} and μ_{ee} ; in particular μ_{gg} can be zero and yet have $\beta \neq 0$ as long as $\mu_{ee} \neq 0$. Accordingly one can have molecules with large β but vanishing μ_{gg} in which case dipole-dipole interactions in the lattice are weak and cannot prevent the noncentrosymmetric lattice configuration to be formed. Such is the case of POM(77) and presently good quality large crystals for this compound have been grown and used in second harmonic generation.

This situation however is not common and one must cope with molecules having large β-values but also appreciable μ_{gg} -values ; the remedy then is either to render the molecule chiral by the adjunction of an appropriate molecular unit, either left- or right- handed, in which case the lattice structure can never be centrosymmetric, this is the case of MAP(78), or and MNA(79) one can judiciously introduce hydrogen bonding between the molecules to overcome the dipole-dipole forces and combine with chirality to favor noncentrosymmetric structures as in the case of NPP(80). Presumably this situation is prevailing also in the urea(81). In all three cases very good quality crystals with large $\chi_E^{(2)}$ have been grown and used in several applications ; urea seems to be the most stable of them. But clearly these approaches can be used for a wide range of organic materials.

Despite a formidable pluridisciplinary effort(70-72) the bulk organic and molecular crystals, with the possible exception of urea(81), have not reached yet the required standards for applications in nonlinear optics and much less so in integrated optics. The reasons are several and can be traced back, as was previously pointed out, to the weakness of the intermolecular forces with respect to the intramolecular ones and to the sensitivity of organic molecules to light. As a consequence the organic crystals have much larger defect density than the covalent or ionic crystals ; also their mechanical and optical resistance is weaker than in the inorganic crystals. The second issue is even more serious as all organic molecules easily undergo bond breaking, bond arrangement or other photochemical reactions once the photon energy or a multiple thereof is close to a photosensitive transition. In addition organic crystals are apt to aging even without exposure to light. Another serious drawback regarding the use of organic crystals in nonlinear integrated optical devices is the difficulties encountered in doping, processing, polishing or interfacing them with techniques currently used in semiconductors, ionic and other inorganic crystals.

4.2.4. *Disordered oriented media.* Crystallinity is not an imperative requirement for second order effects to occur in a material. It is sufficient for the constituent asymmetric molecules to be oriented on the average so that the medium macroscopically lacks inversion symmetry. This implies that the requirement of translational periodicity is irrelevant as long as the molecules are uniformly distributed but with their axis pointing on the average in the same direction. This can also occur in an otherwise centrosymmetric medium, if one uniformly induces oriented noncentrosymmetric molecular complexes. These two approaches have been successfully put into practice in the case of poled polymers(71,74,75,76) and in certain glasses(82,83). While the first class(74-76) was the outcome of a well thought strategy that led to materials and prototype devices that compete or even surpass those based on $LiNbO_3$ the second class was an unexpected(82,83) finding and the origin of the second harmonic generation there is still not well understood and accounted for with the different models that have been proposed. We remind that solid polymers(84) and glasses can be prepared in excellent optical quality and all show a phase transition around a temperature T_g to a plastic phase where collective flexibility and molecular motion at large distances can take place ; below this temperature the medium is rigid and only slight local motion is allowed like in a crystal. T_g can be very high, well above the room temperature.

a. *Poled polymers.* Here organic molecules with large charge transfer type polarizability β are introduced and uniformly dispersed(76,84) into an excellent optical quality amorphous organic polymer. The solution of the polymer and guest molecules is then evaporated by spin coating, film casting or other techniques ; at this stage the guest molecules are randomly oriented. They are oriented by poling, namely by applying a strong static electric field (electrode poling) or coronar discharge (corona poling) at a temperature T close to T_g ; by proper choice of the polymer its T_g can be very high, much higher than the extreme temperatures that a transparent nonlinear material will be expected to operate in practical applications. Several other precautions allow one to obtain excellent optical quality poled polymers, photoresistant and long aging. Concerning the guest molecule relation to the polymer several situations are possible[76,84] :

* guest-host system : here the nonlinear molecule is not attached to the polymer chain and thus can freely rotate close to T_g,
* side-chain (co) polymers : here one end of the nonlinear molecule chemically reacts and is covalently attached or grafted on the polymer chain but its other end is free,
* cross-lined polymers : here both ends of the nonlinear molecule chemically react and are covalently attached with the polymer chains,
* main-chain systems : here the guest molecule is inserted in the chain.

The poling efficiency depends on several factors but the most important one is the ratio $\mu_{gg}E/kT$ where E is the poling electric field ; corona and electrode poling have their own advantages and disadvantages. The techniques are well understood and controlled and allow one to obtain poled polymers with large orientational order for the guest molecules namely large $<\cos^3\theta>$ where θ is the angle between the guest molecule axis and the poling field E direction.

Poled polymers exhibit[76,84] substantial advantages over all crystalline materials regarding most requirements, nonlinearity, optical quality, processability, interfacing, phase matching etc..., except for one : their stability against aging is not yet satisfactorily solved. Clearly this is connected to the excess of T_g over the highest temperature that the material reaches under practical operation. The best results were obtained with the cross linked polymers where aging can be retarded to any desirable length of time but more recently equally good results where also obtained with side-chain polymers. The aging manifests itself with the lowering of $<\cos^3\theta>$ namely the guest molecules lose preferential orientation and disorientation and $\chi^{(2)} \sim N\beta<\cos^3\theta>$ decreases. Such a decrease of $\chi^{(2)}$ always occurs immediately after the poling is interrupted and the material is cooled to the room or the operating temperature ; however after a few hours or days the average orientation parameter $<\cos^3\theta>$ and $\chi^{(2)}$ stabilize to values that are well above those of most crystalline noncentrosymmetric materials inclusive $LiNbO_3$ and can remain so for years with a judicious choice of the guest molecule and the polymer. Another problem that recently surfaced in connection with their implementation in electrooptic waveguided devices is the occurrence[85] of a photorefractive effect which is not well understood and clearly sets limitations in the long term operation since it implies the existence of charged molecular defects. Notwithstanding these and some other drawbacks, poled polymers presently constitute the most serious and advantageous contenders in the competition and development of nonlinear integrated devices.

b. *Glasses*. Glasses and in particular fused silica have now reached outstanding optical quality, more than any other optical material, as their introduction in modern technology becomes increasingly sophisticated and irreversible. These materials were always expected to be centrosymmetric macroscopically and the observation[82] presently of second harmonic generation in Si-Ge glass fibers, but also in other glassy media, came as a surprise and aroused immense interest and expectations[86]. Despite the substantial and understandable effort in delineating the mechanisms that provoke this effect its understanding is still not satisfactory and falls short of our capability to control and exploit it. The proposed models like electric-field induced nonlinearities, the static electric field being provided by a third order rectification process, or the photovoltaic effect based on the interference between the fundamental and harmonic field, or other mechanisms, do not simultaneously account for all observations related to this effect.

The fact is that a glass fiber of particular constitution, for instance Si-Ge, which initially does not show second order generation after irradiation with a laser beam within a certain time lapse and over a certain distance that never exceeds a few tens of cms evolves to a fiber that subsequently allows instantaneous and highly efficient second harmonic generation at the same wavelength as at its preparation stage ; this is because of a built in photoinduced spatial modulation among with noncentrosymmetricity that allows phase matching at a given wavelength. The SH conversion efficiency can be very large more than 20 %. It has not been proved yet that the same fiber can also be used for efficient electrooptic modulation or other applications related to the electrooptic effect.

More recently it was shown[83] that a large second order nonlinearity less than but still comparable to that of $LiNbO_3$ can be induced within a layer near the surface region of commercial fused silica optical plats at a temperature close to the Orbach temperature and poling process in the presence of a static field ; the layer can be of the order of a few μm namely of the order or larger than typical optical wavelenghs.

4.3. THIRD ORDER NONLINEARITIES

The main frequency preserving third order effect is the optical Kerr effect which is related to the third order polarization[9,10,15]

$$\underline{P}_{\omega}^{(3)} = 3\underline{\underline{\underline{\chi}}}^{(3)}(\omega, -\omega', \omega')\,|\underline{E}_{\omega'}|^2\,\underline{E}_{\omega} \tag{103}$$

where ω and ω' can both be optical frequencies but ω' can also be zero in which case the effect reduces to the static or second order Pockels effect also defined in (44b). It is also convenient to introduce the related the intensity dependent indices $\tilde{n}(I) = n_0 + n_2 I$ and $\tilde{\alpha}(I) = \alpha_0 + \alpha_2 I$ defined in (33) and (34), in section 2.

There are several mechanisms[9] that can lead to such photoinduced changes of the refraction or absorption : electronic, vibrational, orientational translational. With the exception of the first one all the others involve ionic or molecular motion and can be substantial in magnitude but, when all factors are taken into consideration, these are not exploitable in applications. Hence the following discussion will only concern the electronic cubic nonlinearities.

In line with the introductory discussion of section 3.2.4 the electron delocalization over several identical units has been identified(70) as the principal material feature that markedly influences the magnitude of the cubic nonlinearities ; this is in contrast to the case of quadratic nonlinearities where the charge asymmetry within a single unit is the material feature that most matters. The main implications are that the structure must be periodic at least over the electron delocalization length and the polarizability additivity which is a real space property is irrelevant for materials with large cubic nonlinearities like the inorganic semiconductors and the linear conjugated polymers. Instead in such materials where the electron states are delocalized band states one can establish(87) an additivity in k-space which involves the contributions to $\chi^{(3)}$ from a small number of critical points in the joint density of states. This allows one to derive scaling laws that strongly depend on the electron charge dimensionality. In such materials with very delocalized electrons the cubic nonlinearities(88,89) can be sensitive to many body effects like the Fermi exclusion principle or change screening.

It has been suggested(9,48,49) that third order nonlinear processes can also take place through two cascading second order processes if the material is noncentrosymmetric. Here the electric field generated by a second order polarization, induced by any pair of incident beams inside the material, interacts with one of the incident beams to create a second order polarization which however is effectively equivalent to a third order one and can be cast in the form (103) and clammed together with this term ; the corresponding effective third order nonlinearity which is now nonlocal and depends on the wavelengths can be comparable or larger than the intrinsic cubic nonlinearity in particular if phase matching is achieved at the intermediate second order process.

In general the values of $\chi^{(3)}$ (ω, -ω', ω') due to the electrons, when ω and ω' are in the transparency region of the material, are very weak to be exploited for applications involving interaction that occur over a short distance as the case is in miniaturized single pass integrated devices. The only alternative is to resonantly enhance the nonlinearities. The use of resonances however introduces two drawbacks which in general can be very severe : linear and/or nonlinear absorption losses on one hand and on the other the nonlinearity recovery time is long. One can resonantly enhance $\chi^{(3)}$ (ω, -ω', ω') by having ω or ω' (or both) close to dipole allowed resonances of the material ; here the absorption losses are linear in the light intensity and nonlinear at high intensities (saturation effect). One can also enhance $\chi^{(3)}$ (ω, -ω', ω') by having $\omega + \omega'$ (or $\omega - \omega'$) close to a two-photon (or Raman) allowed resonance without any of the frequencies ω or ω' being close to a resonance ; here the losses are nonlinear in the light intensity.

These drawbacks related to the resonant enhancement of the cubic nonlinearities can be brought under control to some extent by exploiting(65,89,91) morphological resonances whose position, width, oscillator strength can be externally controlled and tailored to meet certain requirements related to the desired performances of the device. Such morphological resonances can be introduced through confinement of the delocalized electrons, for instance the quantum and dielectric confinements. The new artificial material fabrication techniques allow[68,69] one to control the interfacing and confinement to better than an atomic layer. Several classes of such artificial semiconductor based microstructures are now studied for this purpose and in the case of quantum wells these studies have reached a high degree of sophistication. For applications however the artificial materials that emerge as the most promising[65,91] are the composites obtained by uniformly dispersing semiconductor or even metal nanocrystals in a glass or in a transparent polymer matrix of high optical quality.

Based on the above general remarks the materials that presently show promises in devices that exploit cubic nonlinearities are

* covalent semiconductors,
* composite materials,
* linear conjugated polymers,
* noncentrosymmetric media with large $\chi^{(2)}$.

We shall cursively discuss the main nonlinear mechanisms that are responsible for large cubic nonlinearities and how they relate to the electron distribution and dynamics.

4.3.1. *Semiconductors*. Covalent semiconductors like Ge or GaAs possess[92] the largest nonresonant cubic nonlinearities among all known crystals but these are unexploitable for several reasons, the most severe being the low absorption threshold in these compounds. On the other hand large and exploitable cubic optical nonlinearities can be produced[88,93,94] by generating a finite concentration of electron-hole pairs by resonant excitation above the band gap.

These large resonant nonlinearities have their origin in the band filling mechanism whereby the photocreated electrons and holes quickly thermalize and fill all states in the bottom of the conduction and the top of the valence bands respectively up to levels that depend on the light intensity and pulse or recombination times, excluding thus these states from further occupation because of the Fermi principle. This blocking mechanism[88,94] appears as a repulsion of the states on either side of the forbidden energy gap or equivalently as a blue shift of the absorption threshold ; at very high intensities this mechanism leads to a saturation of the nonlinearity.

This band filling mechanism is not the sole cause of nonlinearities ; it is accompanied[88,89] by additional many-electron effects, like exciton formation and band renormalization, that have a certain impact on the global nonlinearity. These additional effects have received a lot of attention but their impact on device applications is overestimated and to some extent unreliable ; in particular their relative contribution to the total nonlinearity is greatly reduced at room temperature with respect to that of the band filling.

As previously stated the major problem with resonant nonlinearities and in particular the ones in semiconductors is the concommittant absorption losses and the long times required for the photoinduced nonlinearities to decay. The latter are related to the evolution of the electron and hole populations. These can be modified to some extent with appropriate doping or through the quantum confinement by using sophisticated but well established fabrication techniques of artificial materials.

4.3.2. *Composites. Semiconductor and metal nanocrystals in glasses.* A way to enhance the cubic nonlinearities of materials with very delocalized electrons, like metals, semiconductors, or conjugated polymers is to artificially confine the valence electrons in regions much shorter than their natural delocalization length in the bulk, which extends over many unit cells or even to infinity ; its more conspicuous feature is the appearance(65,89,91) of broad but discrete optical resonances whose position, oscillator strength, and dynamics depend on the extension of the artificial confinement and hence can be modified to meet certain requirements. These morphology(91) related resonances result from two types of confinement ; quantum and dielectric. The first one prevails in semiconductor nanocrystals while the second one in metal nanocrystals.

The dielectric confinement resides(95,96,65) in the difference of the dielectric constants between the crystallites and their surrounding transparent medium. Because of this dielectric inhomogeneity the electric field $E_{\omega\ell}$ that effectively polarizes the charges in these crystallites can be substantially different from the macroscopic Maxwell field E_ω in the composite. Under certain simplifying conditions one shows that the relation between these two fields is

$$E_{\omega\ell} = \frac{3\varepsilon_0}{\varepsilon_m(\omega) + 2\varepsilon_0} E_\omega \equiv f_\ell(\omega)\, E_\omega \tag{104}$$

and the effective cubic susceptibility of the composite is(95)

$$\tilde{\chi}^{(3)}(\omega) = p\,|f_\ell(\omega)|^2\, f_\ell(\omega)^2 \chi^{(3)}(\omega) \tag{105}$$

for spherical particles of volume concentration $p<<1$, dielectric constant $\varepsilon_m(\omega)$ and cubic susceptibility $\chi^{(3)}$ embedded in a transparent dielectric of dielectric constant ε_0. To the extent that only $\varepsilon_m(\omega)$ is complex and frequency dependent in the optical range the field E_ℓ in (104) is enhanced close to the resonance ω_s, such that

$$\mathrm{Re}\,\varepsilon_m(\omega_s) + 2\varepsilon_0 = 0 \tag{106}$$

which is the condition for the surface plasmon frequency ; the cubic nonlinearity (105) is enhanced by the fourth power of the same resonance. The position of the resonance is controlled by modifying ε_0 while its width depends on the size of the crystallites.

Careful studies(96,65) of the nonlinearities of composites formed by uniform dispersion of gold particles of different average sizes in glass matrices revealed that the quantum confinement is irrelevant in these compounds and that the nonlinearity $\chi^{(3)}$ of the crystallites results from essentially the saturation of the interband transition and the hot conduction electron population rearrangement.

The quantum confinement occurs(97,91) when the electron and hole envelops are restricted within a region of extension L equal to or smaller than the electron and hole Bohr radii, a_e and a_h respectively ; the latter are defined in the bulk by the condition that the average value of the electron or hole kinetic energy roughly equals that of the potential. The confinement perturbs this balance, since these energies now vary as $(a_c/L)^2$ and a_c/L respectively where $a_c = a_e, a_h$. The characteristic energy of the confinement is the kinetic energy

$$E_c = \frac{1}{2}\frac{e^2}{a_c}\left(\frac{a_c}{L}\right)^2 \tag{107}$$

and as L decreases E_c increases and gradually suppresses the effect of the other interactions. Otherwise stated the charges behave as free within the confinement region which to a good approximation can be considered as a spherical quantum well of infinite potential height. The main consequence is that the initially continuous energy spectrum is replaced(97) by a discrete one with a spacing that varies as (107) ; the widths of these resonances as well as their oscillator strengths can dependent on L. Because of the selection rules that prevail for the transitions between these quantum confined states each crystallite essentially behaves as a quantum two level system and the main nonlinearity results(98) from the saturation of these transitions. The interface and impurity states however can drastically modify this behavior. In this respect the technique employed to fabricate these composites plays a crucial role since it allows to control these states.

These materials are formed(65) by uniformly dispersing semiconductor nanocrystals, such as CdS_xSe_{1-x} in a glass where x varies from 1 to 0, but also other II -VI and I - VIII compounds as well can be used. The technique is a more or less thermally or chemically controlled precipitation in transparent solid matrices like glasses or polymers. It evolves in two steps ; the nucleation and striking stages. In the nucleation stage one first forms widely supersaturated glass melts with uniformly dispersed semiconductor clusters ; at this stage the resulting batch material is transparent as the initial undoped material, the glass on the polymer. In the ensuing striking stage, that occurs close to the Orbach temperature these clusters grow to crystallites by coalescence, namely larger clusters grow further at the expense of the smaller ones, and reach sizes where volume properties overtake surface properties and solid-state-like features are acquired. Several structural techniques have revealed that these crystallites beyond the nucleation stage have the same structural features as the bulk ones. The size distribution can to some extent be controlled by the temperature and duration of the striking process, and this also fixes the color of the doped glass. It has been found that the optical properties of these materials are strongly affected by a photodarkening process that takes place when they are exposed to high fluences[(99)]. More sophisticated growth techniques like growth in zeolites, on polymers or colloids together with controlled doping allow one to produce composites with narrower size distribution and improved characterization of the interface. Similar techniques are used to obtain metal nanocrystals uniformly dispersed in glasses or polymers.

Although the nonlinearities of these artificial composites and their response times can be tailored to arbitrary values by varying several parameters during the fabrication process their figures of merit are not significantly improved over those that prevail in the bulk semiconductors. However many of their other properties like robustness, interfacing, processability... are superior to those of the bulk semiconductors and this has led to intensive studies to implement them in device applications in integrated nonlinear optics ; with narrower size distribution of the crystallites, improvement in the doping process and interface characterization these materials will become the most serious candidates for applications there.

4.3.2. *Linear conjugated polymers*. There is a large class of conjugated polymers that can be obtained by several stereochemical processes. The most studied one is the polydiacetylene class[(100,71,72)]. Here one initially forms good quality crystals with diacetylene monomers R_1 - $C \equiv C$ - $C \equiv C$ - R_2 where R_1 et R_2 are appropriately chosen radicals and then polymerizes them by heat, mechanical shear force or UV radiation and obtains the conjugated polymer crystals which however contain defects of several types irrespective of the employed fabrication technique. The obtained crystals are photochemically rather stable and of good optical quality but occur only in small sizes and still poor mechanical resistance.

The nonresonant cubic nonlinearity along the chain direction in these compounds is among the largest known[(101)] and comparable to that of bulk semiconductors like Ge or GaAs ; in the direction transverse to the chain direction $\chi^{(3)}$ is two or three orders of magnitude lower. The magnitude and anisotropy of $\chi^{(3)}$ is due to the extensive delocalization that the valence π-electrons undergo along the chain direction. Visualizing these chains as unidimensional semiconductors and using a simple tight-binding model to describe their electron states one obtains expressions (see (97a,b)).

$$\chi^{(3)} \sim \left(\frac{E_F}{E_g}\right)^6 \tag{108a}$$

while

$$\chi^{(1)} \sim \left(\frac{E_F}{E_g}\right)^2 \tag{108b}$$

where E_F is the Fermi level and E_g the energy gap ; the parameter E_F/E_g plays a very important role as it measures the optical delocalization length and also the minimum chain length beyond which the nonlinearities are insensitive to the chain length. This is certainly important since these chains are never infinite even in the best quality crystals. The power behavior of the nonlinearity (108a) is intrinsicly related[87] to the unidimensional character of the electron distribution and is also reproduced with more sophisticated descriptions of the electron states. On the other hand the conjugation defects, like the Pople-Wamsley or solitons, seem irrelevant for the magnitude of the nonresonant cubic nonlinearities contrary to certain claims which were never substantiated. As in the case of bulk semiconductors the nonresonant cubic nonlinearities of the conjugated polymers are not exploitable mainly because of the inconvenient transparency range of these compounds.

The origin of the resonant cubic nonlinearity in these one dimensional organic semiconductors is still a question of debate[90,102] as is also the origin and width of the main absorption peak. This controversy is related to the uncertainties that still persist regarding the assignment of the levels in these conjugated chains and the inclusion of the delocalized of π-electron correlation effects ; another is the strong electron-phonon or vibronic coupling in these chains which introduces a polaron-like character in the electron states. Both these features lead to complicated state configurations and nonlinear mechanisms that are only partially supported by experimental findings. Thus state-state interaction mediated by phonons similar to the inverse Raman scattering has been claimed[102] as being the origin of the resonant cubic nonlinearity. Whatever the mechanism may be these π-electron conjugated polymers are photochemically instable[90] and their use in device applications becomes rather questionable.

4.3.4. *Cascading of second-order nonlinearities.* In noncentrosymmetric media the cubic nonlinearity besides the intrinsic contribution also contains a contribution that results by cascading of second-order nonlinearities. This is actually a retardation effect[9] that leads to a nonlocal or wave vector dependent nonlinearity. The origin was briefly described in the beginning of this section. Thus in the case of the optical Kerr effect in a noncentrosymmetric material the third order polarization at frequency ω besides the direct contribution (102) also contains two additional contributions : a contribution that results from a second order polarization induced by a field of frequency -ω and another at 2ω the latter being the field generated by a second order polarization at frequency 2ω induced by the field at frequency ω ; and a contribution that results from a Pockels effect induced by a static electric field generated by an optical rectification process. One can easily show that all these contributions lumped together lead to[9] an effective third order polarization of the same form as (102) but with a nonlinear susceptibility that now depends on the wave vector mismatch at the intermediate second order process.

These cubic nonlinearities can be equal to or larger[9,49] than the intrinsic ones in the nonresonant regime and are related to coherent processes. All noncentrosymmetric materials discussed in relation to the second order nonlinearities can be used here and the advantage with respect to the centrosymmetric ones is evident ; they benefit from the more developed material technology related to noncentrosymmetric materials with large second order nonlinearities and in addition the same class of materials will be used for quadratic and cubic nonlinearity based devices.

4.4. HYBRID NONLINEARITIES

Besides the applications that exploit the photoinduced or all optical modifications of the optical characteristics of a material a whole class of other or similar applications in integrated nonlinear optics can be envisaged that exploit optical parametric effects, where the modifications of the optical characteristics of the material are provoked by an external agent. Here we have in mind modifications and modulations of the characteristics mediated through electrooptic, acoustrooptic or magnetooptic coupling. The first two cases and the related devices and materials are well documented in the literature but much less so the third one and we briefly discuss here certain aspects of interest for integrated devices and the related materials.

Nonlinear magnetooptical effects can be treated within the framework of nonlinear optics in general but in line with the criteria stated in section we shall only consider the main frequency preserving effects namely the photoinduced Faraday rotation[41,103] and gyrotropy. These effects allow a very efficient photoinduced control of the polarization state of an optical field and the development of nonreciprocal optical devices like optical valves and others.

The photoinduced Faraday effect originates from the combined effect of the Faraday rotation and the optical Kerr effect. In an isotropic medium in the presence of a static magnetic field H_o the two eigenmodes of frequency ω in the direction of H_o are the left and right circularly polarized waves with indices n_- and n_+ respectively ; through the optical Kerr effect the latter become intensity dependent for high light intensities. Accordingly the polarization direction of a linearly polarized input wave E of frequency ω after propagation through a length L in such a medium collinear with H_o rotates by an angle

$$\tilde{\theta}_F = \theta_F + \Delta\theta_{NL}(I) \tag{109}$$

where θ_F is usual linear Faraday rotation angle $\theta_F = \omega L(n_- - n_+)/2c$ and $\Delta\theta_{NL}(I)$ is the photoinduced change of the latter and is proportional to the difference of the optical Kerr coefficients for left and right circular polarizations.

These photoinduced changes are in general very weak even in materials with large optical Kerr effect coefficients, like the semiconductors, unless certain provisions are made such that the difference of the optical Kerr coefficients for left and right polarizations is large. This can be done in II-VI semiconductors like CdSe or CdTe doped with magnetic impurities the so-called[(104)] semimagnetic semiconductors. Through the spin exchange interaction of the band and impurity electrons the Landé factor of the band electrons is enhanced by almost two orders of magnitude and similarly the magnetooptical coupling and the Zeeman splitting. Otherwise stated the magnetic impurities act as "local amplifiers" of the static magnetic field. Without optimizing the interaction configuration giant photoinduced Faraday rotations for instance exceeding 90° degrees have been measured in these compounds under resonant conditions around the liquid nitrogen temperature and for magnetic fields of the order of half Tesla ; these performances can be substantially improved to meet the device requirements.

Similar effects are expected without the presence of a magnetic field, in media with rotatory (gyratory) power and here one has a large choice of organic materials but also their severe problems regarding photochemical stability etc.... At presently the photoinduced natural gyrotropy has been studied in a couple of isolated cases[(105)] but the measured rotations were not sufficiently large.

4.5. GENERAL REMARKS

It would be an euphemism to state that there is no consensus yet regarding the class of nonlinear optical materials that will be used in integrated nonlinear optics. It is a fact that the choice of such materials cannot reside only on the figures of merit for the nonlinear operation but must also take into account many other criteria crucial for the design, production and maintenance of the nonlinear devices. Actually in view of the rather limited nonlinear optical performances of the available materials these other criteria will be the most crucial ones in making a choice. And this choice must be made with less ambitious goals than one would wish or had wished as our understanding of nonlinear integrated optics is progressing. There are now certain materials that can be used not only to conceive prototype nonlinear optical devices but also to proceed to their massive development and production and their insertion to large scale technology. In table 2 we summarize some values.

At first hand these concern materials and devices that exploit quadratic (second order) optical processes. $LiNbO_3$ and poled polymers here seem to provide a good starting point for such development. It is doubtful whether organic crystals at this stage can be used to initiate this process. There are still severe problems associated with their stability, long term optical quality ....

The situation is less bright for materials and devices that will exploit cubic (third order) optical processes. Here either the nonlinearities are weak or can be made large but with a high

Table 2 . Indicative* Table of second order nonlinear coefficients

Crystal	Symmetry	n_0	d_{21}	d_{14}	d_{33} (10^{-12}m/V) **	d_{eff}	6	I_{ob} (GW/cm^2) ***	Transm. (μ) ****
$LiNbO_3$	3m	2.232	-2.1	0	- 27	5.1	70	10	0.35 - 5
$BaBO_4$	3m	1.655	-2.3	0	0	1.9	16	14	0.2 - 2.6
$LiIO_3$	6m	1.857	0	0	4.5	1.8	13	2	0.34 - 4
KDP	$\bar{4}2m$	1.493	0	0.37	0	0.35	1	5	0.18 - 1.8
ADP	$\bar{4}2m$	1.509	0	0.47	0	0.39	1.2	6	0.18 - 1.5
$AgGaGe_2$	$\bar{4}2m$	2.594	0	33	0	28	81	0.3	0.78 -18
$ZnGeP_2$	$\bar{4}2m$	3.073	0	69	0	70	292	0.05	0.74 -12
KTP	mm2	1.737	0	0	8.3	3.2	47	15	0.35 - 4.5
$KNbO_3$	m2	2.119	0	0	- 19.5	-11	312	7	0.4 - 5.5
					d_{11}				
Urea					12			5	0.2 - 1.4
MAP		1.508	16.8					3	0.5 - 2.5
POM		1.663		9.2				2	0.5 - 1.7
MNA		2.0			168		1000		0.48 - 2.0
NPP					85			0.05	0.48~ 2.0

* values selected from Ref.69
** divide each value by $4.2 \cdot 10^{-4}$ to convert to esu
*** optical breakdown
**** transmission range

Indicative Table of third order nonlinear coefficient

Material	$\chi^{(3)}$ 10^{12} esu
glass	10^{-4}
NaCl	10^{-2}
CS_2	1
Si	1
GaAs	10
Ge	10^2
PTS	10^2
SDG* (resonant)	$>10^4$

* semiconductor doped glasses

price paid in absorption losses and reduced operation speed. At presently only the composites offer some limited possibilities but their fabrication technique must be substantially improved. The advantage of these materials is that one can artificially modify several of their properties by external control. Another possibility is to exploit cascading processes but here we need a better control of the different factors that affect these processes. The advantage with these processes is that one can use the same class of materials in devices that exploit quadratic and cubic nonlinear optical effects.

But one can still ask the question whether there is room of improvement of the figures merits by an amount that will allow one to use much lower beam intensities and thus avoid some of the problems related to material stability under high fluences. This seems a difficult task as nonlinearities are associated with very loosely bound electrons and hence the material is apt to undergo irreversible photochemical changes. The relation between the nonlinearity and the photochemical stability is still not understood and much work remains to be done there.

REFERENCES

1 - P.A. Franken, A.E. Hill, C.W. Peters and G. Weinreich, Phys. Rev. Lett. 7, 118 (1961)

2 - G. Fowles, Introduction to Modern Optics, Holt, Rinehart and Winston, New York 1975

3 - M. Born and E. Wolf, Principles of Optics, Pergamon Press, Oxford, 1975

4 - N. Bloembergen, Nonlinear Optics, Benjamin, New York 1964

5 - W.K.H. Panofsky, M. Phillips, Classical Electriciy and Magnetism, Addison. Wesley, New York, 1964

6 - J.A. Armstrong, N. Bloembergen, J.Ducuing and P.S. Pershan, Phys. Rev 127, 1918 (1962)

7 - J. Giordemaine, Phys. Rev. Lett. 8, 407 (1962)

8 - R.W. Terhune, P.D. Maker and C.M. Savage, Phys. Rev. Lett. 8, 404 (1962)
P.D. Maker, R.W. Terhune, M. Nisenoff and C.M. Savage, Ibid 8, 21 (1962)

9 - C. Flytzanis, in Quantum Electronics : A Treatise 1A, Eds H. Rabin and C.L. Tang, Academic Press, New York, (1975)

10 - see for instance Y.R.Shen, The Principles of Nonlinear Optics, Wiley, N. Y., 1984

11 - G.B. Whitham, Linear and Nonlinear Waves, Wiley, N. Y., 1974

12 - V.E. Zakharov and A.B. Shabat, Sov. Phys. JETP, 34, 63 (1972)

13 - see for instance A. Hasegawa, Optical Solitons in Fibers, Springer Verlag, Berlin, 1989 ; G.P. Agrawal, Nonlinear Fiber Optics, Academic Press, Boston, 1989 ;

14 - see also A.C. Newell and J.V. Moloney, Nonlinear Optics, Addison-Wesley, New York, 1992

15 - see also R.W. Boyd, Nonlinear Optics, Academic Press, New York 1992

16 - J.M. Manley and H.E. Rowe, Proc. IRE, 44, 904 (1956) ; ibid 47, 2115 (1959)

17 - see for instance J.F. Reintjes, Nonlinear Optical Parametric Processes in Liquids and Gases, Academic Press, Orlando, 1984

18 - H. Rabin and Bey, Phys. Rev. 156, 1010, 1967

19 - see for instance Y.R. Shen, Ann. Rev. Mat. Sci, 16, 69 (1986)

20 - N. Bloembergen, P.S. Pershan, Phys. Rev 128, 606 (1962)

21 - S.A. Akhmanov and R.V. Khokhlov, Sov. Phys. JETP 16, 252 (1962)

22 - see C.L. Tang, in Nonlinear Optical Materials Principles and Applications, Eds V. Digiorgio and C. Flytzanis, North-Holand, Amsterdam 1994

23 - see for instance Optical Phase Conjugation, Ed. R.A.Fisher, Academic Press, Orlando, 1983 ; B.Ya. Zeldovich, N.F. Pilipetsky and V.V. Shkunov, Principles of Phase Conjugation, Springer Verlag, Berlin, 1985

24 - R.W. Hellwarth, J. Opt. Soc. Am. 67, 1 (1977)

25 - F.S. Felber and J.H. Marburger, Appl. Phys. Lett. 28, 231 (1976) ; Phys. Rev. A17, 335 (1978)

26 - A. Szöke, V. Daneu, J. Goldhar and N.A. Kurnit, Appl. Phys. Lett. 15, 376 (1969)

27 - see for instance H.M. Gibbs, Optical Bistability, Academic Press, Orlando, 1985

28 - V.I. Bespalov and V.I. Talanov, JETP Lett 3, 471 (1966) ; R.Y. Chiao, E. Garmire and C.H. Townes, Phys. Rev. Lett. 13, 479 (1964)

29 - S. Maneuf, R. Desailly and C. Froehly, Opt. Comm. 65, (1988) ; Y. Silberberg, Opt. Comm. 15, 1005 (1990) ; J.S. Aitchison, Y. Silberberg, A.M. Weiner, D.E. Laeird, M.K. Oliver, J.L. Jackel, E.M. Vogel and P.W.E. Smith, J. Opt. Soc. Am. B8, 1290 (1991)

30 - G.B. Altshuler, V.S. Ermolaev, K.I. Krylov, A.A. Manenkov and A.M. Prokhorov, J. Opt. Soc. Am. B3, 660 (1986)

31 - N. Kothari and C. Flytzanis, Opt. Lett. 11, 806 (1986) ; 12, 492 (1987)

32 - A. Ishimaru, Wave Propagation and Scattering in Random Media, Academic Press, New York 1978

33 - J.P.E. Taran, H.A. Haus, J.R. Lisfitz, P.L. Kelley, Phys. Rev. 177, 306 (1969)

34 - A. Hasegawa and F. Tappert, Appl. Phys. Lett 23, 142 (1973)

35 - L.F. Mollenauer, R.H. Stolen and J.P. Gordon, Phys. Rev. Lett. 45, 1095 (1980)

36 - A. Hasegawa and J.Brinkman, IEEE.QE-16, 694 (1980)

37 - D.A. Kleinman, Phys. Rev. 126, 1977 (1962)

38 - L.D. Landau and E.M. Lifshitz, Electrodynamics of continuous media, Pergamon Press, New York 19960

39 - see for instance M.E. Lines and A.M. Glass, Principles of Ferroelectrics and Related Materials, Clarendon Press, Oxford 1979

40 - see for instance A. Yariv, Introduction to Optical Electronics, Holt, Rinehart and Winston, New York, 1971

41 - R. Frey, J. Frey, C. Meriaux, C. Flytzanis in Guided Wave Nonlinear Optics, Eds D.B. Ostrowsky and R. Reinisch, Kluwer, NATO ASI Series, 1992

42 - P.N. Butcher, D.Cotter, The Elements of Nonlinear Optics, Cambridge University Press, Cambridge 1991

43 - J. Jerphagnon, Phys. Rev. B2, 1091 (1970)

44 - S.M. Kogan, Sov. Phys. JETP 43, 217 (1963) ; P.J. Price, Phys. Rev. 130, 1792 (1963)

45 - D.A.B.Miller, C.T. Seaton, M.E. Price and S.D. Smith, Phys. Rev. Lett. 47, 197 (1981)

46 - C. Flytzanis, to be published

47 - P.S. Pershan, Phys. Rev. 128, 2903 (1962)

48 - E. Yablonovitch, C. Flytzanis and N. Bloembergen, Phys. Rev. Lett. 29, 865 (1972)

49 - G.I. Stegeman, M. Sheik-Bahae, E. Van Stryland and G. Assanto, Opt. Lett. 18, 13 (1993)

50 - E.M. Purcell, Electricity and Magnetism, Berkeley Physics Course, vol. 2 (1965)

51 - I.L. Fabelinskii, Molecular Scattering of Light Plenum, New York 1968

52 - E. Garmire, F. Pandarese and C.H. Townes, Phys. Rev. Lett. 11, 160 (1963)

53 - C. Flytzanis and J. Ducuing, Phys. Rev. 178, 1218 (1969)

54 - C. Flytzanis and C.L. Tang, Phys. Rev. B4, 2520 (1971)

55 - B. Davydov , L. Derkacheva, L.D. Dubina, V.V Zhabotinskii, M.E. Zolin, L.G. Koreneva and M.A. Samokhina, Sov. Phys. JETP. Lett. 12, 16 (1970)

56 - J.L.Oudar and D.S. Chemla, Opt. Comm. 13, 164 (1975)

57 - B.F. Levine and C.G. Bethea, J. Chem. Phys. 66, 1070 (1977)

58 - K. Rustagi and J. Ducuing, Opt. Comm. 10, 258 (1974)

59 - G.P. Agrawal and C. Flytzanis, Chem. Phys. Lett. 44, 366 (1976),

60 - G.P. Agrawal, Cojan and C. Flytzanis, Phys. Rev. B17, 776 (1978)

61 - See for instance Nonlinear Optics ; Materials and Devices, Eds C. Flytzanis and J.L. Oudar, Springer Verlag, Berlin 1986 ; Nonlinear Optical Materials Principles and Applications, Eds V. Digiorgio and C. Flytzanis, North Holland Co. Amsterdam 1994 (to appear)

62 - See other contributions in this volume : Nonlinear Optical Materials and Devices and their Applications in Information Technology, Eds A. Miller, B. Daino and K. Welford, NATO ASI Series, Kluwer Publ., Dodrecht 1994

63 - See contribution of G. Stegeman in this volume

64 - See for instance Nonlinear Optics in Signal Processing, Eds R.W. Eason and A. Miller, Chapman and Hall, London, 1993

65 - C. Flytzanis, F. Hache, M.C. Klein, D. Ricard and Ph. Roussignol, in Progress in Optics, Vol. XXIX, Ed. E. Wolf, Elsevier, Amsterdam 1991, p. 321-411

66 - D. Miller, Appl. Phys. Lett. 5, 17 (1964)

67 - See for instance C. Kittel, Introduction to Solid State Physics, J. Wiley, New York 1984

68 - See for instance The Physics and Fabrication of Microstructures and Microdevices, Eds M.J. Kelly and C. Weisbuch, Springer Verlag, Berlin 1988 ; Quantum Semiconductor Structures ; Fundamentals and Applications, Academic Press New York 1991

69 - F.C Zumsteg, J.D. Bierlein, T.E. Dier, J. App. Phys. 47, 4986 (1976) ; for a recent review see for instance F. Bordui and M.M. Fejer, Ann. Rev. Mat. Sci. 23, 321 (1993) ; G.I. Chen and G.Z Liu, Ann. Rev. Mat. Sci. 16, 203 (1986)

70 - See for instance Nonlinear Optical Properties of Organic Molecules and Crystals, Vols. 1 and 2, Eds D.S. Chemla and J. Zyss, Academic Press, New York 1987

71 - See for instance Organic Molecules for Nonlinear Optics and Photonics Eds J. Messier, F. Kajzar and P. Prasad, NATO ASI Series Kluwer, Publ. 1991 ;

P.N. Prasad and D.J. Williams, Introduction to Nonlinear Optical Effects in Molecules and Polymers, J. Wiley Interscience, New York 1990

72 - See for instance Principles and Applications of Nonlinear Optical Materials, Eds R.W. Munn and C.N. Ironside, Blackie Academic, London 1993

73 - B.L. Davydov, S.G. Kotovchikov, V.A. Neferlov,Sov. J. Quant. Elect. 7, 129 (1977)

74 - G.R. Meredith, J.G. Van Dusen and D.J. Williams, Macromolecules 15, 1385 (1982)

75 - K.D. Singer, J. E. Sohn, J. Lalama, Appl. Phys. Lett. 49, 248 (1986)

76 - See Möhlmann in Nonlinear Optical Materials : Principles and Applications, Eds V. Digorgio and C. Flytzanis, North-Holland, Amsterdam 1994

77 - J. Zyss, J.F. Nicoud and D.S. Chemla, J. Chem. Phys. 74, 4800 (1981)

78 - J.L. Oudar and R. Hierle, J. Appl. Phys. 48, 2699 (1977)

79 - B.F. Levine, C.G. Bethea, C.D. Thurmond, R.T. Lynch and J.L. Bernstein, J. Appl. Phys. 50, 2523 (1979)

80 - J. Zyss, J.F. Nicoud, and M. Coquillay, J. Chem. Phys. 81, 4160 (1984) ; R. Mase and J. Zyss, Mol. Eng. 1, 141 (1991)

81 - J.M. Halbout, S. Blit, W. Donaldson and C.L. Tang, IEEE J. Quant. Electr. 15, 1176 (1979)

82 - V. Osterberg and W. Margulis, Opt. Lett. 11, 516 (1986)

83 - R.A. Myers, N. Mukherjee and S.R.J. Brueck, Opt. Lett. 16, 1734 (1991)

84 - See contribution of Zerbi in this volume

85 - M.C.J. Donckers, S.M. Silence, C.A. Walsh, F. Hache, D.M. Burlaud, W.E. Moerner and R.J. Twieg, Opt. Lett. 18, 1044 (1993)

86 - R.H. Stolen and H. Tom, Opt. Lett. 12, 587 (1987) ; D.Z. Audersson, V. Mizrahi, J.E. Sipe, ibid. 16, 796 (1991)

87 - see C. Flytzanis in Ref. 70, also M. Cardona and F.H. Pollack in Optoelectronic Materials, Ed G.A. Albers, Plenum, New York 1971

88 - see for instance Optical Nonlinearities and Instabilities in Semiconductors Ed H. Haug, Academic Press, New York 1988

89 - S. Schmitt-Rink, D.S. Chemla and D.A.B. Miller, Adv. Phys. 38, 89 (1989)

90 - B.I Greene, J. Orenstein, S. Schmitt-Rink, Science 247, 679 (1990)

91 - see for instance C. Flytzanis and J. Hutter in Contemporary Nonlinear Optics, Eds G.P. Agrawal and R.W. Boyd, Academic Press, New York 1992

92 - Wynne J.J., Phys. Rev. 178, 1295 (1969)

93 - D. Weaire, B.S. Wherett, D.A.B. Miller and S.D. Smith, Opt. Lett. 4, 331 (1979) Miller, D.A.B. Seaton and S.D. Smith, Phys. Rev. Lett. 47, 197 (1981)

94 - B.S. Wherett, Proc. Roy. Soc (London) A390, 373 (1983)

95 - D. Ricard, Ph. Roussignol and C. Flytzanis, Opt. Lett. 10, 511 (1985), K.C. Rustagi and C. Flytzanis, Appl. Phys. Lett. 9, 344 (1984)

96 - F. Hache, D. Ricard, C. Flytzanis and U. Kreibig, Appl. Phys. Lett. A47, 347 (1988)

97 - Al.L.Efros and A.L Efros, Sov. Phys. Semic. 16, 772 (1982)

98 - D.A.B. Miller, D.S. Chemla and S. Schmitt-Rink, Phys. Rev. B35, 8113 (1987)

99 - Ph. Roussignol, D. Ricard, J. Lukasik and C. Flytzanis, J. Opt. Soc. Am. B4, 5 (1987)

100 - See for instance G. Wegner, Makromol. Chem. 154, 35 (1971)

101 - C. Sauteret, J.P. Hermann, R. Frey, F. Pradère, J. Ducuing, R.R. Chance and R.H. Baughman, Phys. Rev. Lett. 36, 956 (1976)

102 - B.I. Greene et al, Phys. Rev. Lett. 61, 325 (1988), G.J. Blanchard et al, ibid 63, 887 (1989)

103 - J. Frey, R. Frey and C. Flytzanis, Phys. Rev. B45, 4056 (1992)

104 - See for instance J.K. Furdyna, Appl. Phys. 64, R29 (1988)

105 - H. Ashitaka, Y. Yokoh, R. Shimizu, T. Yohozawa, K. Morita, T. Snehiro, Y. Matsumoto, Nonlinear Optics 4, 281 (1993)

NONLINEAR PHENOMENA IN OPTICAL FIBRES

N J DORAN
Electronic Engineering & Applied Physics
Aston University
Aston Triangle
Birmingham B4 7ET
UK

ABSTRACT.

A review of nonlinear optical phenomena in fibres is presented. The chapter briefly describes important linear properties of single mode fibres including dispersion and incorporates a discussion on amplified long haul communication systems. Fundamental nonlinear short pulse effects in fibres are described and applications to all-optical processing presented. Solitons as the stable digital modulation format for long haul systems are described and limitations on the information carrying capacity of soliton systems identified.

1 Optical Fibres and Optical Communications

Optical communications is now the established technology for trunk and long distance telecommunications. At the heart of every optical communications systems is the fibre which is silica based and usually single mode. The phenomenally low loss of the fibre and enormous bandwidth (which still remains largely under utilised) are the key features which enable this technology [1]. In this chapter we shall see that these very same fibres have enormous potential for nonlinear optics [2] and are not just the passive transmitters of optical signals they were originally designed to be. It is however the refinement of the technology of fibre fabrication for low loss communications which makes them particularly suitable as the nonlinear element in many potential applications [3]. Thus we shall start this chapter with a review of fibre properties relevant to their nonlinear applications.

1.1 PROPAGATION IN OPTICAL FIBRES

Fibres are thin cylinders of glass which guide light largely confined in a higher index core surrounded by a lower index cladding. The cores are small with radii comparable with the intended wavelength of operation (usually around 1 μm). Thus their light guiding properties are best described by solving the electromagnetic wave equations. Fibres are primarily designed to passively transmit light and have thus been developed with extremely low loss. The lowest loss

A. Miller et al. (eds.), Nonlinear Optical Materials and Devices for Applications in Information Technology, 75–102.

occurs in the near infra red at around 1.5 μm where commercially available fibre has a loss of 0.2 dB/km. This loss is dominated by a Rayleigh scattering loss (about 90% at this wavelength) and means that 15 km of fibre will transmit 50% of launched power. A loss plot of a "typical" single mode fibre is shown in the Figure 1. The increased loss around 1.4 μm is from a harmonic of an 0-H vibration absorption line, it is thus important that the fibres are fabricated in a "dry" environment.

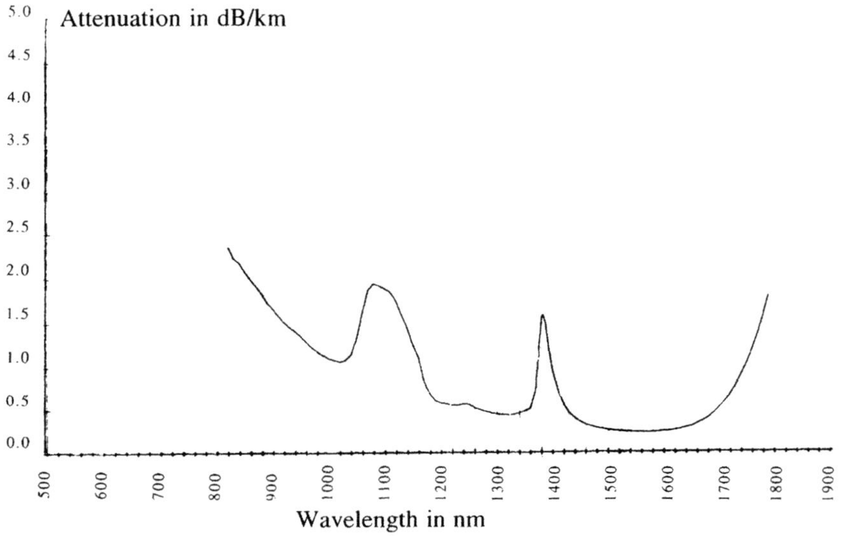

Figure 1. A loss plot of a typical fibre single moded in the 1.3 and 1.5 μm windows.

We now turn to a description of the basic light guiding properties of fibres. A typical fibre consists of a core of refractive index n_1 and a cladding of lower index n_2 ($n_1 - n_2 \sim 10^{-3}$). The modes of the fibre can be determined by solving the radial wave equation

$$\nabla_t \Psi + \left(n^2 k^2 - \beta^2\right)\Psi = 0 \quad (1)$$

where Ψ is one of the six electromagnetic field components and ∇_t the transverse Laplacian. The propagation constant β and the mode field can be obtained by solving the boundary condition problem at the core-cladding interface. The bound solutions in the core are oscillating J Bessel functions and in the cladding K Bessel functions.

The continuity of Ψ results in

$$\Psi(r,\phi) = \begin{cases} \dfrac{J_l(u\frac{r}{a})}{J_l(u)} \cos m\phi & \text{core} \quad r<a \\ \dfrac{K_l(w\frac{r}{a})}{K_l(w)} \cos m\phi & \text{cladd} \quad r>a \end{cases} \tag{2}$$

where $u = \beta^2 - n_1^2 k^2$
$w = n_2^2 k^2 - \beta^2$

For the continuity of $\frac{\partial \Psi}{\partial r}$ $(r = a)$ we obtain

$$\frac{u J_l'(u)}{J_l(u)} = \frac{w K_l'(w)}{K_l(w)} \tag{3}$$

The cut-off of a mode occurs when it is not possible to solve the above equality. Using the V-value which is a useful measure of the degree of guidance sometimes called the normalised frequency

$$V = \frac{2\pi a}{\lambda}\left(n_1^2 - n_2^2\right)^{1/2} \tag{4}$$

The cut-off occurs for

$$\begin{aligned} J_{l-1}(V_c) &= 0 \qquad l \geq 1 \\ J_1(V_c) &= 0 \qquad l = 0 \end{aligned} \tag{5}$$

The modes are usually identified as LP_{lm} where m is the mth mode corresponding to the mth zero of the Bessel function. The fundamental mode LP_{01} has no cut-off at finite wavelength and the next order mode is LP_{11} with a cut-off at V = 2.41. Thus provided $V \leq 2.41$ the fibre will be single moded. In fact LP_{01} is a two fold degenerate mode in a circular fibre.

For nonlinear optical application it is usually important to work with single mode fibres. In addition small field widths will allow greater nonlinearity in the fibre for a given launched power. For communication systems fibres are designed with a V-value around 2. If the V-value is too small, although single mode behaviour is guaranteed, the fibre will be bend sensitive and not useful in practical environments. Thus they are designed to operate near to, but not at, the higher order mode cut-off.

The key characteristic of a single mode fibre is the mode propagation constant obtained as indicated above. This constant will be wavelength dependent and can be expanded around the frequency of interest, ω_o as

$$\beta = \beta_o + \beta'(\omega - \omega_o) + \frac{1}{2}\beta''(\omega - \omega_o)^2 + \ldots\ldots\ldots \tag{6}$$

where

$$\beta' = \frac{d\beta}{d\omega} = \frac{1}{\text{group velocity}} \tag{7}$$

and $\beta'' = \frac{d^2\beta}{d\omega^2}$ is the first order dispersion. This coefficient is quite small in standard fibres around the minimum loss region of 1.5 μm (typically $\beta'' \sim 20ps^2/km$ at 1.55 μm). Nevertheless this dispersion is an important factor in designing communication systems.

Figure 2 shows a measured dispersion in a standard fibre. In the figure the group delay is plotted as a function of wavelength and the dispersion, now group delay dispersion, calculated from that. The dispersion parameter D (in ps/nm/km) illustrated is related to β'' by

$$D = \frac{d\beta'}{d\lambda} = \frac{-2\pi c}{\lambda^2}\beta'' \tag{8}$$

The dispersion is seen to go through zero around 1.3 μm changing sign as a function of wavelength. The effect of dispersion on optical pulses will be to broaden them in time and to introduce a frequency "chirp" across the pulse. In the wavelengths longer than the first order dispersion zero the longer wavelengths travel slower so that the chirp in the pulse has higher frequencies at the leading side of the pulse. This is sometimes called anomalous dispersion and is best described as positive group delay dispersion or negative group velocity dispersion. In wavelengths shorter than λ_o all the statements are reversed and the sign on the dispersion inverted.

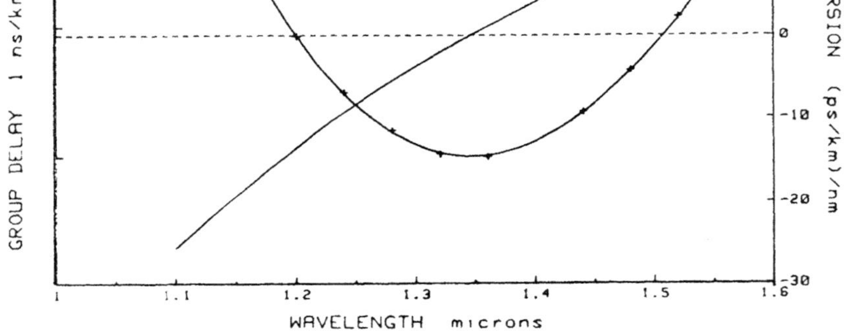

Figure 2. Measured group delay dispersion in a single mode fibre.

The propagation constant β_o can be expressed as an effective index of the mode $\beta = n_{eff} \times \frac{2\pi}{\lambda}$. This effective index ranges between n_1 and n_2. It is possible to alter the value of λ_o by suitable fibre design. In particular it is possible to arrange for λ_o to occur at 1.55 μm the minimum loss in the fibre. This is called dispersion shifting and is important since now the minimum dispersion can be shifted to coincide with the minimum loss of 1.55 μm. The propagation constant switches from that of the higher index core to that of the lower index cladding as the wavelength increases. If the transition is arranged to occur near the material λ_o significant alteration of the effective λ_o can occur.

It is important to know the mode field area in a fibre. This is particularly so for nonlinear studies where the nonlinearity depends on the intensity and thus the field area. The most common method for obtaining the field width is to perform a "best" gaussian fit to the field and to define the $\frac{1}{e}$ width. A well-known approximation to this field width is given by Marcuse as

$$\frac{W}{a} = 0.65 + \frac{1.619}{V^{1.5}} + \frac{2.879}{V^6} \tag{9}$$

This equation usefully illustrates the behaviour of the mode field as a function of core radius. For small a $W \alpha \frac{1}{a^5}$ and thus the field expands rapidly as "a" is decreased; for large a $W \alpha$ a and the field expands to match the waveguide size. There is thus a minimum field width between these two regions. Optimum nonlinearity may thus be obtained by designing fibres with the minimum field area.

For nonlinear studies the appropriate field area is not the above but a nonlinear effective area defined by [2]

$$A_{eff} = \frac{\left[\int\int|\Psi|^2 \, r \, dr \, d\phi\right]^2}{\int\int|\Psi|^4 \, r \, dr \, d\phi} \tag{10}$$

This can be calculated for a particular fibre once the field Ψ has been determined. For typical communications fibres this area is around 50 μm^2 at 1.5 μm.

1.2 COMMUNICATIONS SYSTEMS

Optical telecommunications systems are designed to optimise information flow over long distances. The desired characteristics are high data rate, long distance and flexibility. Systems may be limited by loss, dispersion, amplifier induced noise, nonlinearity or a combination of these. The most common limiting mechanism is fibre loss; the advent of Er fibre amplifiers means that this is now a limit which may be overcome. Dispersion too may be reduced by suitable design and since it is a linear process may be compensated for exactly using a number of schemes [4,5,6,7]. This latter approach is yet to be applied in commercial systems.

In the last 20 years the data rate x distance product has increased exponentially for both experimental and installed systems. The more recent systems employ amplifiers and are limited by nonlinearity and amplifier noise. Amplified transoceanic systems will be installed in 1995/6 and will operate without electronic repeaters for their entire length.

It is illustrative to examine the commercial potential of these amplified transoceanic systems. TAT12/13 will be installed in 1995/6 and will operate at 5 Gbit/s on each of two fibres. The link will form a major highway from the US to Europe and is illustrated in Figure 3. The total system cost is $750m and it might be expected that such a system would have a lifetime of around 25 years. If the system has two active fibres the total capacity is 10 Gbit/s. For an assumed digital telephone channel of 64 kb (this is conservative and allows for no compression techniques) and if we assume a charge of $1/minute (this may be a little above that which many companies charge) per channel the system will generate

$$\frac{10 \times 10^9}{64 \times 10^3} \times 60 = 10^7 \ \$ / \mathrm{hr} \tag{11}$$

It thus takes of order 75 hours for the system to generate revenue equal to its capital cost with the assumption of 100% filling of the capacity. Of course the commercial arithmetic is much more complex. The above analysis does not for example take any account of other costs of connecting an international call nor any maintenance costs. Nevertheless, this does illustrate the high revenue generating potential of such systems. The design and operation of the systems is largely determined by the nonlinearity, dispersion and amplifier noise considerations. Thus the fibre nonlinearity is now a major commercial consideration and we shall see later in the chapter how the next generation of soliton communication may be designed to enhance even further the capacity of such amplified transoceanic communication.

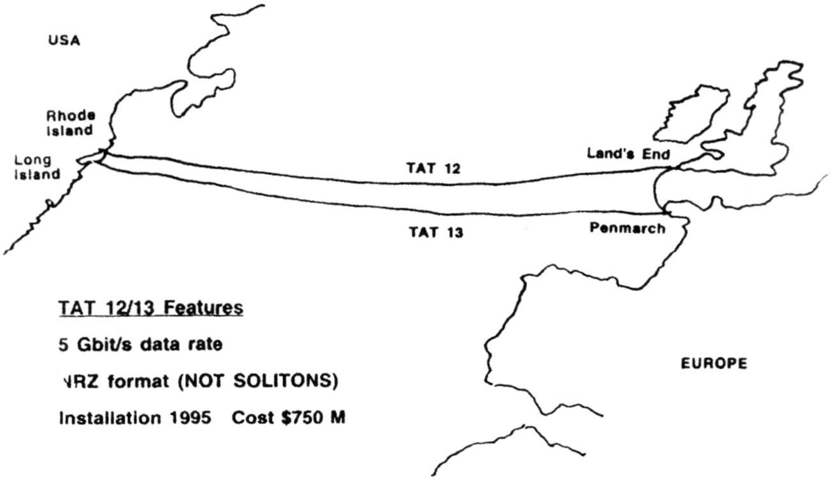

Figure 3. The routes of TAT 12/13. These are all-amplified systems operating at 5 Gbit/s on each of two fibres.

It is interesting to note what the role of the amplifiers is such systems is. The ideal passive lightline would be to have no amplification but sufficient initial power to be detectable at the end of the system. How powerful would the initial source need to be? On a 6000 km system with a loss of 0.2 dB/km it would be necessary to launch 10^{120} photons for just one to emerge at the receiver! This is of course totally unachievable launch energy (more than the energy in the universe) and illustrates dramatically the need for distributed amplification.

2 Nonlinear Optics in Fibres

2.1 WHY SILICA FIBRES?

Optical fibres are made from very pure silica and have exceptionally small nonlinear coefficients. The most important nonlinear coefficient is the nonlinear refractive index, n_2 where the refractive index may be expressed as

$$n = n_o + n_2 I \tag{12}$$

where I is the intensity. The exact value of n_2 and its wavelength dependence is uncertain but it is in the range $3.2 \rightarrow 2.5 \times 10^{-16}$ cm^2/W at 1.5 μm. This is a small value and the refractive index changes induced are always much less than the index difference of the fibre guide.

The effect of this nonlinearity is to modify the propagation constant of the mode, ie

$$\beta = \beta_o + \frac{n_2 P k}{A_{eff}} \tag{13}$$

Thus after a distance z the nonlinear modification to the phase is $\frac{n_2 P k z}{A_{eff}}$.

Because of the low loss in fibres, and also if amplifiers are employed, the distance z may be very long, up to thousands of km. For example for z = 6000 km a π phase change represents 1 part in 10^{13}! Nevertheless, such a small perturbation on the phase velocity will have a profound effect on the optical signal and in particular will induce significant spectral broadening.

Thus, despite the small nonlinearity, significant effects can be expected at modest optical powers because of the extremely long interaction lengths. First we shall consider the case of non amplified fibre transmission and derive simple rules for the thresholds of nonlinear fibre effects. Here we shall consider only χ^3 effects and include the stimulated processes SRS and SBS.

The stimulated processes can be characterised, at their simplest level, as the mixing of two waves of different frequencies one of which is a high power pump. Thus

$$P_1 (L) = P_1 (0) \exp (g P_2 L/A) \tag{14}$$

where P_1 and P_2 are the mixing waves and A the common nonlinear area. Here we assume the pump P_2 is undepleted by this process but we shall consider the effects of loss. Thus

$$P_2 = P_2(0)\, e^{-\alpha z} \qquad \alpha \text{ - loss coefficient} \tag{15}$$

It is useful to define an effective length

$$P(0)\, L_{eff} = \int_0^L P(0)\, e^{-\alpha z}\, dz$$

$$\text{or } L_{eff} = \left(1 - e^{-\alpha L}\right) / \alpha \tag{16}$$

for a loss of 0.2 dB/km ($\alpha = 0.23 \times \gamma$ where γ is in dB/km) for $L \rightarrow \infty$

$$L_{eff} = \frac{1}{\alpha} = \frac{1}{0.046} = 21.7 \text{ km} \tag{17}$$

We shall use this effective length to calculate the SRS and SBS thresholds in fibres. This is then appropriate for very long fibre lengths and assumes no walk-off between P_1 and P_2 and no pump depletion and thus gives a lower limit on the threshold power. For SRS g is measured to be 10^{-13} m/W at the peak of the gain spectrum. If we calculate P from the condition

$$\frac{g\, L\, P}{A_{eff}} = 16 \tag{18}$$

with the above parameters ($A = 50\ \mu m^2$) $\underline{\underline{P = 0.36 \text{ W}}}$.

SBS has a much lower threshold $g = 5 \times 10^{-11}$ m/W and thus $P \approx 7 \times 10^{-2}$ W, however the bandwidths of the two processes are markedly different. For SBS it is ~ 10 MHz and SRS ~ 5 THz. Thus SBS is important for coherent systems and acts as a power limiting factor. SRS has the potential as an amplification process but will generally act as a limitation to the nonlinear refractive index processes.

For nonlinear refractive index we can do a similar calculation for the threshold now defined as the power required for a π phase shift. Thus for $n_2 = 3.2 \times 10^{-16}$, $\lambda = 1.5\ \mu m$, $A = 50\ \mu m^2$

$$\phi = 0.8 \times \pi \times L \times P \tag{19}$$

where L is in km and P in watts. Thus we have the general rule-of-thumb that a π phase shift occurs for ~ 1 Watt.km.

Inserting the effective length 21.7 km the threshold for this nonlinearity

$$P \approx 5.7 \times 10^{-2} \text{ Watts} \tag{20}$$

Thus the nonlinear refractive index threshold is at least an order of magnitude lower than the Raman threshold. The equivalent of this statement is that at least 10 π's of nonlinear phase shift are possible before Raman effects limit the process. This is a lower limit estimate, in most practical situations the effective Raman threshold will be larger due to factors such as pump-signal, walk-off and therefore more than 10 π's of phase shift are possible in most fibre situations.

2.2 NONLINEAR REFRACTIVE INDEX EFFECTS IN FIBRES

Most of the significant nonlinear refractive index effects will involve pulses, this is because high powers are possible with short pulses and that the processing anticipated will act on high data rate signals necessarily constructed for short pulses.

A pulse propagating in a nonlinear medium, where the dominant effect is the nonlinearly induced phase change, will acquire a "chirp" and broaden its spectrum, the temporal width remaining unchanged. This situation can easily be realised in fibres by working close to the zero dispersion wavelength. For a pulse u(t,z) propagating in the z direction, the differential equation description of this effect may be written in a frame moving at the group velocity of the pulse

$$i \frac{du}{dz} = - n_2 k |u|^2 u \tag{21}$$

This has solution

$$u(t,z) = u_o e^{in_2 k|u|^2 z} \tag{22}$$

Thus the nonlinear phase, ϕ, varies with the pulse intensity across the pulse. A local frequency $= - \partial\phi/\partial t$ is thus induced across the pulse proportional to the differential of the intensity. This is illustrated in figure 4 where the pulse, the local frequency and the spectrum are displayed. The chirp is such that lower frequencies (from the central frequency) are induced at the leading edge of the pulse.

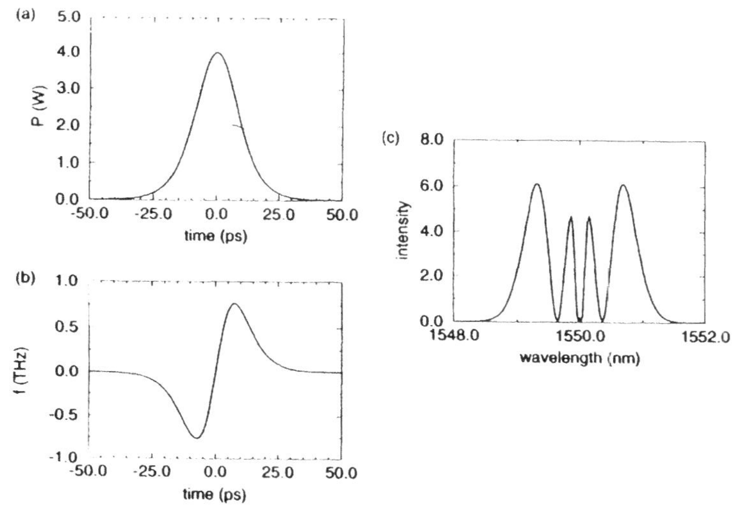

Figure 4. Pulse intensity, frequency chirp and power spectrum of an SPM pulse.

In figure 5 an experimental observation of this effect is shown. A 10 ps pulse from a colour centre laser with 300 m of fibre were used in this observation. A spectral broadening of almost an order of magnitude is observed here confirming the possibility of many π's of phase shift as deduced above.

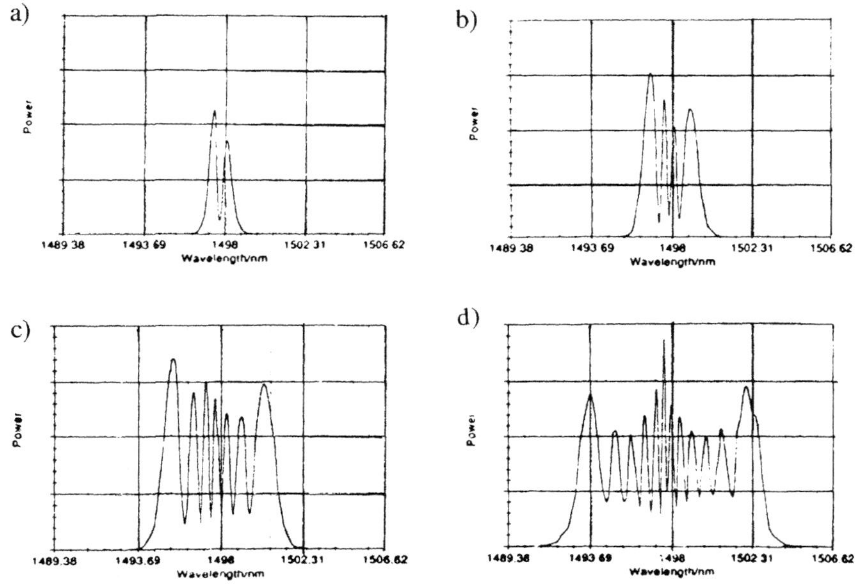

Figure 5. Experimental observation of an SPM broadened spectrum from a 10 ps pulse after 300 m of fibre increasing launched power.

Although fibre loss is small in many cases SPM will occur over long fibre distances and then the effect of fibre loss must be included. Moreover, in amplified systems the power is periodically restored but the effects of SPM will accumulate and the effect of loss between the amplifiers is significant. SPM with loss can be expressed by

$$i \frac{\partial u}{\partial z} = - |u|^2 u - i \Gamma u \tag{23}$$

where Γ (+ve) represents the fibre loss. This equation can be simply solved as follows. Changing the variable $u \rightarrow u' e^{-\Gamma z}$ results in

$$i \frac{\partial u'}{\partial z} = -e^{-2\Gamma z} |u'|^2 u' \tag{24}$$

now assuming a solution of the form $u'(z) = u'_0 e^{i\phi}$ and substituting we obtain

$$- u_o' e^{i\phi} \frac{\partial \phi}{\partial z} = - e^{-2\Gamma z} |u_o'|^2 u_o' e^{i\phi} \tag{25}$$

Integration of this equation gives

$$\phi = |u_o'|^2 \int_o^z e^{-2\Gamma z} = \left(\frac{1 - e^{-2\Gamma z}}{2\Gamma} \right) |u_o'|^2 \tag{26}$$

$$\text{ie} \quad u = u_o e^{-\Gamma z} e^{i\Lambda^2 |u_o|^2} \tag{27}$$

$$\text{with} \quad \Lambda^2 = \left(\frac{1 - e^{-2\Gamma z}}{2\Gamma} \right) \tag{28}$$

Thus an intensity dependent modification of the phase occurs which increases with propagation distance. A pulse will acquire a similar chirp to the lossless case but with a rescaling of the size of the chirp. The result presented here will be appropriate for the amplified transmission systems discussed in section 4.

In addition to SPM nonlinear interaction between different waves in fibres will lead to Cross-Phase-Modulation (XPM) that is the modification of the phase of one wave by the intensity of another wave. In a fibre the two primary cases are where the waves are of different wavelength or of different polarisations. If the two wavelengths are close enough such that other effects can be ignored (pulse walk-off) then XPM can be added to the SPM equation simply by an addition cross term, but now there is a factor of two enhancement.

$$\begin{aligned} i \frac{du}{dz} &= |u|^2 u + 2|v|^2 u \\ i \frac{dv}{dz} &= |v|^2 v + 2|u|^2 v \end{aligned} \tag{29}$$

where u and v are the two waves.

Nondegenerate four-wave mixing can occur also in fibres. We shall consider the simple case of a high power pump mixing with a signal wave at a different frequency generating a new wave at a frequency on the other side of the pump frequency, ie waves u_s, u_p and u_i will be related by $\omega_p - \omega_s = \omega_i - \omega_p$. This will conserve energy provided this condition is obeyed. This also means that the generated wave will be the conjugate of the signal wave ie phase conjugation will be the result but at a new frequency displaced from the pump frequency an equal amount to the difference of the pump and signal.

Phase matching will require

$$k_i = -k_s + k_p + k_p \tag{30}$$

energy conservation requires

$$\omega_i = -\omega_s + \omega_p + \omega_p \tag{31}$$

Phase matching in a single mode fibre can be achieved by operating near the zero dispersion. The propagation constant β of the mode replaces the k-vector in the above equality. The propagation constant can be expanded around ω_p.

$$\beta = \beta_o + \frac{d\beta}{d\omega}\Delta\omega + \frac{1}{2}\frac{d^2\beta}{d\omega^2}(\Delta\omega)^2 + \ldots.. \tag{32}$$

For $\beta = \beta_i$ $\qquad \omega_i = \omega_p + \Delta\omega$
$\beta = \beta_s$ $\qquad \omega_s = \omega_p - \Delta\omega$

the expansion is good provided $\Delta\omega$ is small. If ω_p is the zero dispersion frequency $\frac{d^2\beta}{d\omega^2} = 0$ and the phase matching condition can be written

$$\beta_i = -\beta_s + 2\beta_p \tag{33}$$

This equation is satisfied up to the term $\frac{d^4\beta}{d\omega^4}(\Delta\omega)^4$ so that provided the pump is at the zero first order dispersion and the other frequencies are close, good phase matching may be achieved.

This type of phase conjugation has an important application in communication systems. If phase conjugation of a signal is performed at the mid-point of a transmission line the effects of dispersion, which causes pulse broadening, are reversed ie midpoint phase conjugation provides complete compensation for dispersion [7]. This is then one of the clear applications for fibre nonlinearity in communications systems.

2.3 SOLITONS: BASIC PRINCIPLES

We now consider the combined effects of SPM and dispersion on pulse propagation in a fibre. Dispersion leads to chirp formation and temporal broadening of a pulse. SPM leads to chirp formation and spectral broadening. If these two effects are of a similar size and occur simultaneously they will balance provided the chirps are of the opposite sense. We can combine the effects by combining the differential equations describing them - a complete proof of this assertion is not presented here and requires a complete formulation of the problem starting with Maxwell's equation -

$$i\frac{\partial u}{\partial z} + \frac{1}{2}\left|\ddot{\beta}\right|\frac{\partial^2 u}{\partial t^2} + n_2 k\,|u|^2\,u = 0 \tag{34}$$

The first two terms alone describe a dispersing pulse in a frame moving at the group velocity. The third term is that giving rise to SPM. Equation 34 is the so-called Nonlinear Schrödinger equation (NLS) which is more conventionally written in normalised units as

$$i\frac{\partial u}{\partial z} + \frac{1}{2}\frac{\partial^2 u}{\partial t^2} + |u|^2 u = 0 \tag{35}$$

This equation applies in an optical fibre in the negative group velocity dispersion regime, ie for wavelengths longer than the dispersion zero. In the positive group velocity dispersion regime a minus sign must be used before the second term.

The NLS has well-known soliton solutions of which the most simplest form is

$$u(z, t) = e^{i\,z/2}\,\text{sech}\, t \tag{36}$$

This pulse thus propagates without change of shape and only accumulates a uniform phase shift proportional to the distance. This nondispersing pulse is the one sought for communications since it is invariant in both temporal and spectral width.

It is useful now to relate this normalised solution to the real world values. The parameters which specify a fibre are its dispersion D in ps/nm/km related to $\ddot{\beta}$ by $D = -\frac{2\pi c}{\lambda^2}\ddot{\beta}$, the nonlinear mode area A_{eff} and the nonlinear coefficient n_2. For a fibre with $A_{eff} = 50\ \mu m^2$ at 1.55 μm we can write simple formulae for the pulse peak power and energy in terms of the pulse duration τ (FWHM) in picoseconds

$$P(W) = 1.34 \times \frac{D}{\tau^2}$$
$$E(pJ) = 1.53\,\frac{D}{\tau} \tag{37}$$

Thus we see that shorter pulses require larger peak power and energy for a given dispersion. Thus to generate a soliton in a particular fibre what is required is a pulse of the correct peak power and shape as given by the above formulae. However, the soliton is the stable feature of nonlinear pulse propagation and all that is required is that a pulse energy close to the ideal value is chosen. For example, if a pulse Asecht is launched into a fibre (where A = 1 corresponds to the exact pulse energy) a soliton of width $\tau = \frac{1}{2\left(A - \frac{1}{2}\right)}$ and amplitude $2\left(A - \frac{1}{2}\right)$ will be generated. The remaining energy is shed from the pulse on propagation as a dispersive wave often called radiation since it radiates in time away from the pulse.

A similar effect occurs when the initial pulse shape is not ideal, that is the propagation in the fibre converts the pulse to the stable soliton with some of the initial energy shed by radiation. Figure 6 illustrates this point showing an initial distorted pulse and a pulse after considerable propagation in the fibre. There is one other important parameter characterising the soliton; the so called soliton period z_0. This is defined as the length for a soliton to acquire a phase of $\pi/4$, ie in normalised distances this is the distance $z = \pi/2$. This clearly gives the length scale over which the soliton phase evolves and would probably be more appropriate if defined as the distance for a 2π phase change which would be 8 x z_0 as defined above. In real units

$$z_o = 0.42 \frac{\tau^2}{D} \tag{38}$$

Thus short pulses have short soliton periods. This soliton period is also closely related to the nonlinear length defined in 2.3.

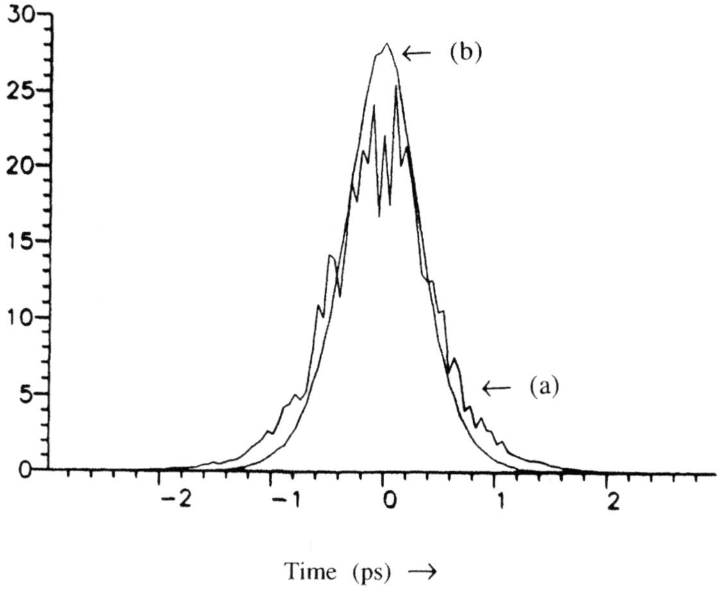

Figure 6. Initial "noisy" pulse (a) and pulse after considerable propagation (b) in a lossless NLS.

There are some important soliton features which we now discuss. Probably the most important feature of solitons is that they do not mutually interact. This statement requires some clarification in the optical pulse context. Consider first, solitons of different wavelength. If these are propagated in a fibre they will have different group velocities (the group velocity is unaltered by the nonlinearity) and if launched separately in time will pass through each other unnoticed. This is illustrated in figure 7. However, although no energy is exchanged between the solitons they impart a phase shift on each other. This interaction is notable because the pulses are in essence nonlinear but they are in this context behaving as linear pulses.

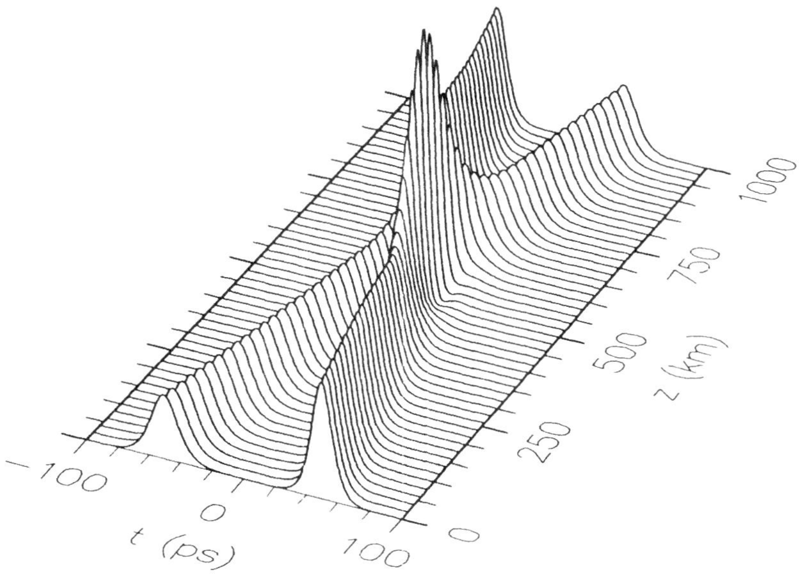

Figure 7. Crossing of two solitons of different frequencies and amplitudes.

When two (or more) solitons are launched with the same frequency they naturally have the same group velocity and so there is no tendency for them to pass through each other. It might then be expected that if they are launched separately in time they would remain so. This is not in fact the case. Consider two pulses separated in time by ΔT. The initial condition can be written allowing for a relative phase difference

$$u(z=0,\ t) = \operatorname{sech}\left(t + \frac{\Delta T}{2}\right) + e^{i\phi} \operatorname{sech}\left(t - \frac{\Delta T}{2}\right) \tag{39}$$

On propagation an interaction occurs which depends on the phase ϕ. If $\phi = 0$ (ie in phase) the solitons "attract", coalesce and then re-emerge. This happens periodically and so it can be said that there is no permanent exchange of energy and this does not violate the principle of non-mutual interaction. If $\phi = \pi$ the pulses will repel and never cross. This effect is illustrated in figure 8. Soliton-soliton interaction as illustrated here is an important constraint in designing soliton communications systems.

The collapse length can be simply expressed in terms of the solitons' initial separation ΔT in normalised units

$$Z_L = \frac{\pi}{2} e^{\frac{\Delta T}{2}} \tag{40}$$

Solitons will suffer attenuation on propagation in the fibre. However, if the loss is small the soliton will respond to the effect of loss by broadening but still remaining with the condition for soliton formation. Loss can be considered as adding a term $i\Gamma u$ to the NLS analogous to the effect of SPM with loss. The soliton solution is now

$$u(z,t) = e^{-\Gamma z} e^{i\left(\frac{1 - e^{-2\Gamma z}}{4\Gamma}\right)} \operatorname{sech}\left(e^{-\Gamma z} t\right) \tag{41}$$

Thus the pulse broadens in width exponentially. The above formulation is only correct where the loss length is long compared to the soliton period.

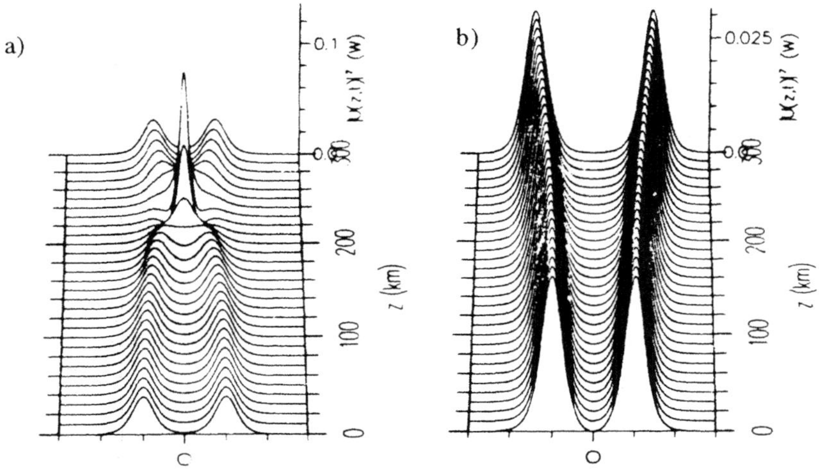

Figure 8. Soliton-soliton interaction (a) initial in phase pulses (b) initial out-of-phase pulses.

3 Nonlinear Effects Applied to Optical Processing

The nonlinear intensity dependence of refractive index can be used in some very simple fibre configurations to perform useful ultrafast processing functions. The functions which can be performed include demultiplexing, modulation, clock recovery and saturable absorption (soliton filtering). There are many device configurations which are SPM, XPM and 4WM coupled with soliton effects for good pulse processing effects. Here we will concentrate on probably the most commonly used element, the Nonlinear Optical Loop Mirror (NOLM) [8,9]. This device consists of a four port coupler with the two "output" ports connected by a length of fibre in which the nonlinearity occurs. This is illustrated in figure 9. The nonlinear function is achieved by breaking the symmetry between the clockwise and counter clockwise propagation in the loop. One of the simplest symmetry breaking procedures is to operate with an unbalanced coupler. If the coupler power splitting ratio is expressed as α: $1 - \alpha$ the transmitted intensity P_t is expressed in terms of the launched power P_i as

$$P_t = P_i \left(1 - 2\alpha(1 - \alpha)\left(1 + \cos\left[(1 - 2\alpha) n_2 k L P_i\right]\right)\right) \tag{42}$$

Thus a periodic nonlinear response to input power is obtained. The particular advantageous features of the NOLM are:

(1) assured co-incidence of pulse at coupling point.

(2) environmentally stable with long fibre loop lengths.

(3) ultrafast nonlinearity enables operation with ultrashort pulses.

(4) multifunctional (eg demultiplexing, modulation, saturable absorption).

(5) polarisation control allows operation as a high power transmitter or high power reflector.

(6) compatible with other linear and nonlinear fibre elements.

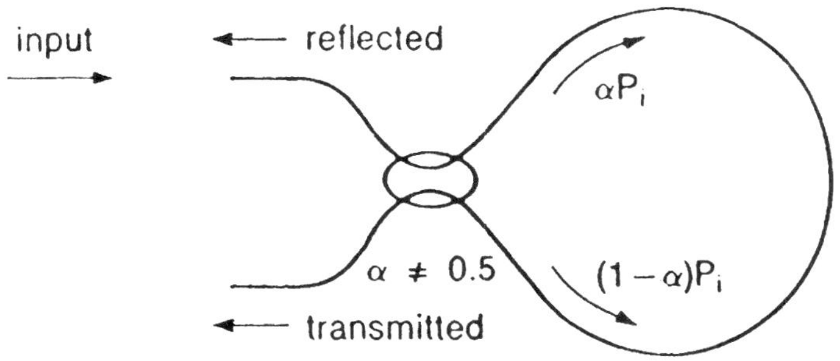

Figure 9. The Nonlinear Optical Loop Mirror.

The first application to consider for the NOLM is as a saturable absorber. Simply it will transmit high power (up to the first maximum) and reflect low power. Thus an incoming pulse with a low power pedestal will be separated into a transmitted pulse with a reduced pedestal and a low power reflected wave. This property makes a NOLM an obvious candidate for a modelocking element in a laser cavity. However, since the response of the loop is phase sensitive the uniform nonlinear phase acquired by a soliton (equation 36) makes the pulse discrimination feature most effective when appropriate dispersion as well as nonlinearity is incorporated in the nonlinear loop.

NOLM's have been used to generate ultrashort soliton pulses in doped fibre laser cavities [10]. The most common arrangement is the figure-of-eight configuration [11,12] illustrated in figure 10. In this case the amplifying fibre is incorporated in the NOLM which is often then referred to as a NALM (nonlinear amplifying loop mirror) [13], the amplifier not only amplifies the circulating pulses but if placed assymetrically in the loop also acts as the symmetry breaking element. In the absence of additional control features, solitons of fixed size but at random relative positions are generated. Very short sub-ps pulses may be generated this way but incorporation of suitable control is necessary to stabilise the temporal position of the pulses.

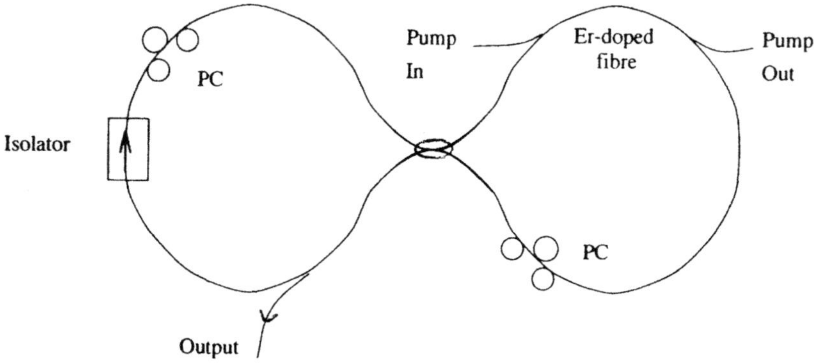

Figure 10. Figure-of-Eight Er Fibre Laser.

The saturable absorber properties of NOLM's make them suitable control elements in long transmission lines. Recently it has been shown that short solitons ~ 1.5 ps may be stabilised in long periodically amplified systems using NOLMs [14]. We shall return to solitons in long amplified transmission lines in the next section.

NOLM's have a simple application as all optical modulators. The general principle is that data on one optical stream may be imposed on another via cross phase modulation within the loop. The principle may be understood by considering the two waves at different wavelengths although polarisation can be used as the distinguishing feature. If the modulated stream is injected to follow just one path around the loop and the wave to be modulated is split symmetrically by the loop then pulses in the direction coincident with the data have a nonlinear phase imparted by the presence of data pulses. If the loop is polarisation biased to be a nonlinear transmitter a data sequence identical to the original is transmitted at the new wavelength. The original data stream exchanges, in principle, no energy with the other wave and is thus left unaffected. An impression of the data is thus imparted on the unmodulated stream.

This principle has been recently incorporated in an all-fibre all-optical regenerator [15]. In this experiment all the features of regeneration are performed (ie amplification, retiming, reshaping) in the optical domain. The experiment has two stages. In the first a doped fibre ring laser is modelocked by cross phase modulation from the signal data sequence. Although the data contains "ones" and "zeros" the cavity arrangement of the laser will generate a modelocked sequence of pulses at the repetition rate of the data sequence. The cross phase modulation ensures phase modulation at the required frequency. In the second stage the data pulses are used in the NOLM arrangement described above to modulate the extracted clock pulse train. The wavelength of the output from the laser is different from the original incoming data so that wavelength translation is achieved in addition to regeneration. The experiment in ref. 15 shows that the regenerated data may be retimed and amplified to reduce jitter.

The particular process described here functions efficiently if solitons or soliton-like pulses are used to encode the data. Thus all-optical processing is particularly compatible with solitons and we turn in the next section towards practical design of soliton communications systems.

4 Soliton Communications Systems

4.1 SOLITONS IN LONG DISTANCE AMPLIFIED TRANSMISSION - FUNDAMENTAL FEATURES

In this section we shall be concerned with the application of solitons for high data rate (several Gbit/s) very long haul amplified links. As identified in Section 1 there is a new generation of all amplified links which will be installed in 1995/96 across the Atlantic and Pacific linking North America with Europe and Asia respectively. These links are up to 10,000 km long and will employ only direct amplification. This first generation of such systems will employ NRZ modulation format and operate at 5 Gbit/s on a single fibre. Solitons hold the potential for increasing this data rate still further and extending the system to greater distance within the general principle of periodic lumped amplification.

Solitons are non-dispersive in both the temporal and frequency domain and this is, of course, the attraction for communications links. However, in a fibre link the soliton must withstand large periodic changes in the pulse energy. With lumped amplifiers the solitons will be subject to attenuation of 10 dB or more for amplifier spacing of 50 km. It has been shown that the essential soliton features are retained provided that the amplifier spacing is small compared to the soliton period and that the initial soliton amplitude is such that on average over a single span the amplitude is equal to the lossless soliton. This is called the average soliton principle [16] or guiding centre soliton [17,18]. It is important to note that this principle states that the stability is ensured independent of the size of the required amplification.

The above is the primary consideration which must be included in designing a soliton system. For example, if the soliton pulse duration employed is 20 ps the soliton period z_0 is 168/D km. It is relevant to note here that it is the full soliton period, that distance over which a lossless soliton's phase changes by 2π, which is the parameter here. This is 8 times z_0 the conventionally named soliton period. To ensure that the condition is well fulfilled it is often taken that $z_a \leq \frac{8z_0}{10}$. Thus for D = 1 ps/nm/km (using our 20 ps solitons) it is sufficient to ensure that the amplifier spacing is considerably less than 135 km. For standard fibre where D $\approx$ 17 ps/nm/km we would require an amplifier spacing much less than 8 km. Thus soliton systems employing these type of pulse durations are only really practical for dispersion shifted (low dispersion) fibres.

4.2 SOLITON SYSTEM DESIGN FEATURES

Apart from average soliton consideration described above there are other important considerations specific to solitons which must be included in the design of the system. The most important of which are:

(1) soliton-soliton interaction.
(2) amplifier noise induced timing jitter (Gordon-Haus jitter) [19].
(3) signal to amplifier induced noise limit.
(4) multiplexing potential.

The first of these arises from the effect of launching adjacent solitons of the same frequency and amplitude. In figure 8 we show the effect on two solitons, initially in phase, out of phase and with slightly different amplitudes. This is a lossless calculation but shows the "attraction" for in phase solitons and repulsion for out-of-phase solitons. This interaction depends exponentially on the initial soliton separation, the distance to coallescence of in phase pulses L_c in normalised units is simply expressed as

$$L_c = \frac{\pi}{4} e^{\Delta T/2} \tag{43}$$

where ΔT is the initial separation. To avoid this effect it is necessary to work with relatively large ΔT's. This means that soliton systems require a large mark-to-space ratio usually in the range 5:1 to 10:1.

The second effect is a little more complex and involves the noise induced by the amplifiers. Signal-spontaneous beat noise will induce a random frequency fluctuation (jitter) in the individual soliton. This frequency jitter has mean zero but finite variance $\langle \Omega^2 \rangle$. The frequency jitter is induced at each amplifier. The subsequent propagation will translate this frequency jitter into a timing jitter on arrival at the end of the system. The temporal variance may be simply expressed as

$$\langle \delta t^2 \rangle = \left(\frac{\lambda}{2\pi c} \right)^2 \sum_{j=1}^{N} D_j^2 \langle \Omega^2 \rangle \tag{44}$$

where j numbers the amplifiers (which are taken to be identical) and D is the average dispersion in the fibre. This jitter is known as Gordon-Haus jitter. A complete formula for this effect including the gain of the amplifiers, G, and the soliton pulse width τ is

$$\langle \delta t^2 \rangle = \frac{L_s^3 (G - 1) h \mu D n_2}{9 \tau A_{eff} z_a \Lambda^2 (G)} \tag{45}$$

where L_s is the total system length and $\Lambda^2 (G) = G \ln G / (G - 1)$.

The important feature of equation 45 is the strong dependence on overall system length, ie L_s^3. The jitter also increases with the gain G in the amplifiers and therefore indicates that large amplifier spacing is undesirable. By employing appropriate statistics it is straightforward to calculate for a required error rate the jitter limit for a specific data rate as a function of amplifier spacing.

We are now in a position to construct a design diagram for a soliton system with specified system length and chosen average dispersion. For this example we shall take D = 1 ps/nm/km and a total system length of 10000 km and a nonlinear area of 50 μm^2. These are the only parameters we need to specify for this exercise. We will however also define the acceptable error rate (10^{-12} will be taken here) and the SNR which we will take as 23dB. These are very conservative limits, but this is a simplified design and there are potentially other detrimental effects neglected here (the obvious example is that we are ignoring any imperfections in the source).

Two design diagrams are shown in figure 11. It is sufficient to take just the pulse duration and the amplifier spacing and to generate specific diagrams for the desired data rates (2.5 and 5 Gbit/s are displayed). The upper limit (on pulse duration) is provided by the SNR and soliton-soliton (SS) limits; lower limits are determined by the Average Soliton (AS) and Gordon-Haus (GH) jitter. For this example there is a clear window of operation at 2.5 Gbit/s using pulse durations of ~ 15 ps and amplifier spacing of up to 40 km. There is apparently no operating window for 5 Gbit/s. For all but the SS limit we could generalise the pulse duration axis to be τ/D so thus if D is taken as 0.5 ps/nm/km the pulse duration would be twice that shown in the diagram. We have been very conservative in the SNR limit which we could increase by a factor of 2 and still operate the system but with smaller margin, in that case there may be a small operating region for amplifier spacing up to 30 km for 5 Gbit/s.

It is clear from the diagrams that the GH limit is the primary obstacle to high data rate operation. Much effort has been devoted to the elimination or at least reduction of this limit. These efforts are best described by the term soliton control. The intention is to control the soliton's position in time in order to prevent errors which occur when a soliton drifts into a wrong time slot.

Solitons may be controlled by a variety of techniques but probably the best known is the use of distributed band pass filters along the transmission line [20,21]. The principle is that the filters are individually weak, that is their bandwidth is much greater than the soliton's bandwidth, but that there are many of them, usually there is a filter at each amplification stage. The concatenated effect of all the filters would be enough to effectively destroy a linear pulse of the soliton's bandwidth but the soliton "recovers" its integrity between each stage allowing an undestorted transmission path for the soliton but controlling the central frequency of the soliton. The soliton can pass through a transmission line which has an effective bandwidth much less than the individual soliton's bandwidth. This is a remarkable effect and utilises usefully the inherent nonlinearity of the soliton.

With this approach it is possible to reduce the G-H jitter standard deviation by a factor of ~ 2. If the filters are individually too strong the gain in the amplifiers has to be increased and the noise at the central frequency will be increased and thus destroy the operation of the system.

A more effective way to use filters is to slide progressively the central frequency of the filters across the system [22]. This sliding-frequency-filter scheme can almost completely eliminate jitter and thus permit soliton data rates of 10 or 20 Gbit/s.

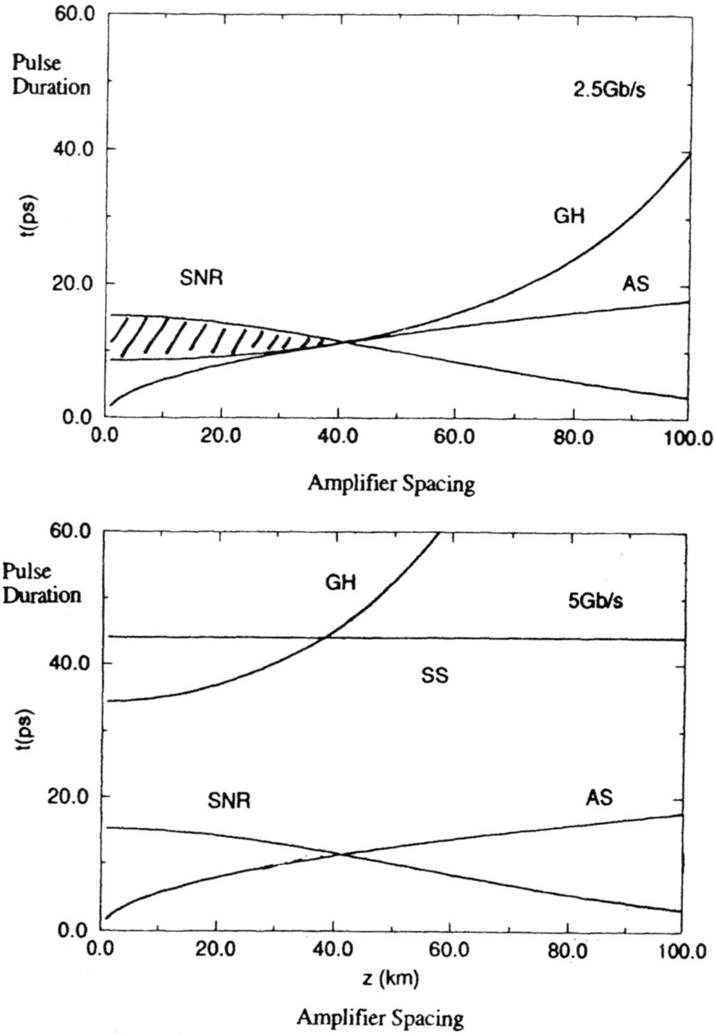

Figure 11. Design diagrams for 10,000 km soliton systems with no control, 10^{12} BER and 23dB SNR, D = 1 ps/nm/km. The operating window (shaded) exists below the SNR and SS curves and above all the others. (SNR - signal-to-noise ratio requirement, GH - Gordon-Haus limit, AS - average soliton limit and SS - soliton interaction limit).

Table 1 shows a review of soliton transmission experiments using both recirculating loop configurations and in-line results. The data rates and distances achieved are still rapidly increasing. The prospects of multiplexing, both using polarisation and wavelength have not yet been fully investigated or exploited. It is probable that using these techniques 40 Gbit/s over intercontinental distances will be achieved in the next few years. The ultimate limits on soliton capacity have not yet been determined in particular for shorter (up to 1000 km) length systems where data rates approaching 100 Gbit/s may be possible.

Year	Reference	Transmission Path Length	Bit Rate	Amplifiers & Spacing
1988	Mollenauer [23]	4000 km (42 km loop)	100 MHz	Raman
1990	Iwatsuki [24]	23 km	2.8 Gb/s	Raman
	Nakazawa [25]	100 km	3.2-5 Gb/s	Er 25 km
	Suzuki [26]	250 km	5 Gb/s	Er 25 km
	Suzuki [27]	250 km	5 Gb/s	Er 25 km
	Iwatsuki [28]	23 km	5 Gb/s	Raman
	Nakazawa [29]	200 km	10-20 Gb/s	Er 25 km
	Mollenauer [30]	10,000 km (75 km loop)	200 MHz	Er 25 km
	Iwatsuki [31]	70 km	20 Gb/s	Raman
1991	Wickens [32,33]	205 km	20 Gb/s	Er 51.2 km
	Mollenauer [34]	12,000 km (75 km loop)	2.4 Gb/s	Er 25 km
	Ellis [35]	350 km	5 Gb/s	Er 70 km
	Nakazawa* [36]	10^6 km (510 km loop)	10 Gb/s	Er 50 km
	Olsson [37]	9000 km (120 km loop)	2 x 2 Gb/s WDM	Er 30 km
	Spirit [38]	28 km	10 Gb/s	Dist. Er
	Yamada [39]	1000 km	10 Gb/s	Er 50 km
1992	Mollenauer [40]	15,000 km (80 km loop)	5 Gb/s	Er 26 km
	Mollenauer [40]	11,000 km (80 km loop)	2 x 5 Gb/s WDM	Er 26 km
	Nakazawa [41]	1,200 km	10 Gb/s	Er
	Nakazawa [42]	1,020 km	20 Gb/s	Er 50 km
	Taga [43]	6,000 km (99 km loop)	2.5 Gb/s	Er 33 km
	Nakazawa [44]	350 km	20 Gb/s	Er 25 km
	Iwatsuki [45]	65 km	40 Gb/s	Er 16 km
	Mollenauer [46]	14,000 km (81 km loop)	2.5 Gb/s	Er 27 km
	Taga [47]	3,000 km	5 Gb/s	Er 33 km
1993	Nakazawa* [48]	10^6 km (500 km loop)	10 Gb/s	Er 50 km
	Mollenauer† [49]	20,000 km (77 km loop)	10 Gb/s	Er 26 km
	Mollenauer [49]	13,000 km (77 km loop)	2 x 10 Gb/s WDM	Er 26 km
	Ellis [50]	205 km	40 Gb/s	Er 50 km
	Nakazawa [51]	2000 km	20 Gb/s	Er 50 km
	Nakazawa [52]	1000 km	40 Gb/s	Er 50 km
	Suzuki [53]	13,100 km (90 km loop)	5 Gb/s	Er 30 km

* Modulator Used
† Sliding Filters Used

Table 1

Acknowledgement

The author wishes to thank W Forysiak, N J Smith and F Knox for help in the preparation of the text and diagrams.

References

[1] For introduction to optical fiber technology:
Keiser, G. (1991) "*Optical fiber communications*", McGraw-Hill.
Senior, J. M. (1992) "*Optical fiber communications: principles and practice*", Prentice-Hall.

[2] For overview of nonlinear fiber optics:
Agrawal, G. (1989) "*Nonlinear fiber optics*", Academic Press.

[3] For overview of nonlinear fiber devices:
Islam, M. N. (1992) "*Ultrafast fiber switching devices and systems*", C.U.P.

[4] Heidemann, R., Wedding, B. and Veith, G. (1993) "10 Gb/s transmission and beyond", *Proceedings of the IEEE*, 81:1558-1567.

[5] Wedding, B. (1992) "New method for optical transmission beyond dispersion limit", *Electronics Letters*, 28:1298-1300.

[6] Dugan, J. M., Price, A. J., Ramadan, M., Wolf, D. L., Murphy, E. F., Antos, A. J., Smith, D. K. and Hall, D. W. (1992) "All-optical fiber based 1550 nm dispersion in a 10 Gb/s, 150 km transmission experiment over 1310 nm optimised fiber", *Proceedings OFC'92*, paper PD14.

[7] Murata, S., Tomita, A., Shimizu, J. and Suzuki, A. (1991) "THz Optical-Frequency Conversion of 1 Gb/s-Signals using highly nondegenerate four-wave mixing in an InGaAsP semiconductor laser", *IEEE Photonics Technology Letters*, 3:1021-1023.

[8] Doran, N. J. and Wood, D. (1988) "A nonlinear optical loop mirror", *Optics Letters*, 13:56-58.

[9] Nayar, B. K., Blow, K. J. and Doran, N. J. (1991) "All-optical switching in nonlinear fibre loop mirror devices", *Optical Computing and Processing*, 1:81-89.

[10] Bulushev, A. G., Dianov, E. M. and Okhotnikov, O. G. (1990) "Passive mode locking of a laser with a nonlinear fiber reflector", *Optics Letters*, 15:968-970.

[11] Duling III, I. N. (1991) "All-fiber ring soliton laser mode locked with a nonlinear mirror", *Optics Letters*, 16:539-541.

[12] Richardson, D. J., Laming, R. I., Payne, D. N., Matsas, V. and Phillips, M. W. (1991) "Self-starting, passively mode-locked erbium fibre ring laser based on the amplifying Sagnac switch", *Electronics Letters*, 27:542-544.

[13] Fermann, M. E., Haberl, F., Hofer, M. and Hochreiter, H. (1990) "Nonlinear amplifying loop mirror", *Optics Letters*, 15:752-754.

[14] Smith, N. J. (1994), PhD Thesis, University of Strathclyde.

[15] Smith, K. and Lucek, J. K. (1993) "All-optical signal regenerator",*Optics Letters*, 18:1226-1228.

[16] Blow, K. J. and Doran, N. J. (1991) "Average soliton dynamics and the operation of soliton systems with lumped amplifiers", *IEEE Photonics Technology Letters*, 3:369-371.

[17] Hasegawa, A. and Kodama, Y. (1990) "Guiding-center soliton in optical fibres", *Optics Letters*, 15:1443-1445.

[18] Hasegawa, A. and Kodama, Y. (1991) "Guiding-center soliton", *Physical Review Letters*, 66:161-164

[19] Gordon, J. P. and Haus, H. A. (1986) "Random walk of coherently amplified solitons in optical fiber transmission", *Optics Letters*, 11:665-667.

[20] Mecozzi, A., Moores, J. D., Haus, H. A., and Lai, Y. (1991) "Soliton transmission control", *Optics Letters*, 16:1841-1843.

[21] Kodama, Y. and Hasegawa, A. (1992) "Generation of asymptotically stable solitons and suppression of the Gordon-Haus effect", *Optics Letters*, 17:31-34.

[22] Mollenauer, L. F., Gordon, J. P. and Evangelides, S. G. (1992) "The sliding frequency filter: an improved form of soliton jitter control", *Optics Letters*, 17:1575-1577.

[23] Mollenauer, L. F. and Smith, K. (1988) "Demonstration of soliton transmission over more than 4000km in fiber with loss periodically compensated by Raman gain", *Optics Letters*, 13:675-677.

[24] Iwatsuki, K., Nishi, S., Saruwatari M. and Shimizu, M. (1990) "2.8 Gb/s optical soliton transmission employing all laser diodes", *Electronics Letters*, 26:1-2.

[25] Nakazawa, M., Suzuki, K. and Kimura, Y. (1990) "3.2-5 Gb/s 100 km error free soliton transmissions with erbium amplifiers and repeaters", *IEEE Photonics Technology Letters*, 2:216-219.

[26] Suzuki, K., Nakazawa, M., Yamada, E. and Kimura, Y. (1990) "5 Gbs^{-1}, 250 km error-free soliton transmission with Er^{+++}, doped fiber amplifiers and repeaters", *Electronics Letters*, 26:551-553.

[27] Suzuki, K. and Nakazawa, M. (1990) "Automatic optical soliton control using cascaded Er^{3+}, doped fiber amplifiers", *Electronics Letters*, 26:1032-1034.

[28] Iwatsuki, K., Nishi, S., Saruwatari, M. and Nakazawa, K. (1990) "5 Gbs^{-1}, optical soliton transmission experiment using Raman amplification for fibre-loss compensation", *IEEE Photonics Technology Letters*, 2:507-509.

[29] Nakazawa, M,. Suzuki, K., Yamada, E. and Kimura, Y. (1990) "20 Gbit/s soliton transmission over 200 km using erbium doped fibre repeaters", *Electronics Letters*, 26:1592-1593.

[30] Mollenauer, L. F., Neubelt, M. J., Evangelides, S. G., Gordon, J. P., Simpson, J. R. and Cohen, L. G. (1990) "Experimental study of soliton transmission over more than 10,000 km in dispersion shifted fiber", *Optics Letters*, 15:1203-1205.

[31] Iwatsuki, K., Suzuki, K., Nishi, S., Saruwatari, M. and Nakazawa, K. (1990) "20 Gb/s optical soliton data transmission over 70 km using distributed fiber amplifiers", *IEEE Photonics Technology Letters*, 2:905-907.

[32] Wickens, G. E., Spirit, D. M. and Blank, L. C. (1991) "20 Gbit/s, 205 km optical time division multiplexed transmission system", *Electronics Letters*, 27:973-974.

[33] Wickens, G. E., Spirit, D. M. and Blank, L. C. (1992) "Nonlinear transmission of 20 Gbit/s optical time-division-multiplexed data, over 205 km of dispersion shifted fibre", *Electronics Letters*, 28:117-118.

[34] Mollenauer, L. F., Nyman, B. M., Neubelt, M. J., Raybon, G. and Evangelides, S. G. (1991) "Demonstration of soliton transmission at 2.4 Gbit/s over 12,000 km", *Electronics Letters*, 27:178-179.

[35] Ellis, A. D., Cox, J. D., Bird, D., Regnault, J., Wright, J. V. and Stallard, W. A. (1991) "5 Gbit/s soliton propagation over 350 km with large periodic dispersion coefficient perturbations using erbium doped fibre amplifier repeaters", *Electronics Letters*, 27:878-880.

[36] Nakazawa, M., Yamada, E., Kubota, H. and Suzuki, K. (1991) "10 Gb/s soliton data transmission over one million kilometers", *Electronics Letters*, 27:1270-1272.

[37] Olsson, N. A., Andrekson, P. A., Simpson, J. R., Tanbun-ek, T., Logan, R. A. and Wecht, K. W. (1991) "Bit-error-rate investigations of two-channel soliton propagation over more than 10000 km", *Electronics Letters*, 27:695-697.

[38] Spirit, D. M., Marshall, I. W., Constantine, P. D., Williams, D. L., Davey, S. T. and Ainslie, B. J. (1991) "Nonlinear, dispersion-free 10 GHz optical pulse train transmission in a distributed erbium-doped fibre", *Electronics Letters*, 27:222-224.

[39] Yamada, E., Suzuki, K. and Nakazawa, M. (1991) "10 Gbit/s single-pass soliton transmission over 1000 km", *Electronics Letters*, 27:1289-1291.

[40] Mollenauer, L. F., Lichtman, E., Harvey, G. T., Neubelt, M. J. and Nyman, B. M. (1992) "Demonstration of error-free soliton transmission over more than 15,000 km at 5 Gbs^{-1}, single channel, and over more than 11,000 km at 10 Gbs^{-1} in two channel wdm", *Electronics Letters*, 28:792-794.

[41] Nakazawa, M., Suzuki, K., Yamada, E., Kubota, H. and Kimura, Y. (1992) "10 Gbs^{-1}, 1200 km error-free soliton data transmission using erbium doped fibre amplifiers", *Electronics Letters*, 28:817-818.

[42] Nakazawa, M., Suzuki, K. and Yamada, E. (1992) "20 Gbit/s - 1020 km penalty-free soliton data transmission using erbium-doped fibre amplifiers", *Electronics Letters*, 28:1046-1047.

[43] Taga, H., Suzuki, M., Tanaka, H., Yoshida, Y., Edagawa, N., Yamamoto, S. and Wakabayashi H. (1992) "Bit error rate measurement of 2.5 Gbit/s data modulated solitons generated by InGaAsP EA modulator using a circulating loop", *Electronics Letters*, 28:1280-1281.

[44] Nakazawa, M., Suzuki, K., Yamada, E. and Kubota, H. (1991) "Observation of nonlinear interactions in 20 Gbit/s soliton transmission over 500 km using erbium-doped fibre amplifiers", *Electronics Letters*, 27:1662-1663.

[45] Iwatsuki, K., Suzuki, K., Nishi, S. and Saruwatari M. (1992) "40 Gbit/s optical soliton transmission over 65 km", *Electronics Letters*, 28:1821-1822.

[46] Mollenauer, L. F., Neubelt, M. J., Honer, M., Lichtman, E., Evangelides, S. G. and Nyman, B. M. (1991) "Demonstration of error-free soliton transmission at 2.5 Gbit/s over more than 14,000 km", *Electronics Letters*, 27:2055-2056.

[47] Taga, H., Suzuki, M., Edagawa, N., Yoshida, Y., Yamamoto, S., Akiba, S. and Wakabyashi H. (1992) "5 Gb/s optical soliton transmission experiment over 3000 km employing 91 cascaded Er doped fibre amplifier repeaters", *Electronics Letters*, 28:2247-2248.

[48] Nakazawa, M., Suzuki, K., Yamada, E., Kubota, H., Kimura, Y. and Takayen, M. (1993) "Experimental demonstration of soliton data transmission over unlimited distances with soliton control in time and frequency domains", *Electronics Letters*, 29:729-730.

[49] Mollenauer, L. F., Lichtman, E., Neubelt, M. J. and Harvey, G. T. (1993) "Demonstration, using sliding-frequency guiding filters, of error free soliton transmission over more than 20 mm at 10 Gbit/s, single channel, and over more than 13 Mm at 20 Gbit/s in a two-channel", *Electronics Letters*, 29:910-911.

[50] Ellis, A. D., Widdowson, T., Shan, X., Wickens, G. E. and Spirit, D. M. (1993) "Transmission of true single polarisation 40 Gbit/s soliton data signal over 205 km using a stabilised erbium ring laser and 40 GHz electronic timing recovery", *Electronics Letters*, 29:990-991.

[51] Nakazawa, M., Suzuki, K., Yamada, E., Kubota, H. and Kimura, Y. (1993) "Straight-line soliton data transmission over 2000 km at 20 Gbit/s and 1000 km at 40 Gbit/s using erbium doped fibre amplifiers", *Electronics Letters*, 29:1474-1476.

[52] Suzuki, M., Taga, H., Edagawa, N., Tanaka, H., Yamamoto, S. and Akiba, S. (1993) "Experimental investigation of Gordon-Haus limit on soliton transmission by using optical short pulses generated by an InGaAsP electroabsorption modulator", *Electronics Letters*, 29:1643-1644.

GIANT OPTICAL NONLINEARITIES IN NEMATIC LIQUID CRYSTALS

E. SANTAMATO
Dipartimento di Scienze Fisiche
Universita di Napoli
Pad.20. Mostra d'Oltremare
80125 Napoli
Italy

1. Introduction

The advent of coherent high-intensity laser sources has made apparent that the light itself can change the optical properties of the medium. As a result, a new field in physics originated: nonlinear optics. Nonlinear optical phenomena are so important for both fundamental and practical reasons that nowadays the main part of laser physics concerns nonlinear optics. Many devices and powerful techniques based on harmonic generation, stimulated scattering of light, wavemixing, optical phase conjugation, self focusing and other nonlinear optical effects have been developed. Usual materials, however, have a very small optical nonlinearity, so that very powerful laser pulses are required. To obtain appreciable nonlinear effects using cw laser sources, a medium is needed having exceptionally high nonlinear optical response. The development of new materials showing such huge optical nonlinearity is still in progress. In the case of Kerr-like nonlinearity, the nonlinear optical response of the material is measured by the change (δn of its refractive index n produced by the light itself. In most materials one has

$$\delta n = n_2 I \tag{1}$$

where I is the intensity of the laser beam and n_2 is a parameter characterizing the optical nonlinearity of the medium. This index change produces a corresponding change

$$\delta\psi = \frac{2\pi}{\lambda}\int_0^L \delta n \, dz = \frac{2\pi L}{\lambda}\langle n_2\rangle I \tag{2}$$

in the phase of the optical beam of wavelength λ, when it traverses a layer of thickness L. In Eq.(2) we introduced the Kerr index, $\langle n_2\rangle = L^{-1}\int_0^L n_2(z)dz$ averaged over the layer thickness. The change $\delta\psi$ can be easily measured, e.g. with interferometric techniques.

A. Miller et al. (eds.), Nonlinear Optical Materials and Devices for Applications in Information Technology, 103–139.

Besides a large n_2, in most practical applications one requires also fast response time. Unfortunately these two requirements cannot be met simultaneously, since highly nonlinear media have usually very long response time, as shown in Fig. 1. Liquid crystals have nonlinear Kerr index n_2 in the range of $n_2 \cong 10^{-3}$ cm^2/W more than 10^8 times the n_2 of a typical organic liquid as CS_2. The only materials having comparable nonlinear response are photorefractive materials. The response time, however, is very long (seconds or more) in both cases.

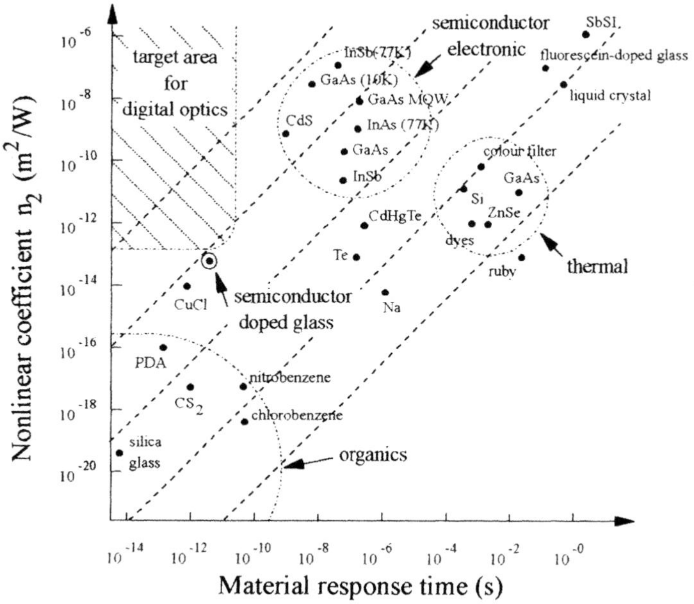

Figure 1. Survey of nonlinear optical materials by Kerr coefficient n_2 and dominant material response time.

The discovery of the extraordinarily large optical nonlinearity of liquid crystals has made possible to observe several nonlinear optical processes, which cannot be even imagined in usual materials, as self oscillation, optical bistability, self-induced stimulated light scattering, etc., by using laser powers in the milliwatt range [1].

Beside high optical nonlinearity, liquid crystals have other very useful properties which render these materials very appealing for some applications. They are fluid, so they can be easily inserted in microscopic devices or mixed with other substances to form new composite materials having peculiar optical and thermo-mechanical properties or even dispersed in the form of microscopic droplets in a transparent polymeric matrix. They are strongly affected by external electric or magnetic fields, providing excellent materials for electro-optical devices and, finally, they have low cost. Shortcomings are the long response time, the relatively large light scattering and the fact that the liquid crystalline phase exists only in a limited temperature range. For example *5CB*, a compound widely used in optics, is nematic in the range 24 to 35°C. The nematic temperature range can be increased by conveniently mixing different liquid crystals. A commonly used mixture is *E7*, from *BDH*, whose nematic range is from -10 to 60°C.

2. The liquid crystal mesophases

Liquid crystals are composed of large organic molecules with typical cylindrical structure. In many respects, one can regard such materials as made of microscopic rigid rods, that are free to move along the fluid but can have some tendency to remain aligned along a common average direction. At high temperatures the molecules are completely randomly oriented and the substance behaves as an isotropic ordinary liquid. At lower temperatures new phases become stable, where the tendency of molecules to align prevails and the fluid becomes anisotropic. In these phases (usually called mesophases) the fluid shows a lower symmetry than in the isotropic phase, i.e. is more ordered. The degree of order of the mesophases is usually characterized by an order parameter, which is nonzero in the considered phase, but vanishes, by symmetry, in the isotropic phase.

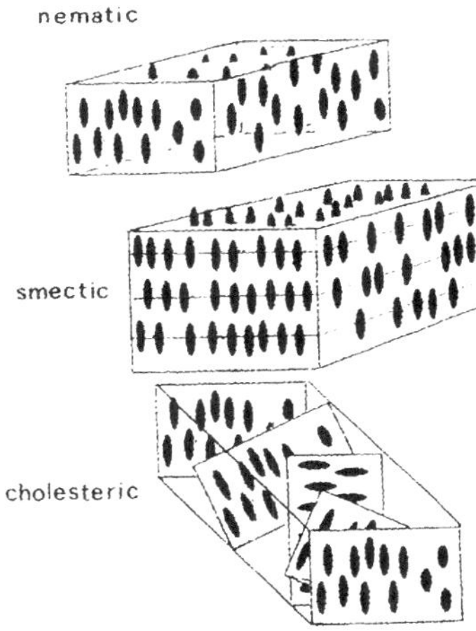

Figure 2. The three mesophases: (a) nematic; (b) cholesteric; (c) smectic.

Macroscopic samples may exist in three typical distinct mesophases as a function of the temperature: the nematic, skeptic and cholesteric mesophases, Fig.2. In the nematic phase the molecules have their centers of mass at random in the fluid but they are directionally correlated to point, on the average, along a common direction. In the smectic phases (there are actually eight known smectic phases) the molecule center of mass are ordered along parallel layers (the smectic planes), but still the molecules tend to be aligned in each layer. The cholesteric phase is similar to the nematic one on the microscopic scale, but over distances of the order of one micron the average direction of the molecules tends to assume a characteristic helical structure, having its pitch strongly dependent on temperature [2]. Cholesterics are evidently not centrosymmetric and show optical activity. Moreover, they are uniaxial and not polar. Also nematics, having the same local structure of cholesterics, are uniaxial and not polar, but, being centrosymmetric, cannot form helical structures nor have optical activity. Cholesteric like structures can be obtained nevertheless also innematics by adding some small quantities of chiral molecules. Smectics can be polar and hence ferroelectric. Moreover, since they show two preferred directions (the normal to the smectic layers and the molecular orientation in each layer) they can be biaxial. In the present work we will deal principally with nematics because of their simple structure and also because they are the most widely studied in the field of nonlinear optics.

3. Microscopic approach

The order of the molecules in the nematic phase can be quantitatively described by giving the probability $dP = f(\theta,\phi)d\Omega$ of finding a fixed molecule (assimilated to a rigid rod) in the solid angle $d\Omega = \sin\theta\, d\theta\, d\phi$ forming the angles (θ, ϕ) with respect to the average direction $\hat{n}$ (that we take as polar z-axis of a large number of molecules in the considered volume dV. The average direction $\hat{\boldsymbol{n}}$ of the molecules contained in dV is named the molecular director. In the case of the nematic phase the following symmetry properties of the distribution function $f(\theta,\phi)$ apply:

1. $f(\theta,\phi)$ has cylindrical symmetry, i.e. $f(\theta,\phi) = f(\theta)$ is independent of ϕ;
2. $f(\theta) = f(\pi-\theta)$, i.e. $\hat{\boldsymbol{n}}$ and $-\hat{\boldsymbol{n}}$ are physically equivalent.

The last property is due to the fact that nematic fluids are not polar. The alignment is often characterized not by the whole distribution function $f(\theta,\phi)$ but by some numerical parameters related to it. Denoting with $\hat{\boldsymbol{l}}$ the unit vector along the molecule, we cannot use the parameter, $\langle \hat{l}\cdot\hat{n}\rangle = \langle\cos\theta\rangle$, because it is zero and there is no average dipole. We must use then the average quadrupole tensor given by

$$Q_{ij} = \langle q_{ij}\rangle = \tfrac{1}{2}\langle 3\hat{l}_i\hat{l}_j - \delta_{ij}\rangle . \qquad (i, j = x, y, z) \qquad (3)$$

Because of property (1) of $f(\theta,\phi)$ we have $\langle \hat{l}_x\hat{l}_y\rangle = \langle \hat{l}_x\hat{l}_z\rangle = \langle \hat{l}_y\hat{l}_z\rangle = 0$, $\langle \hat{l}_x^{\,2}\rangle = \langle \hat{l}_y^{\,2}\rangle = \tfrac{1}{3}(1-S)$, $\langle \hat{l}_z^{\,2}\rangle = \tfrac{1}{3}(1+2S)$, and instead of the whole tensor Q_{ij} we can use the quadrupole momentum $S \equiv Q_{ij}\, n_i\, n_j$, defined as

$$S = \langle P_2(\cos\theta)\rangle = \tfrac{1}{2}\langle 3\cos^2\theta - 1\rangle , \qquad (4)$$

where $P_2\,(\cos\,\theta)$ is the Legendre polynomial. The constant in front of S is chosen so that if all molecules are exactly aligned along the common direction $\hat{\boldsymbol{n}}$ $(\theta = 0)$ we have $S = 1$. For random orientation, instead, we get $\langle\cos^2\theta\rangle = \tfrac{1}{3}$ and $S = 0$. Thus S is a measure of the alignment. In the isotropic phase the molecules are randomly oriented and $S = 0$; in the nematic phase they are more or less well aligned along the director $\hat{\boldsymbol{n}}$, depending on the temperature. Thus, we expect that in the nematic phase S is less than 1 and a function of temperature T. A detailed study of the isotropic-nematic phase transition is beyond the scope of this work, but it is found that S is a universal function of the reduced temperature $\tau = T\,V^2 / (T_c\,V_c^2)$, where T_c is the critical temperature of the nematic-to-isotropic phase transition, and V and V_c are the volumes at T and T_c, respectively. For most nematics the order parameter can be approximated as

$$S = (1 - 0.98\tau)^{0.22}. \tag{5}$$

Since nematics are locally uniaxial, the tensor order parameter assumes the typical form

$$Q_{ij} = Q(T)\left(n_i n_j - \tfrac{1}{3}\delta_{ij}\right) \tag{6}$$

with $Q(T) = \frac{3}{2} S(T)$

The order parameter can be measured from the macroscopic tensor properties of the liquid crystalline material. For example, denoting with χ_{ij} the magnetic susceptibility, we can define a tensor order parameter as

$$Q_{ij} = G\left(\chi_{ij} - \frac{1}{3}\delta_{ij} Tr[\chi]\right) \tag{7}$$

where $Tr[\chi] = \sum \chi_{ii}$ and G is a constant. The normalization constant G can be chosen at will, but it is often convenient to define G by setting $Q_{zz}=1$ in a sample fully oriented along the z-axis. The choice of the magnetic susceptibility as order parameter is arbitrary. Any tensorial quantity could have been used as well. For example, in optics, the natural choice is the dielectric constant ε_{ij} at the proper optical wavelength. The advantage of using the magnetic susceptibility is that it can be directly related to the molecular properties. In fact, if $\alpha_{\parallel}$ and $\alpha_{\perp}$ are the magnetic susceptibilities of one molecule, respectively parallel and perpendicular to the rod axis, we have [3]

$$\chi_{\parallel} - \chi_{\perp} = c(\alpha_{\parallel} - \alpha_{\perp})S(T), \tag{8}$$

where $\chi_{\parallel}$ and $\chi_{\perp}$ are the principal values of the tensor χ_{ij} and c is a constant. The strong local field effects prevent to obtain such simple relationship in the case of the dielectric constant. Since nematics are uniaxial, the optical dielectric constant is related to the molecular director $\hat{\boldsymbol{n}}$ by

$$\varepsilon_{ij} = \varepsilon_0 \delta_{ij} + \varepsilon_a n_i n_j \tag{9}$$

where $\varepsilon_a=\varepsilon_e-\varepsilon_0$, $\varepsilon_0=n_0^2$, $\varepsilon_e=n_e^2$. n_0 and n_e being the ordinary and the extraordinary refractive indices of the material. Nematics have large optical birefringence, typical values being n_0=1.5 and n_e=1.7.

The tendency of molecules to align together can be accounted for by introducing a phenomenological interaction potential $U(\hat{l}_1,\hat{l}_2)$ depending on the directions $\hat{l}_1$ and $\hat{l}_2$ of the two interacting rods. The statistical mechanics calculations were carried out by Maier and Saupe [4] assuming the interaction only due to van der Waals forces. In the mean field approximation, the potential energy of a molecule directed along $\hat{l}_1$ is obtained by averaging $U(\hat{l}_1,\hat{l}_2)$ over all directions $\hat{l}_2$. One obtains $U(\hat{l})=uSq_{ij}n_in_j=uSP_2(\cos\theta)$, where S is the scalar order parameter and u is a constant[1]. Then, according to general principles of statistical mechanics, the distribution function at equilibrium is given by

$$f(\theta)=Ae^{-\frac{uSP_2(\cos\theta)}{k_BT}} \tag{10}$$

where k_B is Boltzmann's constant and A is a normalization factor.

Evaluating the order parameter S using this distribution function, one obtains a consistency implicit relation, that can be solved yielding S as a function of temperature. When k_BT/u is small, there are two solutions. The trivial solution S=0 is associated with the isotropic phase, while the other corresponds to the nematic mesophase. When T is below the critical temperature T_c given by $(k_BT_c)/u(T_c)$ = 4.55 the nematic phase is the stable one. At $T=T_c$ there is a first-order phase transition. The order parameter just below T_c is $S_c \equiv S(T_c)$ = 0.44. If we assume with Maier and Saupe that u is independent of temperature, the order parameter becomes a universal function of T/T_c. In despite of the very rough assumptions of the Maier and Saupe molecular model, the agreement with experiment is qualitatively good.

4. Macroscopic approach

In the absence of external constraints or applied fields the average molecular direction $\hat{\boldsymbol{n}}$ tends to be uniform at equilibrium. External fields may reorient the liquid crystal molecules so that $\hat{\boldsymbol{n}}$ varies from point to point in the sample. When the external constraints are removed, elastic forces appear to restore the equilibrium. These elastic forces can be derived in the framework of a macroscopic approach where it is assumed that the spatial variations of the tensor order parameter occur over a length scale much larger that the molecular scale. Experimental observations show that in the distorted nematic the local tensor properties are those of a uniaxial crystal, the magnitude of the anisotropy being the same as in the undistorted sample. Only the orientation of the local symmetry axis $\hat{\boldsymbol{n}}(\boldsymbol{r})$ at point $\boldsymbol{r}$ is rotated by the external field. We can therefore assume

[1] If steric interactions are considered, u becomes a complicated function of temperature.

that in the distorted nematic the order parameter is still given by Eq.(6), but $\hat{\boldsymbol{n}}_l$ with replaced by $\hat{\boldsymbol{n}}_l(\boldsymbol{r})$. The distorted state is then entirely described by the vector field $\hat{\boldsymbol{n}}(\boldsymbol{r})$. The director $\hat{\boldsymbol{n}}(\boldsymbol{r})$ varies slowly and smoothly with $\boldsymbol{r}$ *except* possibly on a few singular points or lines (defects). This description was initiated by Oseen [5]. The theory is usually expressed in terms of the distortion free energy density F_d, i.e. the energy (or virtual work) per unit volume associated with a distortion $\delta\hat{\boldsymbol{n}}$ of $\hat{\boldsymbol{n}}$. The requirements on F_d are

1. F_d must be invariant by rotations;

2. F_d must be even function of $\hat{\boldsymbol{n}}$ ($\hat{\boldsymbol{n}}$ and $-\hat{\boldsymbol{n}}$ are physically equivalent);

3. there are no terms linear in the spatial derivatives of $\hat{\boldsymbol{n}}$.

Using these properties and retaining only terms up to the second order in the space derivatives of $\hat{\boldsymbol{n}}$, it can be shown that F_d must have the general form

$$F_d = \frac{1}{2}\left[k_{11}(div\hat{\boldsymbol{n}})^2 + k_{22}(\hat{\boldsymbol{n}}\cdot rot\hat{\boldsymbol{n}})^2 + k_{33}(\hat{\boldsymbol{n}}\times rot\hat{\boldsymbol{n}})^2\right] \tag{11}$$

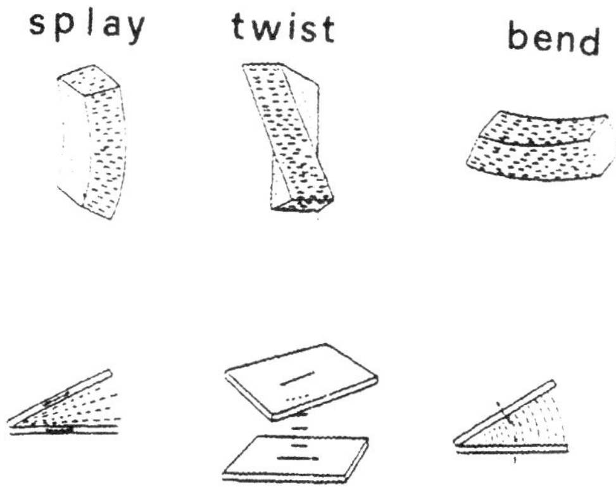

Figure 3 The three basic deformations: splay, twist, bend.

The positive constants k_{11}, k_{22} and k_{33} are named Frank's elastic constants and are related to the three basic types of deformations displayed in Fig.3, i.e. splay, twist and bend, respectively. The order of magnitude of the nematic elastic constants is u/a, where u is a topic interaction energy between molecules and a is a molecular dimension. Taking $U \approx k_BT \approx 0.026$eV and $a \approx 10\AA$ we expect $k \approx 3.10^{-7}$ dyne. Franck's elastic constants decrease strongly as T increases, being roughly proportional to the square of the scalar order parameter S [6]. Although all k_{ii} are of the same order of magnitude, the twist constant k_{22} is usually smaller than the other two. Nevertheless, in order to obtain simpler equations, the so-called one-constant approximation $k_{11} = k_{22} = k_{33} = K$ is widely used. In this approximation the distortion free energy reduces to

$$F_d = \frac{1}{2}k\left[(div\hat{\boldsymbol{n}})^2 + (rot\hat{\boldsymbol{n}})^2\right]. \tag{12}$$

In the absence of external fields the conditions for equilibrium in the bulk of the material can be obtained by minimizing the total distortion energy $\Im_d = \int_V F_d dV$ with respect to all variation of the director $\hat{\boldsymbol{n}}$ with the constraint $\hat{\boldsymbol{n}}^2=1$. This yields the Euler-Lagrange equations

$$h_i \equiv \partial_j\left(\frac{\partial F_d}{\partial\partial_j n_i}\right) - \frac{\partial F_d}{\partial n_i} = \mu(\boldsymbol{r})n_i \tag{13}$$

where $\mu(\mathbf{r})$ is a Lagrange multiplier. The vector $\mathbf{h}$ is named the molecular field and it is given by $\boldsymbol{h}=\boldsymbol{h}_S+\boldsymbol{h}_T+\boldsymbol{h}_B$, where

$$\begin{aligned} \boldsymbol{h}_S &= k_{11} grad(div\hat{\boldsymbol{n}}) \\ \boldsymbol{h}_T &= -k_{22}\left[A rot\hat{\boldsymbol{n}} + rot(A\hat{\boldsymbol{n}})\right] \\ \boldsymbol{h}_B &= k_{33}\left[\boldsymbol{B}\times rot\hat{\boldsymbol{n}} + rot(\hat{\boldsymbol{n}}\times\boldsymbol{B})\right] \end{aligned} \tag{14}$$

with $A = \hat{\boldsymbol{n}}\cdot rot\hat{\boldsymbol{n}}$ and $\boldsymbol{B} = \hat{\boldsymbol{n}}\times rot\hat{\boldsymbol{n}}$. Equations (13) simply state that at equilibrium $\boldsymbol{h}$ and $\hat{\boldsymbol{n}}$ are parallel. In the one constant approximation $\boldsymbol{h}$ reduces simply to

$$\boldsymbol{h} = K\nabla^2\hat{\boldsymbol{n}}. \tag{15}$$

The Euler-Lagrange equations in their full form are too complicated for practical use. In most cases, however, the deformation is plane, i.e. $\hat{\boldsymbol{n}}$ depends, at equilibrium, on one coordinate, that we take as z. Expressing the director $\hat{\boldsymbol{n}}$ in terms of polar angles (θ, ϕ), F_d can be then rewritten as

$$F_d = \frac{1}{2}\left(k_{22}\sin^2\theta + k_{33}\cos^2\theta\right)\sin^2\theta\left(\frac{d\phi}{dz}\right)^2 + \frac{1}{2}\left(k_{11}\sin^2\theta + k_{33}\cos^2\theta\right)\left(\frac{d\theta}{dz}\right)^2. \tag{16}$$

5. Anchoring forces

From Eq.(16) it is evident that $\Im_d$ affords an absolute minimum for a uniformly aligned sample, with θ=const. and ϕ=const. This is certainly true for a sample free of any external constraint. Constraints can arise either from externally applied fields or from boundaries. Boundary effects can be accounted for by adding suitable phenomenological surface terms to the total free energy $\Im_d$. In most practical cases, however, surface forces at boundaries are much stronger than bulk elastic forces and their only effect is to impose a well-defined direction to the director $\hat{\boldsymbol{n}}$ at the surface. This is the case of strong anchoring. In this case the equilibrium distribution of $\hat{\boldsymbol{n}}$ in the sample is obtained by minimizing $\Im_d$ with respect all fields $\hat{\boldsymbol{n}}(\boldsymbol{r})$ that assume given values at the walls containing the liquid crystal. For weak anchoring, we must minimize the sum of the bulk distortion free energy $\Im_d$ and the surface free energy. The anchoring depends in general on the nature of the external medium and on the conditions of the surface. If the external medium has some preferential direction, the director $\hat{\boldsymbol{n}}$ at the surface is usually parallel to the wall along this easy direction. This kind of alignment is said to be planar. In practice, planar alignment is obtained either by rubbing a glass plane surface (eventually coated with appropriate surfactants as polyvinyl alcohol) in one direction or by evaporating on the glass a thin film (3-30nm) of inorganic materials as SiO_x at oblique incidence. The anchoring obtained by rubbing is usually weaker than the one obtained by oblique thin film evaporation. Moreover, in the last case, depending on the evaporation angle, the director may result obliquely tilted with respect to the wall surface. If the external surface has no preferred direction one observes either a conical alignment with $\hat{\boldsymbol{n}}$ oblique at the surface (this happens for instance at free air interface of some liquid crystals) or a homeotropic alignment, where $\hat{\boldsymbol{n}}$ is perpendicular to the wall. Homeotropic alignment is usually obtained by coating the glass surface enclosing the sample with appropriate organic surfactants as *HTAB* or *DMOAP* [7]. If we consider a nematic film enclosed between two plane walls at $z = 0$ and $z = L$, the appropriate boundary conditions are

$$\theta(0) = \theta(L) = 0 \tag{17}$$

for homeotropic strong anchoring,

$$\theta(0) = \theta(L) = \frac{\pi}{2} \tag{18}$$

$$\phi(0) = \phi(L) = \phi_0 \tag{19}$$

for planar strong anchoring along the easy direction at angle ϕ_0 with respect to the x-axis (without loss of generality we can set $\phi_0 = 0$. In the case of weak anchoring, the boundary conditions can be written as

$$\frac{\partial F_d}{\partial(\partial_z\theta)} + \frac{\partial F_s}{\partial\theta} = 0 \tag{20}$$

at z=0

$$\frac{\partial F_d}{\partial(\partial_z\phi)} + \frac{\partial F_s}{\partial\phi} = 0 \tag{21}$$

and

$$\frac{\partial F_d}{\partial(\partial_z\theta)} - \frac{\partial F_s}{\partial\theta} = 0$$

at z=L (22)

$$\frac{\partial F_d}{\partial(\partial_z \phi)} - \frac{\partial F_s}{\partial \phi} = 0,$$

where $F_s(\theta, \phi)$ is the surface free energy density. In general, the Euler-Lagrange equations obtained from F_d must be completed with proper boundary conditions chosen from Eqs.(17-22).

It should be noted that having a uniform anchoring at the walls is essential to have good optical quality, because it is the surface anchoring which provides the initial alignment in the bulk. If the anchoring is poor, so is the bulk alignment and light scattering losses are greatly increased, lowering the overall transparency of the medium. In most cases both surfaces of the liquid crystal film are coated in the same way, so that a homeotropic or planar surface anchoring induces a similar alignment in the bulk. Sometimes, however, the film surfaces are coated differently, one for homeotropic and the other for planar alignment. Then the director $\hat{\boldsymbol{n}}$ is no longer uniformly oriented in the bulk, but it changes from the homeotropic to the planar alignment in passing from one wall to the other. Such an alignment is said to be hybrid. A not uniform orientation of the molecular director in the bulk can also be obtained by coating both walls for planar alignment but with different rubbing directions thus forming what is called a twisted nematic cell. Hybrid and twisted nematic cells have special properties that are often used in the applications.

6. Dynamically effects

By minimizing $\Im_d$ we obtain only the equilibrium distribution of $\hat{\boldsymbol{n}}$ in the sample. No information is obtained on the motion of the liquid crystal. From the hydrodynamical point of view a liquid crystal is a fluid having internal rotational degrees of freedom, represented by the field $\hat{\boldsymbol{n}}(\boldsymbol{r},t)$. The other hydrodynamical variable is the fluid velocity field $v(\boldsymbol{r},t)$. These two fields are generally coupled, yielding the so-called backflow effects. In some geometries and in particular in all known experiments of nonlinear optics with liquid crystals, the backflow effects are made negligible, so they will be neglected. Out of equilibrium the motion of the director $\hat{\boldsymbol{n}}$ is entirely dominated by viscous forces exchanged with the fluid (inertial terms are completely negligible). A macroscopic derivation of viscous forces in liquid crystals was made by Ericksen, Leslie, and Parodi (ELP) [8]. The theory is very complicated: it can be shown that to describe these forces one needs five independent viscosity coefficients. Nevertheless the drastic simplification of one viscous constant γ_1 is often made. In this case the viscous forces are accounted for by adding to the free energy the dissipation function

$$\Re = \gamma_1 \int_V \frac{\partial \hat{\boldsymbol{n}}}{\partial t} dV. \tag{23}$$

In the case of a plane non stationary deformation given by the director angles $[\theta(z,t), \phi(z,t)]$, the use of Eq. (23) adds the viscous terms

$$-\gamma_1 \frac{\partial \theta}{\partial t} \quad \text{and} \quad -\gamma_1 \sin^2\theta \frac{\partial \phi}{\partial t} \tag{24}$$

to the Euler-Lagrange equations obtained from Eq.(16) for θ and ϕ, respectively [see Eqs.(30) below].

7. External fields

The most useful property of liquid crystals is that they can be reoriented by external fields. The possibility of reorienting liquid crystals by applying static magnetic or electric fields was known since long time [9]. The reorientation with optical fields was discovered only in 1980 by Zel'dovich and Tabiryan and by Zolot'ko et al [10] and later quantitatively discussed by Durbin, Arakelian, and Shen in 1981 [11].

The effect of the external field can be accounted for by adding to the elastic free energy the contribution due to the field. Nematics are usually diamagnetic and not polar, so in the case of static electric and magnetic external fields we must simply add to $\Im_d$ the work

$$W = \frac{1}{4}\int_V \left(\int_0^D \boldsymbol{E} \cdot d\boldsymbol{D} + \int_0^B \boldsymbol{H} \cdot d\boldsymbol{B} \right) dV = \frac{1}{8\pi}\int_V (\boldsymbol{E} \cdot \boldsymbol{D} + \boldsymbol{H} \cdot \boldsymbol{B}) dV \tag{25}$$

needed to induce the polarization in the medium. The total free energy $\Im=\Im_d+W$ depends on $\hat{\boldsymbol{n}}$ implicitly, because of the constitutive relations [cfr. Eqs.(8) and (9)]

$$\begin{aligned} \boldsymbol{D} &= \varepsilon_0 \boldsymbol{E} + \varepsilon_a (\hat{\boldsymbol{n}} \cdot \boldsymbol{E}) \boldsymbol{E} \\ \boldsymbol{B} &= (1 + 4\pi\chi_\perp) \boldsymbol{H} + 4\pi\chi_a (\hat{\boldsymbol{n}} \cdot \boldsymbol{H}) \boldsymbol{H} \end{aligned} \tag{26}$$

where $\varepsilon_a=\varepsilon_e-\varepsilon_0$ and $\chi_a=\chi_\parallel-\chi_\perp$ are the dielectric and magnetic anisotropies, respectively[2]. Inserting Eqs.(26) into Eq.(25) the contribution to the free energy due to the external fields can be written as

$$W = \int_V \left\{ \frac{1}{2}\left[\chi_\perp H^2 + \chi_a (\hat{\boldsymbol{n}} \cdot \boldsymbol{H})^2 \right] + \frac{1}{8\pi}\left[\varepsilon_0 E^2 + \varepsilon_a (\hat{\boldsymbol{n}} \cdot \boldsymbol{E})^2 \right] \right\} dV. \tag{27}$$

In order to formulate a variational principle from which the equilibrium director distribution can be obtained by varying $\hat{\boldsymbol{n}}$ at constant fields $\boldsymbol{E}$ and $\boldsymbol{H}$, we need, in place of $\Im$, the new thermodynamic potential $\tilde{\Im}$ given by [12]

$$\tilde{\Im} = \Im - \frac{1}{4\pi}\int_V (\boldsymbol{E} \cdot \boldsymbol{D} + \boldsymbol{H} \cdot \boldsymbol{B}) dV = \Im_d - W. \tag{28}$$

[2] In actual experiments with electric fields, alternate fields are used (frequency $\approx$1kHz) to avoid conduction effects. The dielectric constant must be therefore evaluated at the frequency used in the experiment.

A straightforward calculation based on the variational principle $\delta\Im=0$ and on the dissipation function in Eq.(23) shows that the dynamically equations can be written out in the form of a balance between viscous, elastic, magnetic and electric torques per unit volume acting on the director $\hat{\boldsymbol{n}}$, viz.

$$\tau_v + \tau_d + \tau_m + \tau_e = 0. \tag{29}$$

In the case of plane distortion, the equations for $\theta(z,t)$ and $\phi(z,t)$ are

$$-\gamma_1 \frac{\partial\theta}{\partial t} + \left[k_{33} - (k_{33} - k_{11})\sin^2\theta\right]\left(\frac{\partial^2\theta}{\partial z^2}\right) - (k_{33} - k_{11})\sin\theta\cos\theta\left(\frac{\partial\theta}{\partial z}\right)^2 -$$

$$\left[k_{33} - 2(k_{33} - k_{22})\sin^2\theta\right]\sin\theta\cos\theta\left(\frac{\partial\phi}{\partial z}\right)^2 + (\hat{\boldsymbol{n}} \times \tau_{ext})_z = 0$$

$$-\gamma_1 \sin^2\theta \frac{\partial\phi}{\partial t} + \frac{\partial}{\partial z}\left[\left(k_{22}\sin^2\theta + k_{33}\cos^2\theta\right)\frac{\partial\phi}{\partial z}\right] + (\tau_{ext})_z = 0 \tag{30}$$

where $\boldsymbol{\tau}_{ext}=\boldsymbol{\tau}_m+\boldsymbol{\tau}_e$ is the total torque due to external fields. The torques $\boldsymbol{\tau}_m$ and $\boldsymbol{\tau}_e$ are given by

$$\tau_m = \frac{1}{4}\boldsymbol{B} \times \boldsymbol{H} = \chi_a(\hat{\boldsymbol{n}} \cdot \boldsymbol{H})(\hat{\boldsymbol{n}} \times \boldsymbol{H})$$

$$\tau_e = \frac{1}{4}\boldsymbol{D} \times \boldsymbol{E} = \frac{\varepsilon_a}{4\pi}(\hat{\boldsymbol{n}} \cdot \boldsymbol{E})(\hat{\boldsymbol{n}} \times \boldsymbol{E}). \tag{31}$$

Once the external fields $\boldsymbol{H}$ and $\boldsymbol{E}$ are given, Eqs.(30) are a set of nonlinear partial parabolic differential equations that can be solved with appropriate initial and boundary conditions. At steady state $\partial\theta/\partial t = \partial\phi/\partial t = 0$ and Eqs.(30) reduce to a set of ordinary differential equations yielding the equilibrium distribution of $\theta(z)$ and $\phi(z)$.

Appreciable reorientation of the sample is achieved when the external torque is comparable with the elastic one: this happens when $\chi_a H^2 \approx K/L^2$ or $\varepsilon_a E^2 \approx 4\pi\ K/L^2$, where K is the elastic constant and L the sample thickness. Taking $K\approx10^{-6}$ dyne, $\chi_a\approx10^{-7}$, $\varepsilon_a\approx0.1$ and $L\approx30\mu$m, we get $H\approx10^3$ gauss and $E\approx10^3$ V/cm.

8. The Fréedericksz effect

Consider a nematic film of thickness L enclosed between glass walls coated for homeotropic alignment along z in a magnetic field $\boldsymbol{H}$ directed along x. Since $\boldsymbol{H}$ and $\hat{\boldsymbol{n}}$ are perpendicular, no torque is exerted by the magnetic field and the anchoring forces at the walls maintain the homeotropic alignment. Any fluctuation of $\hat{\boldsymbol{n}}$ off the z-axis, however, tends to be enhanced by the magnetic torque. If the applied field becomes larger than a critical threshold value H_c the surface anchoring forces are overcome by the magnetic field and the sample is reoriented in the xz-plane towards $\boldsymbol{H}$. One says that the sample has undergone a Fr'eedericksz transition. Since at the threshold the angle θ is almost zero, we may evaluate the threshold field by linearizing Eqs.(30) for small θ and setting $\phi = 0$. The result is

$$H_c = \left(\frac{k_{33}}{\chi_a} \right)^{\frac{1}{2}} \frac{\pi}{L}. \tag{32}$$

The Freedericksz transition is second order, in the sense that the equilibrium reorientation $\theta_m(H)$ inside the sample (θ_m is usually evaluated at the center of the nematic film) is a continuous function of the applied field $\boldsymbol{H}$, as shown in Fig.4.

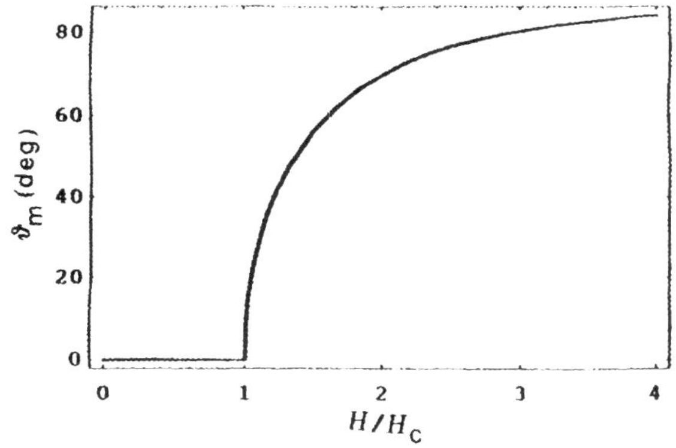

Figure 4 θ_m at the center of the sample as a function of the normalized field strength H/H_c in the magnetic field-induced Freedericksz transition.

The determination of the Freedericksz threshold H_c may be conveniently used to measure the elastic constant k_{33}. In general all experimental geometries where $\hat{\boldsymbol{n}}$ is initially perpendicular to the applied field at any point in the sample have a threshold behavior. By changing the geometry, different elastic constants can be determined. Similar results hold true in the case of the electric field, but the solutions of the relevant equations is slightly complicated by the fact that the distortion of the director field reacts to the electric field distribution inside the sample.

It should be noted that also the state with $\hat{\boldsymbol{n}}$ parallel to the applied field has zero torque. It can be shown, however, that increasing the field this state becomes more and more stable as long as χ_a or ε_a are positive. Freedericksz transition cannot occur. The situation is reversed for media having negative dielectric anisotropy ε_a ($\chi_a > 0$ in all known nematics): the state $\hat{\boldsymbol{n}} \parallel \boldsymbol{E}$ is now unstable and shows the transition, while the other state is always stable. Negative ε_a are obtained in practice by driving some nematic liquid crystals as MBBA with electric fields having appropriate frequency. Finally, states where the director $\hat{n}$ is oblique with respect to the external field have no threshold behavior and reorientation is induced for all values of the field.

Liquid crystals are usually highly birefringent (typical values of the ordinary and extraordinary refractive indices are $n_0 \approx 1.5$ and $n_e \approx 1.7$, respectively). When they are reoriented by the external fields the birefringence changes as well and may strongly affect the polarization of a beam passing through the film. Many electro-optical devices and displays are based on this peculiar property of liquid crystals. The response time of such devices is quite long ($\approx$ 0.1-1s) when nematics are used. Response times of the order of 10μs or less can be obtained by using smectic-C ferroelectric liquid crystals [13].

As examples of electro-optical devices based on liquid crystals we briefly consider here the Light Valve (*LV*) and the Polymer Dispersed Liquid Crystal (*PDLC*) panel.

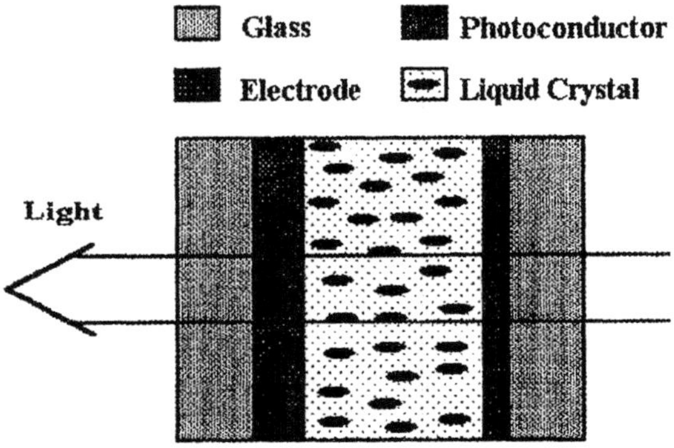

Figure 5 Light Valve (after Ref.[1]).

A scheme of light valve is shown in Fig.5. The liquid crystal film and a weakly absorbing photosemiconductor are sandwiched between transparent electrodes to which a voltage is applied by a radiofrequency generator. The illumination of the photoconductive layer changes its resistivity and hence the electric field seen by the liquid crystal increases locally, producing reorientation. The phase of the beam traversing the *LV* is changed accordingly. In the first approximation the induced optical phase change is proportional to the local intensity of the incoming beam. The overall effect is changing any amplitude modulation (AM) of the incident wave into a corresponding phase modulation (PM) with very high sensitivity. Phase changes of $\approx 2\pi$ can be obtained with light intensity $I \approx 10^{-5}$ - 10^{-6} W/cm^2. Phase holograms and phase conjugation were indeed obtained by using a *LV* at incident power of 10μW and intensity of 10 μW/cm^2 [14]. Low power optical bistability with a *LV* was also reported [15].

Polymer Dispersed Liquid Crystals are usually made by phase separation from a mixture of nematic liquid crystal and epoxy resin. In this way microdroplets of liquid crystal are formed, randomly dispersed in the polymeric matrix [16]. The polymeric matrix is chosen so that its refractive index is very close to the ordinary index of the liquid crystal. The final sample is then enclosed between conductive glasses. In the absence of external voltage, the sample appears opaque, because the molecular directors in different microdroplets have random initial directions and, due to the refractive index mismatch with the epoxy matrix, the incoming light is strongly scattered around. With the voltage present, the droplets reorient along the field so that the light beam sees the liquid crystal ordinary refractive index, which is very close to the one of the polymer. The light scattering is then strongly reduced and the film becomes transparent. When the voltage is removed, the previous disordered state is restored and the film returns opaque. The switching time is usually of the order of 10 - 100ms. The great advantage of such devices is that they can have very large area ($\approx 1\text{m}^2$) so that panels having adjustable transparency can be made with this technique.

9. The optical reorientation

The reorientation of nematic liquid crystals by laser beams presents both analogies and differences with the static field case. The analogies come out from the fact that the torque $\boldsymbol{\tau}_e$, being proportional to the square of the electric field strength [see Eqs.(31)], is not zero when averaged over the period of the optical field. Considering a monochromatic optical wave of frequency ω and using the complex notation, we may write the optical torque acting on $\hat{\boldsymbol{n}}$ as

$$\tau_o = \frac{1}{8\pi}\Re\left[\boldsymbol{D}^* \times \boldsymbol{E}\right] = \frac{\varepsilon_a}{8\pi}\Re\left[\left(\hat{\boldsymbol{n}}\cdot\boldsymbol{E}^*\right)\left(\hat{\boldsymbol{n}}\times\boldsymbol{E}\right)\right]. \tag{33}$$

We see from this equation that a linearly polarized optical field is equivalent to a static field having amplitude $|E|^2/2$. Then, for instance, in the case of a linearly polarized laser beam impinging at normal incidence on a homeotropically aligned nematic film, we expect to find a threshold in the intensity I of the incident laser beam to induce reorientation. This was indeed found in the experiment and shown in Fig.6. The phenomenon is called Optical Freedericksz Transition (OFT). The absence of the threshold at oblique incidence is also evident in the figure, in qualitative agreement with the observations made with static fields. The OFT is second-order, as in the static case, and has no hysteresis. First-order reorientational transitions can be achieved by adding to the optical field a suitable static bias, e.g. a constant magnetic field [17]. In this case weak hysteresis loops have been observed, as shown in Fig.7. Much stronger optical bistability and hysteresis loops can be obtained by doping the nematic with small amounts of chiral agents [18]. An example is shown in Fig.9.

The optical reorientation of liquid crystals differs however from the static-field case in some fundamental respects:

1. the optical radiation is usually focused onto the sample, so that the region where the reorientation occurs is very small ($\approx 100\mu\text{m}$) as compared to the sample size (but often of the same order of the film thickness);

2. when two beams are superimposed in the sample, they can act incoherently, i.e. their intensities (and hence torques) add and not their fields.

3. the light may be elliptically polarized.

Points (1) and (2) will be briefly considered in further sections. For now we consider only point (3), which is particularly important from the fundamental point of view.

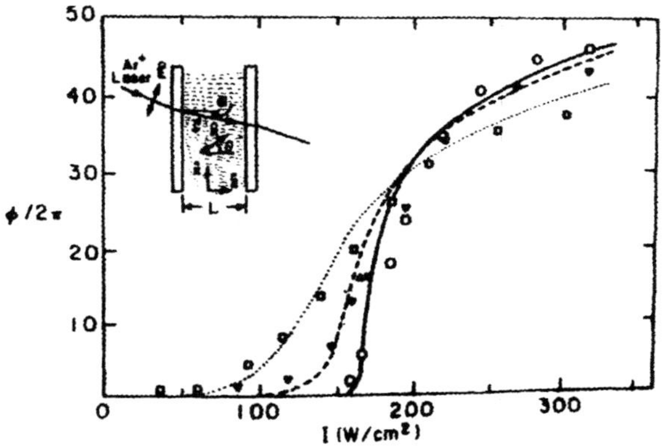

Figure 6 Induced phase shift $\pi/2$ as a function of the light intensity I in the Optical Freedericksz Transition. Normal incidence: circles. Oblique incidence: triangles and squares (after Ref.[11]).

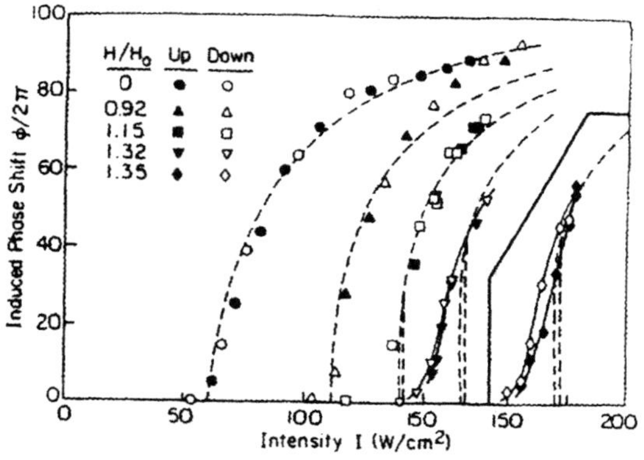

Figure 7 Optical Freedericksz transition biased by a magnetic field. The insert shows a weak hysteresis loop (after Ref. [17]).

Elliptically polarized light can reorient a nematic film. But the process is now essentially different from the static case, because elliptical light carries intrinsic angular momentum. The nematic material is a fluid having internal orientational degrees of freedom parameterized $\hat{\boldsymbol{n}}$. The fluid internal degrees of freedom couple directly with the internal degrees of freedom of light, i.e. its polarization state or, say, its angular momentum. The angular momentum lost by light is transferred to the medium and produces a corresponding torque on the director $\hat{\boldsymbol{n}}$. As we shall see later, in some cases the angular momentum transfer can put the director $\hat{\boldsymbol{n}}$ into continuous rotation, realizing a process of Self-Induced Stimulated Light Scattering (SISLS) [19].

In general, the way the light reorients the medium depends on its polarization state. In turn, the light polarization state depends on the sample local birefringence, i.e. on the sample orientation. We need therefore a set of coupled equations describing both the liquid crystal orientation and the light polarization state simultaneously. It should be noted that liquid crystals being highly transparent media, only the beam polarization is affected by molecular reorientation, while its intensity I remains practically unchanged in traversing the medium [3].

In principle, to describe the propagation of a light beam in a nematic medium we must solve Maxwell's equations for the optical field and the torque equations for the molecular director field simultaneously. In the case of monochromatic light of frequency $\omega=2\pi/\lambda$ *both* Maxwell's and director equations can be obtained by minimizing the thermodynamic potential

$$\tilde{\mathfrak{I}} = \mathfrak{I}_d - \frac{1}{16\pi}\int_V \left(\boldsymbol{D}\cdot\boldsymbol{E}^* - \boldsymbol{H}\cdot\boldsymbol{H}^*\right)dV \tag{34}$$

with $\boldsymbol{D}$ given by the first of Eqs.(26) and

$$\boldsymbol{H} = -\frac{i}{k_0}\,rot\boldsymbol{E} \qquad k_0 = \frac{\omega}{c}. \tag{35}$$

When $\tilde{\mathfrak{I}}$ is minimized with respect to the fields $\boldsymbol{E}(\boldsymbol{r})$ and $\hat{\boldsymbol{n}}(\boldsymbol{r})$ we get Maxwell's equations

$$rot\ rot\boldsymbol{E} = k_0^2\boldsymbol{D} \tag{36}$$

and the torque equations

$$\hat{\boldsymbol{n}}\times\boldsymbol{h}+\tau_o = 0 \tag{37}$$

with τ_o given by Eq.(33) and $\boldsymbol{h}$ by Eqs.(14). These equations describe steady states only. Appropriate viscous torques must be added in the dynamic case. Equations (36) and (37), however, are too complicated and are never used in practice, even in the simple case of plane wave solutions (all fields depend on one coordinate). Approximate methods have been then introduced to

---

[3] Nematic films are usually very thin ($\approx 100\ \mu$m) and self-focusing *inside* the medium is negligible.

describe the propagation of a plane wave in stratified anisotropic transparent media based essentially on the Geometric Optics Approximation [20].

10. The Geometric Optics Approximation

The Geometric Optics Approximation (*GOA*) is widely used to describe the light propagation in inhomogeneous anisotropic media as liquid crystals. The approximation is usually very good, because the director field $\hat{\boldsymbol{n}}(\boldsymbol{r})$ varies very slowly over distances of the order of the optical wavelength λ. Even in the *GOA*, however, the light propagation problem is still very complicated, due to the birefringence of the medium. A great simplification is obtained by assuming plane wave solutions [$\boldsymbol{E}=\boldsymbol{E}(z)$, $\hat{\boldsymbol{n}}=\hat{\boldsymbol{n}}(z)$]. All previously mentioned works on *GOA* [20] use the plane wave approximation. The final goal should be obtaining a set of coupled equations describing the director field and the light polarization, the latter parameterized, e.g., by Jones' or Stokes' vectors. The light intensity should enter the equations as a control parameter. Unfortunately this goal has not been reached yet, even in the plane wave approximation, except in the case of normal incidence [21] or in geometries where the light polarization does not change, as in the OFT. Since both cases have practical relevance, we will consider them separately.

10.1. PLANE WAVE, NORMAL INCIDENCE, ELLIPTICAL POLARIZATION

In this case, the problem can be still handled in terms of a suitable thermodynamic potential, whose density acts as a Lagrangian for the director polar angles θ and ϕ and for the angle ψ yielding the orientation of the light polarization ellipse. Although a Lagrangian approach is possible, a Hamiltonian approach has some advantages. Following Ref. [21], the director equations and the equations governing the light polarization are derived as Hamilton's equations associated to the Hamiltonian function

$$H(p_\theta,\theta;p_\phi,\phi;-l_z,\psi)=H_d(p_\theta,\theta;p_\phi,\phi)+w(\theta,\phi;-l_z,\psi) \tag{38}$$

where

$$H(p_\theta,\theta;p_\phi,\phi)=\frac{1}{2}\left[\left(k_{22}\sin^2\theta+k_{33}\cos^2\theta\right)\sin^2\theta\right]^{-1}p_\phi^2+\frac{1}{2}\left[k_{11}\sin^2\theta+k_{33}\cos^2\theta\right]^{-1}p_\theta^2 \tag{39}$$

is the elastic deformation contribution and w is given by

$$w=\frac{I}{2c}\left[\bar{n}(\theta)-n_0\right]\left\{1+\sqrt{1-e^2}\cos\left[2(\psi-\phi)\right]\right\}. \tag{40}$$

with $e=-(w/I)l_z$ and

$$\bar{n}(\theta)=\frac{n_e n_0}{\sqrt{n_e^2\cos^2\theta+n_0^2\sin^2\theta}}. \tag{41}$$

The quantities ψ, e and l_z turn out to have simple physical meaning: ψ is the angle formed by the major axis of the light polarization ellipse with the laboratory frame x-axis, e is the polarization ellipticity and finally l_z is the angular momentum flux carried by the beam along z. The relevant thermodynamic potential can be recovered by switching back to the Lagrangian formulation, using θ, ϕ and ψ as generalized coordinates.

The equations for θ and ϕ are obtained by adding to the left-hand side of Eq.(30) the following optical contributions

$$\tau_\phi\equiv\frac{I}{c}\left[\bar{n}(\theta)-n_0\right]\sqrt{1-e^2}\sin\left[2(\psi-\phi)\right] \tag{42}$$

and

$$\tau_\theta\equiv\frac{I}{2c}\bar{n}'(\theta)\left\{1+\sqrt{1-e^2}\cos\left[2(\psi-\phi)\right]\right\} \tag{43}$$

where $\bar{n}'(\theta)=d\bar{n}/d\theta$. The equations for the light polarization can be written in terms of Stokes' unit vector $\hat{s}$ on the Poincar'e sphere as

$$\frac{\partial\hat{s}}{\partial z}=\Omega\times\hat{s} \tag{44}$$

where

$$\Omega=\frac{2\pi}{\lambda}\left[\bar{n}(\theta)-n_0\right](\cos 2\phi,\sin 2\phi,0). \tag{45}$$

The Hamiltonian (38) depends on the difference ψ-ϕ and hence it is invariant with respect to the transformation $\psi\to\psi+\delta$, $\phi\to\phi+\delta$. By Noether's theorem we have therefore

$$p_\phi-l_z=\left(k_{22}\sin^2\theta+k_{33}\cos^2\theta\right)\sin^2\theta\frac{d\phi}{dz}+\left(\frac{I}{\omega}\right)e=const. \tag{46}$$

This conservation law has a simple physical meaning. The momentum p_ϕ, in fact, is the negative of the angular momentum flux carried by the elastic forces in the nematic along z [22]. Relation (46) expresses therefore the conservation of the total (elastic + optical) angular momentum flux along the beam propagation direction. In the case of a nematic film of thickness L with homeotropic strong anchoring at the wall [$\theta(0)=\theta(L)=0$], the conservation law (46) assumes, at steady state, the simple form

$$\Delta e = e(L) - e(0) = 0. \tag{47}$$

We have then the following alternative: either the ellipticity e of the light emerging from the sample is the same as in the input beam and the system reaches a steady state or the beam ellipticity changes and the director continuously moves without reaching equilibrium[4]. The reorientation induced in a homeotropic nematic film by an elliptically polarized laser beam at normal incidence was studied carefully in Ref. [24]. Depending on the intensity and ellipticity of the beam, various dynamically regimes were found as director oscillations, nutations and eventually steady states.

The two cases of circular and linear polarization have received particular attention. In the case of circular polarization a uniform precession of $\hat{\boldsymbol{n}}$ about the beam direction was observed [25]. The transition from the undistorted state to the rotating state is first-order and accompanied by hysteresis. The angular momentum conservation law in this case is given by

$$\left(\frac{I}{\omega}\right)\Delta e = \gamma_1 \int_0^L \sin^2\theta \frac{\partial\phi}{\partial t} dz. \tag{48}$$

For a uniform precession $\partial\phi/\partial t = \Omega =$ const. In this case we have pure Self Induced Stimulated Light Scattering [19]. Photons of energy $\hbar\omega$ and angular momentum z-component $\hbar$ are scattered into photons of energy $\hbar\omega'$ and angular momentum z-component $-\hbar$. If $\dot{n} = I/(\hbar\omega)$ is the photon flux in the beam, the torque transferred to the medium is $2\dot{n}\hbar$ and the work done per unit time is $2\dot{n}\hbar\,\Omega$. This work is dissipated by viscous forces in the fluid and must be provided by the optical field. The photon frequency is then shifted towards the red according to the energy conservation law $\dot{n}\hbar(\omega' - \omega) = -2\dot{n}\hbar\Omega$, from which we get the frequency shift $\omega' - \omega = -2\Omega$. Since energy can be only dissipated, the photon frequency shift is always towards the red. The light beyond the sample is therefore a coherent superposition of unscattered right-handed circularly polarized photons with frequency ω and scattered left-handed circularly polarized photons with frequency $\omega - 2\Omega$. We may also say that the light emerging from the sample has an elliptical polarization continuously rotating with angular frequency Ω. The scattering differs from other stimulated processes as Brillouin or Raman stimulated light scattering because it is self-induced, in the sense that the photon frequency shift Ω is determined by the process itself and depends, in fact, by the light intensity. The angular velocity Ω and the ellipticity change Δe in the beam can be measured, so that the validity of relation(48) was experimentally checked [25].

---

[4] Checking Eq.(47) is not easy in practice, because of the deviations from the plane wave assumption due to the finite beam cross section [23].

The case of linear polarization has been studied by many authors because it corresponds to the Optical Freedericksz Transition (OFT) [10,11]. In this case the director and light polarization equations have the solution $\phi = 0$, $\psi = 0$, $e = 0$, and $\theta = \theta(z,t)$. The equation for $\theta(z,t)$ is

$$-\gamma_1 \frac{\partial\theta}{\partial t} + \left[k_{33} - (k_{33} - k_{11})\sin^2\theta\right]\left(\frac{\partial^2\theta}{\partial z^2}\right) - (k_{33} - k_{11})\sin\theta\cos\theta\left(\frac{\partial\theta}{\partial z}\right)^2 +$$

$$\left(\frac{I n_e n_o}{c}\right)\frac{(n_e^2 - n_o^2)\sin\theta\cos\theta}{(n_e^2\cos^2\theta + n_o^2\sin^2\theta)^{\frac{3}{2}}} = 0. \tag{49}$$

For homeotropic strong anchoring the initial and boundary conditions are $\theta(z,0) = 0$ and $\theta(0,t) = \theta(L,t) = 0$. The trivial solution $\theta(z,t) = 0$ becomes unstable when the light intensity I reaches the threshold value I_{th} for the OFT, given by

$$I_{th} = \frac{n_e^2 c k_{33}}{n_o(n_e^2 - n_o^2)}\left(\frac{\pi}{L}\right)^2. \tag{50}$$

Equation (50) is obtained by linearizing Eq.(49) at steady state ($\partial\theta/\partial t = 0$) and looking for a solution having the form

$$\theta \approx \theta_m \sin\left(\frac{\pi z}{L}\right) \tag{51}$$

As shown by Eq.(50), the threshold intensity scales with the sample thickness as L^{-2}. For a typical nematic sample having $L \approx 100\mu m$, $I_{th} \approx 1 kW/cm^2$. Such intensities can be easily obtained by focusing onto the nematic film a commercial cw laser beam.
The approach to the distorted steady state for $I > I_{th}$ can be obtained also from Eq.(49) by expanding it up to the second order in the small angle θ and looking for an approximate solution still of the form (51) with time dependent amplitude $\theta_m = \theta_m(t)$ [26]. The time constant τ_{on} to buildup the distortion depends on the light intensity I and it is given by

$$\tau_{on} = \frac{\gamma_1}{k_{33}}\left(\frac{I_{th}}{I - I_{th}}\right)\left(\frac{\pi}{L}\right)^2. \tag{52}$$

The time constant τ_{on} diverges as $I \to I_{th}$ from above. This time slowing down is characteristic of pretransitional phenomena near the critical point of a second-order phase transition. The relaxation time τ_{off} towards the equilibrium undistorted state at laser switching-off is given by

$$\tau_{off} = \frac{\gamma_1}{k_{33}}\left(\frac{\pi}{L}\right)^2. \tag{53}$$

The measurement of τ_{off} can be used to obtain the value of the viscoelastic ratio γ_1/k_{33} [27]. Typical values of τ_{off} are in the range 1-2s. Similar low response times are typical of all reorientational phenomena in liquid crystals and are the price to pay to have giant optical nonlinearity. Of course τ_{off} can be shortened by choosing materials with low viscosity and high elastic constant and using very thin films. In any case it is very hard to obtain response time below a few milliseconds.

10.2. PLANE WAVE, OBLIQUE INCIDENCE, LINEAR POLARIZATON

The problem of laser-induced molecular reorientation in nematics at oblique incidence has not been completely solved yet, even in the case of linearly polarized plane wave. The *GOA* solution is known only in the simplest case of linear polarization in the incidence plane, where a pure extraordinary wave is excited in the medium and the wave maintains its linear polarization. The *GOA* optical fields in this case have been evaluated as [28]

$$E_x = A\left(\varepsilon_{zz} - s^2\right)^{\frac{1}{4}} e^{i\psi(z)}$$

$$E_z = -A\frac{\varepsilon_{xz}\left(\varepsilon_{zz} - s^2\right)^{\frac{1}{2}} + sn_o n_e}{\varepsilon_{zz}\left(\varepsilon_{zz} - s^2\right)^{\frac{1}{4}}} e^{i\psi(z)} \tag{54}$$

where

$$\psi(z) = \frac{2\pi}{\lambda}\int_0^z \frac{-s\varepsilon_{xz}(u) + \left[\varepsilon_0\varepsilon_e\left(\varepsilon_{zz} - s^2\right)\right]^{\frac{1}{2}}}{\varepsilon_{zz}(u)} du \tag{55}$$

and $s = \sin\alpha$, α being the incidence angle. The constant A is related to the average z-component of Poynting's vector $I \equiv \langle S_z \rangle$ by

$$I = \frac{cn_o n_e}{8\pi}|A|^2. \tag{56}$$

The molecular director $\hat{\boldsymbol{n}}$ stays in the incidence (x,z)-plane and hence has components $\hat{\boldsymbol{n}} = (\sin\theta, 0, \cos\theta)$. The equation for θ is the same as Eq.(49) with the optical torque last term replaced by [29]

$$\tau_0 = \frac{(n_e^2 - n_o^2)I}{cG_+^2}\left[n_o n_e \frac{G_+ - 2s^2}{(G_+ - s^2)^{\frac{1}{2}}} \sin\theta\cos\theta - sG_- \right], \tag{57}$$

where

$$G_\pm = n_e^2 \cos^2\theta \pm n_o^2 \sin\theta \tag{58}$$

In the case of normal incidence ($s = 0$), Eq.(49) is recovered. At oblique incidence the optical torque (57) at $\theta = 0$ reduces to

$$\tau_0(0) = -\frac{(n_e^2 - n_o^2)Is}{cn_e^2} \tag{59}$$

which is different from zero and proportional to the intensity I. No intensity threshold is therefore expected to reorient a homeotropic sample. For small I we have in fact the small angle approximate steady state solution

$$\theta(z) \approx \frac{(n_e^2 - n_o^2)Is}{2cn_e^2 k_{33}} z(L - z). \tag{60}$$

In the same approximation the phase change suffered by the wave in traversing the medium is given by

$$\psi = \psi_0 + \beta\langle\theta\rangle, \tag{61}$$

where $\langle\theta\rangle = L^{-1}\int_0^L \theta dz$ and

$$\psi_0 = \frac{2\pi L}{\lambda}\left[\frac{n_o^2(n_e^2 - s^2)}{n_e^2}\right]^{\frac{1}{2}} \tag{62}$$

$$\beta = \left[\frac{2\pi(n_e^2 - n_o^2)L}{n_e^2\lambda}\right]s. \tag{63}$$

The first term ψ_0 is due to the natural birefringence of the homeotropic film, while the second term is due to the laser induced molecular distortion. Inserting the θ-profile (60), we obtain the nonlinear contribution to the wave phase change as

$$\delta\psi = \beta \frac{\left(n_e^2 - n_o^2\right)L^2}{12cn_e^2k_{33}} I_0 \sin\alpha\cos\alpha. \tag{64}$$

where we posed $s = \sin\alpha$ and $I = I_0 \cos\alpha$, I_0 being the light intensity measured in a section normal to the incident beam. Comparing this result with Eq.(2) and using Eq.(63) we get the effective average nonlinear optical Kerr index $\langle n_2 \rangle$ of the film as

$$\langle n_2 \rangle = \frac{\left(n_e^2 - n_o^2\right)^2 L^2}{12cn_e^4k_{33}} \sin^2\alpha\cos\alpha \tag{65}$$

The medium behaves therefore as a true optical Kerr medium having effective average nonlinear Kerr index given by Eq.(65). Inserting typical values for nematic materials (n_e = 1.7, n_o = 1.5, k_{33} = 3 10^{-6} dyne, c = 3 10^{10} cm/s, L = 100μm, $\alpha = 45^0$), we get $\langle n_2 \rangle \approx 10^{-4}$ cm^2/Watt. Such Giant Optical Nonlinearity (*GON*) is due to the cooperative character of molecular reorientation. In general, *GON* is obtained in experimental geometries where the optical torque is non zero inside the initially prepared sample. A case is the homeotropic (or planar) alignment at oblique incidence, another could be the hybrid alignment (one wall homeotropic, one wall planar) at normal incidence [30].

The case of pure ordinary wave in homeotropic nematic film (light polarization normal to the incidence plane) was also studied experimentally [31], but a suitable theoretical model is still lacking. The observed behavior is quite different from the case of the extraordinary wave. For ordinary wave and small incidence angle ($\alpha \approx 3^0$- 4^0), in fact, above a critical threshold for the light intensity, the molecular director starts to oscillate and cannot reach a steady state. For higher intensity, the observed director motion seems to assume the character of deterministic chaos [32]. We may have a qualitative insight about the phenomenon, observing that, above the threshold, the optical torque tends to move $\hat{\boldsymbol{n}}$ out of the incidence plane, so that the light polarization inside the film becomes elliptical. Then *SISLS* takes place and angular momentum is transferred to the medium, that starts to rotate about the beam propagation direction. The process is similar to the analogous oscillations observed in the case of elliptical polarization at normal incidence [24]. The angular momentum conservation cannot be preserved in steady states and the system is forced to move continuously. The oscillation process stops for higher incidence angles where a steady distorted state is reached. Why this happens is not clear yet.

11. Thermal Indexing

Although liquid crystals are highly transparent to visible light, a small fraction of the incident power is absorbed and the sample temperature is slightly increased, producing thermal indexing, i.e. a thermal gradient of the refractive index [1]. Both ordinary and extraordinary indices are affected by thermal indexing. The indices n_o and n_e are strongly affected by the order parameter S and collapse to the same value n_{iso} in the isotropic phase, as shown in Fig.8. The derivatives

dn_o/dT and dn_e/dT have opposite sign and become very large near the critical temperature T_c. In general $| dn_e/dT | > dn_o/dT$. The refractive index in the isotropic phase is roughly given by $(n_e+2n_o)/3$ and it is therefore intermediate between n_e and n_o. Depending on the experimental geometry, there is a complicated interplay between optical reorientation and thermal indexing. Let us consider, for example, a linearly polarized laser beam impinging at normal incidence onto a homeotropic cell, as in the standard OFT. Initially, the wave sees the index n_o and, since $dn_o/dT > 0$, the thermal indexing produces a phase change in the wave having the same sign as the optical reorientation. Once the sample is reoriented, however, the wave sees a mixture of n_o and n_e, depending on the director tilt angle θ [see Eq.(41)] and therefore, at some time, the thermal indexing is dominated by $dn_e/dT < 0$ and changes in sign. The existence of thermal indexing near the threshold for the OFT prevents usually to get reliable measurements of I_{th}. In fact a nonlinear behavior of the homeotropic cell due to thermal indexing was reported also for $I < I_{th}$ [33].

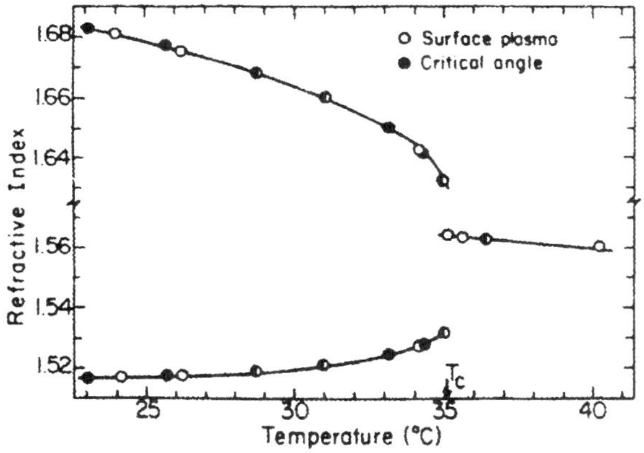

Figure 8 Ordinary and extraordinary refractive indices of nematic *5CB* as functions of temperature (after Ref. 57).

It is usually difficult to obtain a simple relationship between the thermal indexing and the beam intensity, because experimental parameters as the cell size and the thermal conductivity of the walls determine the environmental heat exchanges. Anyway the following simple estimation holds true for the temperature raising δT produced by a beam of intensity I:

$$\delta T \approx \frac{\alpha I}{c_p \sigma}\left(\frac{L}{\pi}\right)^2, \tag{66}$$

where α is the absorption coefficient, σ is the thermal diffusivity and c_p the specific heat per unit volume of the sample. Using typical values as $c_p \approx 10^7$ erg/(^{0}C cm^3), $\sigma \approx 10^{-3}$ cm^2/s, $\alpha \approx 0.1$ cm^{-1}, $L \approx 100\mu$m, we get $\delta T \approx 5^0$C, for $I \approx 500$kW/cm^2. The laser induced refractive index change due to thermal indexing is given therefore by $\delta n_o = (dn_o/dT)\delta T$ and $\delta n_e = (dn_e/dT)\delta T$ for the ordinary and extraordinary indices, respectively. Both changes are proportional to the light intensity I producing optical Kerr nonlinearity.

Although reorientational effects in liquid crystals are much stronger than thermal ones, distinguishing between the two it is often not easy. Thermal effects are weaker but faster than reorientational effects. The thermal response time is approximately given by

$$\tau_T \approx \frac{L^2}{\sigma\pi^2}. \qquad (67)$$

With the previous values we get $\tau_T \approx 10^{-2}$s, i.e. two order of magnitude less than orientational response times.

Light absorption in liquid crystals may lead to phenomena other than thermal indexing which may produce a change in the refractive index. An example are photostimulated chemical reactions [34] or changes in the elastic constant anisotropy [35]. These, however, are minor effects and in general they are observable only very close to the critical temperature T_c.

Nonlinear optical effects due to laser local heating can be observed in all liquid crystals, including Polymer Dispersed Liquid Crystal [36]. Thermal effects can be greatly enhanced by adding small amounts of organic dyes to the pure nematic. It should be stressed, however, that some dyes (as antraquinone dyes) may interact with the nematic host when they are excited by the incoming radiation, so that the effective optical torque may be increased over more than two order of magnitude [37]. In some cases the sign of the optical torque is even reversed, changing the Kerr medium from self focusing into self-defocusing [38]. This may be important for applications, because negative optical Kerr media are very unusual. It is worth noting that the increase of the nonlinear optical response is not accompanied here by a lowering of the response time. The mechanism producing this remarkable effect is yet not known, but all experiment seem to confirm that it should be of molecular origin.

12. Optical Bistability

The high orientational or thermal optical nonlinearity of nematicliquid crystals render these materials particularly suitable to obtain Optical Bistability. Optical Bistability (*OB*) is the property of some optical media or devices of having two (or more) stable states characterized by different optical parameters as transmission, reflection, birefringence, etc. [47]. Good optically bistable devices should have fast commuting time between the two states and good contrast, i.e. the two states must differ significantly in some macroscopic property. The *OB* is said to be intrinsic if it is produced by some mechanism internal to the medium itself and not e.g. by using external electronic bistable amplificators. The *OB* is said to be all-optical if no electronic at all is used in the device and the switching is produced by the incoming light itself. The usual way to obtain all-optical *OB* is inserting a highly nonlinear Kerr optical medium into a Fabry-Perot cavity. The cavity provides the feedback needed to have *OB*. Any Kerr medium is good, in principle, provided his Kerr nonlinearity is high enough. Liquid crystals, of course, are particularly convenient. Their nonlinearity is so large, however, that multiple stable states were indeed obtained, characterized by different hysteresis loops [48]. At higher intensity, beyond a critical threshold, continuous oscillations of the light transmitted through the Fabry-Perot were observed. This "flip-flop" behavior was interpreted as the result of two competing mechanisms having very different response times, actually reorientation and thermal indexing [48]. More recent experiments have shown that the time oscillations observed when a liquid crystal is put into a resonator may be also accompanied by complicated transverse spatial structures [49].

As previously mentioned, intrinsic *OB* without feedback cavity can be obtained biasing with a magnetic field the OFT in nematics films [17] or even without external fields by using chiral doped nematic cells [18]. In the last case the *OB* hysteresis loops are very large and the contrast is very high. An example of bistability loop in the OFT in *E7* doped with small amount of cholesteric *C15* is shown in Fig.9. Another way to obtain large *OB* in nematic films is to drive the cell with two beam at oblique incidence [50]. With the same twobeam geometry is also possible to modulate a strong beam by using a very weak beam, obtaining a sort of "transistor" action [51]. It should be stressed here that in these experiment the two beam were made not interfere into the film. This is a crucial difference with respect to standard wave mixing experiments.

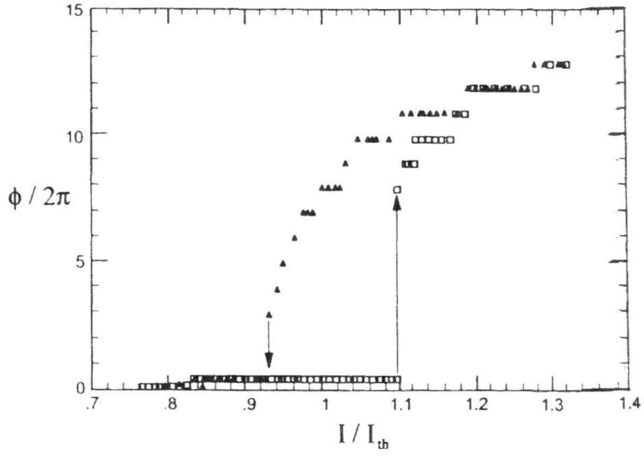

Figure 9 Large bistability loop in chiral-doped nematic. No bias field is needed. The induced optical phase shift is plotted as a function of the reduced intensity I/I_{th}.

We notice, however, that although very interesting from the fundamental point of view, optical bistable devices based on nematic liquid crystals suffer the drastic limitation of having an exceedingly long response time.

13. Transverse Effects

We have hitherto considered the plane wave approximation. The effects due to the finite laser beam profile are however very important in many experimental situations. Among the transverse effects arising when a laser beam propagates into a nematic film, we shall consider here only the self-phase modulation, the index grating, and the phenomenon of spontaneous pattern formation. It should be stressed that all these phenomena are not specific of liquid crystals, but could be observed in any medium having the optical Kerr nonlinearity. In liquid crystals they are particularly impressive, however, due to the presence of *GON*.

Before starting to study these phenomena it is worth spending a few words about the effect of the finite beam cross section on the threshold intensity I_{th} for the OFT [see Eq.(50)].

13.1 OPTICAL FREEDERICKSZ TRANSITION IN NARROW BEAMS

The departure from the plane wave approximation may be of some relevance in the OFT, because strongly focused laser beams are often used in the experiments. As pointed out in Sec. 9 the laser beam is usually focused onto a small spot having approximately the same linear dimensions of the nematic film thickness ($\approx 100\mu$m or less). The problem has been treated with some extent by N.V.Tabiryan, A.V.Sukhov, and B.Ya.Zel'dovick in Chapter 5 of their review paper [1]. The effect of the finite beam cross section is in general to increase the threshold I_{th} of a factor roughly given by ($1+L^2/w^2$), where L is the film thickness and w a measure of the beam spot size at the sample. The threshold increasing can be understood by considering that the liquid crystal molecules are reoriented in a region $\approx w$, thus a term proportional to K/w^2 (K is an elastic constant) is added to the elastic energy $\approx K/L^2$ of the distorted film. The detailed analysis, however, depends on the actual beam profile. In general, we may say that the plane wave approximation is good when the beam spot size w is much larger than the film thickness L. The case of oblique incidence was treated in some detail in Ref. [41]. As expected, there is no threshold intensity in the oblique incidence case. The laser-induced transverse index profile is not a Gaussian, in general, as the beam intensity profile, but a smooth bell shaped function with somewhat larger halfwidth.

13.2. SELF-PHASE MODULATION

It is perhaps the most spectacular effect observed when a laser beam traverses a liquid crystal film. In the far field beyond the sample a series of concentric rings appear having a divergence more than two orders of magnitude larger than the divergence of the incoming beam. The number of rings increases with the beam intensity. Up to 100 rings can be easily observed. An example of ring pattern is shown in Fig.10a. The origin of the phenomenon can be easily understood as follows [39]. The phase of the beam emerging from the sample is transversely modulated according to

$$\delta\psi(r) = \frac{2\pi}{\lambda}\int_0^L \delta n(r,z)dz \tag{68}$$

where r is the coordinate transverse to the beam and δn is the nonlinear index change due to the laser beam. Finding the exact nonlinear index profile $\delta n(r,z)$ is almost impossible, but we know that for a Gaussian beam profile, δn, and hence $\delta\psi$ should have a bell shaped profile as well. We may take, for example, a Gaussian profile

$$\delta\psi(r) = \delta\psi_0 e^{-2\frac{r^2}{w^2}}, \tag{69}$$

but the qualitative results will be equally true for any smooth bell shaped profile. The nonlinear phase shift (69) introduces a transverse component $k_\perp \approx d\delta\psi/dr$ in the wavevector of the outcoming wave, so that some rays are deflected out off the beam direction. We may assume that, in the first approximation, the ray emerging at distance r with respect to the beam axis is deviated at an angle $\alpha(r) \approx k_\perp(r)\lambda/(2\pi)$. The outer ring of the far field pattern corresponds therefore to the divergence angle α_{max} given by

$$\alpha_{max} = \frac{\lambda}{2\pi}\left(\frac{d\delta\psi}{dr}\right)_{max} \tag{70}$$

where the derivative $(d\delta\psi/dr)_{max}$ is to be evaluated at the flex of the Gaussian profile (69).

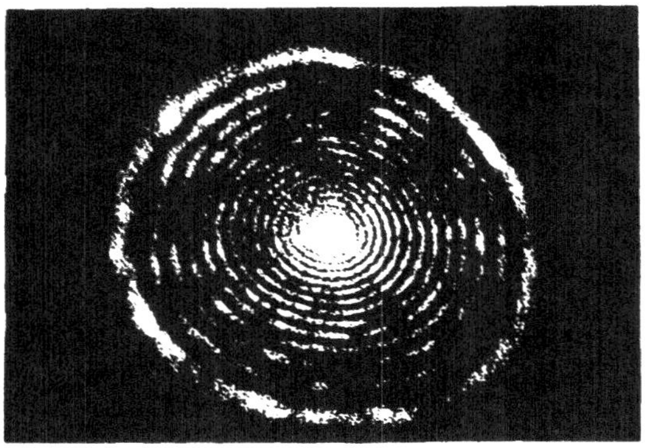

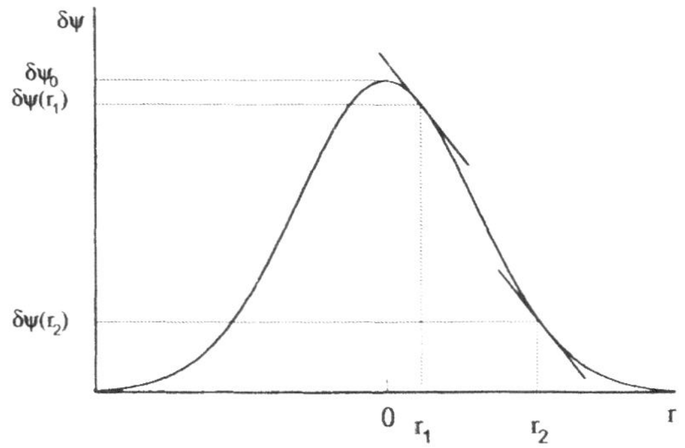

Fig.10(a) Picture of typical far field diffraction ring pattern due to self-phase modulation in a nematic liquid crystal film. (b) Profile of the laser-induced phase shift. Light diffracted at r_1 and r_2 has the same wavevector and interferes in the far field producing rings.

Looking at Fig.10b, where the phase profile is depicted, we see that for each point r_1 on the profile there is another r_2 having the same slope. The rays passing through r_1 and r_2 are deflected along the same direction and can interfere in the far field. The interference is not complete, because the two rays have different intensities, but nevertheless maxima and minima are obtained in the far field as long as $\delta\psi(r_1) - \delta\psi(r_2) \approx m\pi$, with integer m. This interference produces the observed rings. When N rings are present, we have therefore

$$N \approx \frac{\delta\psi_0}{2\pi}. \tag{71}$$

Combining Eqs.(69), (70), and (71), we obtain

$$\alpha_{\max} \approx \frac{2\lambda}{w\sqrt{e}}N. \tag{72}$$

Setting $\lambda \approx 0.5\mu m$, $w \approx 50\mu m$, we get $\alpha_{\max} \approx 0.01N$. For $N \approx 100$ observed rings, we find $\alpha_{\max} \approx 1$ rad, as measured. A more rigorous treatment based on Fresnel diffraction integral can be found in Ref. [40].

13.3. INDEX GRATING

Both thermal and reorientational effects in geometries where no threshold occurs, lead to a huge Kerr-like optical nonlinearity in liquid crystal film. Usual nonlinear optical effects based on Kerr effect ($\chi^{(3)}$-effects) can then be reproduced in liquid crystals [42]. Moreover, due to the relatively low response time of liquid crystals, transient nonlinear effects can be studied with great accuracy. Most of $\chi^{(3)}$-effects are based on the idea of creating a suitable index grating in the Kerr material, by interference of two (or more) beams. If the interfering beams have the same frequency, the induced grating is stationary. The grating amplitude is proportional to the geometric mean of the intensities of the two beams and its spatial period is $\Lambda = 2\pi/\Delta k$ where Δk is the wavevector mismatch between the interfering waves. Both thermal and reorientational gratings can be produced in liquid crystals. The reorientational grating problem was considered by various authors [43]. The starting point are the director torque equations with the optical intensity I given by

$$I(x) = I_1 + I_2 + 2\sqrt{I_1 I_2}\cos qx \tag{73}$$

where $q = \Delta k_x$ is the interference term between the two waves in the x-direction, transverse to the beam axis. In most cases the laser-induced molecular distortion is very small and a linearized approach can be used. When the grating is created, a term $\approx K/\Lambda^2$ is added to the elastic free energy, corresponding to the work made against the elastic forces to buildup the grating. Short period gratings require large energy at disposal. As a consequence, the grating depth decreases roughly as Λ^{-2} as the grating period is decreased. In practice, for a nematic film of thickness L, the grating depth decreases sharply when $\Lambda \leq L$. Nevertheless the index modulation may be large enough to produce strong diffraction of the incident beam over many diffraction orders, as shown in Fig.11. If the beams have frequencies differing by an amount $\Delta\omega$, the grating is moving like a plane wave proportional to exp $i(\Delta k_x - \Delta\omega t)$, so that a corresponding frequency shift in the diffracted beams is produced. In these cases, we may think also of the process as a stimulated light scattering process where photons pass from one beam to the other, the excess energy $\hbar\Delta\omega$ being leaven into (or taken from) the medium in the form of a "material wave" of frequency $\Delta\omega$. This material wave, corresponds, in the reorientational case, to a collective motion of the molecular director.

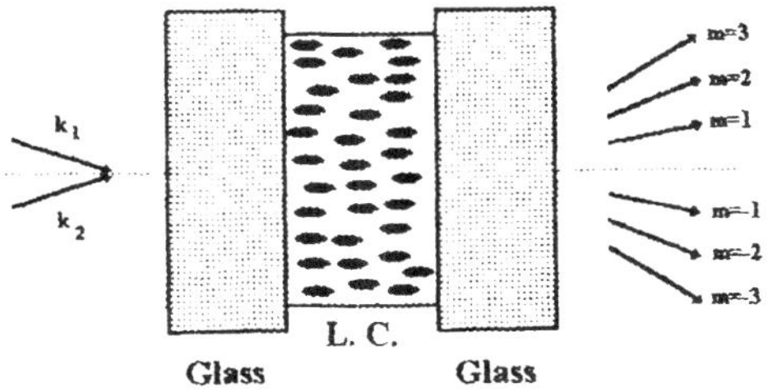

Figure 11 Experimental geometry for degenerate wave mixing in a homeotropic nematic film. Various diffraction orders are shown.

As a result of the laser-induced refractive grating, a beam traversing the medium can experience gain (taking energy from another strongest beam), loss (giving energy to another beam), diffraction over many orders, phase conjugated reflection, frequency broadening, multistable transmission, optical wave mixing, optical switching, and many other interesting effects. A complete account of all nonlinear optical experiments made with liquid crystal is impossible. We quote here only a few experiments about wave mixing [44], beam amplification [45], and stimulated scattering [46], and will refer to the literature for further references [1].

13.4. PATTERN FORMATION

Among transverse effects in optical Kerr media, spontaneous pattern formation has received recently great attention. Potential applications are image recognition and optical computing, but there is also a fundamental interest in the phenomenon, because regular structures are formed spontaneously from chaos in dissipative systems. We already mentioned that transverse structures in the laser beam are formed in highly nonlinear Fabry-Perot cavities [49]. Even more complicate spatial structures as defects or vortex are obtained when a degenerate four wave mixing with gain is realized in a highly nonlinear Kerr material (actually a BSO photorefractive crystal) inserted in a ring cavity [52].

The nonlinear Fabry-Perot scheme, however, is somewhat complicate, because of the complexity of the mechanism providing the optica feedback and require a very careful alignment optimization. For this reasons simpler systems, based on one back reflecting mirror or even without external optical feedback at all are of great interest. Not trivial spatial structures (two-dimensional dark-solitons) with no external feedback were observed in sodium vapors near the D_2 resonance, where the gas behaves as a negative (defocusing) highly nonlinear optical Kerr medium [53]. These structures, however, do not form spontaneously from noise, but must be initialized by some "seed", made, e.g., by making the incident laser beam to pass through a wire mesh. The possibility of forming highly structured hexagonal patterns from noise using just a single back reflecting mirror was proved both theoretically [54] and experimentally using a hybrid [55] or tilted homeotropic [56] liquid crystal cell in order to have *GON*. An example of the observed patterns is in Fig.12. Since the subject is of current interest, we give here some details about the theoretical model.

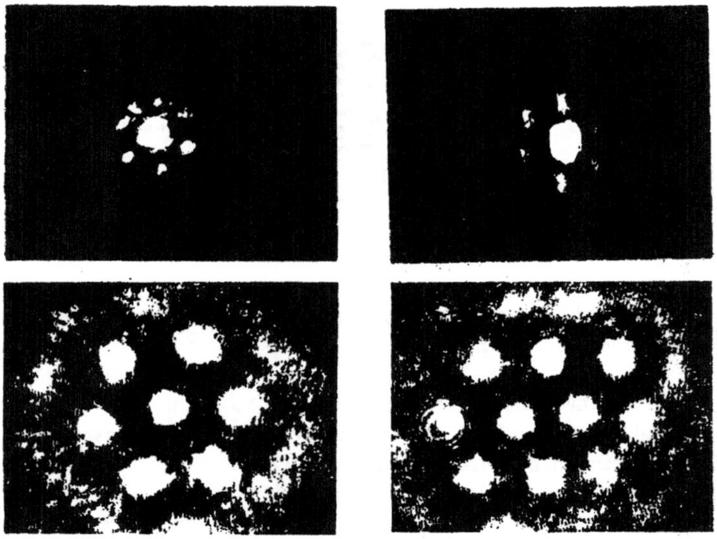

Figure 12 Pictures of the observed hexagonal patterns for various mirror-to-sample distance d. (a1-a2) far field; (b1-b2) near field. The liquid crystal was homeotropic nematic *E7*, 50μm thick (after Ref. [56]).

The equation governing the phenomenon can be obtained easily by director torque equation. Consider, for example, the case of *GON* in a homeotropic nematic film at oblique incidence. The director tilt angle θ is very small, so we can linearize the torque Eq.(49) with the optical torque last term replaced by $\tau_o(0)$ given by Eq.(59). Moreover, since we are interested in the transverse variations of the director field, we must restore in the elastic part of the torque the terms containing the derivatives of θ with respect to the x and y coordinates. In the one elastic constant approximation we get

$$-\gamma_1\frac{\partial\theta}{\partial t}+K\Delta_\perp\theta+K\frac{\partial^2\theta}{\partial z^2}+\tau_o(0)=0, \tag{74}$$

where we split out $\Delta_\perp = \partial^2/\partial x^2 + \partial^2/\partial y^2$ the transverse part of the Laplacian. The nonlinear phase $\phi(x,y,t)$ change suffered by the wave in traversing the medium is given by Eq.(61) and hence it is proportional to $\langle\theta\rangle$. Then a simple integration of Eq.(74) along z yields the following equation for the transverse phase change

$$-\gamma_1 \frac{\partial\phi}{\partial t} + K\Delta_\perp\phi - \frac{12K}{L^2}\phi + \tau_o(0) = 0. \tag{75}$$

In obtaining Eq.(75) we assumed, for small θ, the θ-profile $\theta = 6\langle\theta\rangle z(L-z)/L^2$, [see Eq.(60)]. The final step is writing out the intensity I appearing in the optical torque Eq.(59) as the sum $I(x,y,t) = I_0(x,y) + I_1(x,y,t)$ of the intensities I_0 and I_1 of the incident and back reflected beam, respectively[5]. The intensity profile $I_0(x,y)$ of the incident beam is known (e.g. a Gaussian profile), but the profile I_1 depends on the unknown phase change ϕ, due to the free propagation to and from the back reflecting mirror. Indeed, the optical field at the exit face of the film is $\sqrt{I_0(x,y)}e^{i\phi(x,y,t)}$, so that the reflected intensity will be given by the Fresnel diffraction integral

$$I_1(x,y,t) = \left| \frac{1}{2\lambda d} \iint du dv \sqrt{I_0(u,v)} e^{i\phi(u,v,t)} e^{\frac{i\pi\left[(x-u)^2+(y-v)^2\right]}{2\lambda d}} \right|^2 \tag{76}$$

where d is the mirror-to-sample distance. When Eq.(76) is inserted back in the torque $\tau_o(0)$ appearing in Eq.(75) we are left with an integro-differential equation in the unknown function $\phi(x,y,t)$. This was precisely the equation solved numerically by D'Alessandro and Firth [54] showing the occurrence of hexagonal patterned solutions. Although the derivation of Firth's integro-differential equation is based on simple principles, a far more reaching physical insight about the phenomenon can be obtained in terms of Talbot's effect [56]. The effect is well known. Consider a wave whose transverse profile is phase modulated with the spatial wavenumber q at the plane $z = 0$. Upon free propagation over a distance $z_T/4$, where $z_T = 8\pi^2/(\lambda q^2)$ is the Talbot length, the modulation is converted from PM to AM (with the same wavenumber). The wave is again PM modulated at the plane $z = z_T/2$, and AM at $z = 3z_T/4$. Any two successive PM or AM modulations are out of phase, so that at the Talbot distance $z = z_T$ the initial PM is reproduced. Then, for a fixed distance d between the Kerr cell and the back reflecting mirror, one obtains a resonant enhancement of the spatial transverse mode having wavenumber q whenever $2d = \left(\frac{1}{4} + n\right) z_T$, $n=0,1,\ldots$ In fact, in that case, an initially PM modulated wave returns AM modulated on the cell, while the Kerr optical nonlinearity of the medium converts the AM back into PM, in phase with the initial one. The spatial modes having wavenumbers $q_n = \left(\pi/\sqrt{\lambda d}\right)(1+4n)^{\frac{1}{2}}$ are therefore selectively enhanced and phase locked together. The

[5] We assume that the intensities I_0 and I_1 add incoherently and that their transverse profiles are unaffected by the transparent film.

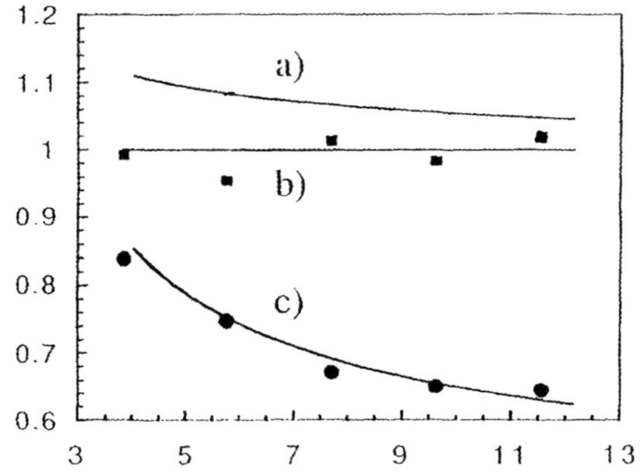

Figure 13 Hexagon lattice constant and threshold intensity for pattern formation as functions of the reduced mirror-to-sample distance $\pi\, \lambda\, d\, /\, L^2$. Curve (a): ratio Λ/Λ_T as obtained from Firth's linearized theory. Curve (b): experimental data (squares) for Λ/Λ_T and curve $\Lambda/\Lambda_T = 1$. Curve (c): Plot of $\left(\pi^2 / 12\right)\left(\beta \sin\alpha / n_o\right)\left(I_{patt} / I_{th}\right)$, where I_{patt} is the threshold for pattern formation as obtained from Firth's linearized theory, I_{th} is the threshold for the optical Freedericksz transition at normal incidence [Eq.(51)] and β is given by Eq.(64). Circles are the experimental data. The incidence angle was α=45°.

phenomenon is qualitatively analogous to the onset of periodic pulse modulation in mode-locked lasers. A periodic structure with strong regular peaks is therefore created. A more refined analysis shows that hexagonal patterns are preferred. The Talbot model implies that the hexagon pitch Λ should be given by $\Lambda = \Lambda_T = 2\sqrt{\lambda d}$. This is actually the case, as clearly shown in Fig.13, where the ratio Λ/Λ_T is plotted for different mirror-to-sample distances.

References

[1] Some reviews on nonlinear optics in liquid crystals have been appeared recently in the literature: N.V.Tabiryan, A.V.Sukhov, and B.Zel'dovich, 'The orientational optical nonlinearity of liquid crystals', Mol. Cryst. Liq. Cryst., **136**, 1-140 (1986); I.C.Khoo, 'Nonlinear optics of liquid crystals', Progress in Optics, XXVI, 107-161 (1988); F.Simoni, 'Nonlinear optical phenomena in nematics', in Physics of liquid crystals, Gordon & Breach, N.Y. 1989;

[2] Classical books where all mesophases are considered in some detail is P.G. de Gennes, 'The physics of liquid crystals', Clarendon Press, Oxford, 1974; S.Chandrasekhar, 'Liquid Crystals', Cambridge Univ. Press, 1977;

[3] A.Saupe and G.Englert, Mol. Cryst. **1**, 503 (1966);

[4] W.Maier and A.Saupe, Z. Naturf. **13**, 564 (1958); **14**, 882 (1959); **15**, 287 (1960);

[5] C.W.Oseen, Trans. Faraday Soc. **29**, 883 (1933);

[6] A.Saupe, Z. Naturf. **15**, 810 (1960);

[7] B.Bahadur, Mol.Cryst.Liq.Cryst. **301**, 1 (1984);

[8] J.L.Ericksen, Archs. Ration. Mech. Anal. **4**, 231 (1960); Phys. Fluids **9**, 1205 (1966); F.M.Leslie, Archs. Ration. Mech. Anal., **28**, 265 (1968); O.Parodi, J. Physique, **31**, 581 (1970);

[9] V.Fréederickz and V.Zolina, Trans. Farad. Soc., **29**, 919 (1933);

[10] B.Y.Zel'dovich and N.V.Tabiryan, Sov. J. Quant. El. **10**, 440 (1980); A.S.Zolot'ko, V.F.Kitaeva, N.Kroo, N.N.Sobolev, and L.Csillag, JEPT Lett., **32**, 158 (1980);

[11] S.D.Durbin, S.M.Arakelian, and Y.R.Shen, Phys. Rev. Lett. **47**, 1411 (1981);

[12] L.Landau and E.Lifchitz, 'Electrodynamics od continuous media', Pergamon Press, 1960;

[13] N.A.Clark, S.T.Lagerwall, 'Ferroelectrics', **59**, pp.25-67 (1984); M.A.Handschy and N.A.Clark, Ferroelectrics **59**, pp.69-116 (1984)

[14] O.V.Gabyrian, I.N.Kompanets, A.V.Parfenov, N.F.Pilipetskij, V.V.Shkunov, A.N.Sudarkin, A.V.Sukhov, N.V.Tabiryan, A.A.Vasiliev, B.Ya.Zel'dovich, Opt. Comm. **38**, 64 (1981);

[15] I.N.Kompanets, A.V.Parfenov, Ju.M.Popov, Opt. Comm. **36**, 415 (1981);

[16] S.Zumer and J.W.Doane, Phys. Rev. A **34**, 3373 (1986); B.G.Wu, J.L. West, and J.W.Doane, J. Appl. Phys. **62**, 3925 (1987); G.P.Montgomery, Jr and N.A.Vaz, Appl. Opt. **26**, 738 (1987);

[17] A.J.Karn, S.M.Arakelian, Y.R.Shen, H.L.Ong, Phys. Rev. Lett. **57**, 448 (1986);

[18] G. Abbate, P. Maddalena, L. Marrucci, L. Saetta, A.Ferraiuolo, S.Ferraiuolo, E. Santamato, Mol. Cryst. Liq. Cryst. **223**, 11 (1992); G. Abbate, A.Ferraiuolo, P. Maddalena, L. Marrucci, E.Santamato, Liq. Cryst. (1993), in press.

[19] E.Santamato, M.Romagnoli, M.Settembre, B.Daino, Y.R.Shen, Phys. Rev. Lett. **61**, 113 (1988);

[20] H.Kubo and R.Nagata, J. Opt. Soc. Am. **73**, 1719 (1983); H.L.Ong and R.B.Mayer, J. Opt. Soc. Am. **2**, 198 (1985); E.Santamato and Y.R.Shen, J. Opt. Soc. Am. A **4**, 356 (1987); C.Oldano, Phys. Rev. A **40**, 6014 (1990); P.Allia, .Arlone, C.Oldano, and L.Trossi, Mol. Cryst. Liq. Cryst. **179**, 253 (1990);

[21] E.Santamato, G.Abbate, P.Maddalena, Phys. Rev. A **38**, 4323 (1988);
[22] C.Mauguin, Phys. Z. **12**, 1011 (1911);
[23] A.S.Zolot'ko and V.F.Kitaeva, Sov. Phys. JEPT. **64**, 76 (1986);
[24] E.Santamato, G.Abbate, P.Maddalena, L.Marrucci, Y.R.Shen, Phys. Rev. Lett. **64**, 1377 (1990);
[25] E.Santamato, B.Daino, M.Romagnoli, M.Settembre, Y.R.Shen, Phys. Rev. Lett. **57**, 2423 (1986);
[26] H.L.Ong, Phys. Rev. A **28**, 2393 (1983);
[27] K.Skarp, S.T.Lagerwall, B.Stebler, Mol. Cryst. Liq. Cryst. **60**, 215 (1980); S.T.Wu, A.M.Lackner, U.Efron, Appl. Opt. **26**, 3441 (1987); S.T.Wu, C.S.Wu, Phys. Rev. A, **42**, 2219 (1990);
[28] B.Ya.Zel'dovich and N.V.Tabiryan, Sov. Phys. JEPT **55**, 656 (1987);
[29] E.Santamato, G.Abbate, R.Calaselice, P.Maddalena, and A.Sasso, Phys. Rev. **37**, 1375 (1988);
[30] F.Simoni and R.Bartolino, Opt. Comm. **53**, 210 (1985);
[31] B.Ya.Zel'dovich, S.K.Merzlikin, N.F.Pilipetskij, A.V.Sukhov, and N.V.Tabiryan, JEPT Lett. **37**, 676 (1983);
[32] G.Cipparrone, V.Carbone, C.Versace, C.Umeton, R.Bartolino, and F.Simoni, private communication, to appear in Phys. Rev. E;
[33] I.C.Khoo, Phys. Rev. A **23**, 2077 (1981);
[34] S.G.Odulov, Yu.A.Reznikov, M.S.Soskin, and A.I.Khizhnyak, Sov. Phys. JEPT **58**, 1154 (1983);
[35] B.Ya.Zel'dovich and N.V.Tabiryan, Sov. Phys. JEPT **63**, 80 (1986);
[36] F.Simoni, G.Cipparone, C.Umeton, G.Arabia, G.Chidichimo, Appl. Phys. Lett. **54**, 896 (1989);
[37] I.Janossy, A.D.Lloyd, and B.S.Wherret, Mol. Cryst. Liq. Cryst. **179**, 1 (1990); I.Janossy, L.Csillag, and A.D.Lloyd, Phys. Rev. A **44**, 8410 (1991);
[38] I.Janossy and T. Kosa. Opt. Lett. **17**, 1183 (1992);
[39] S.D.Durbin, S.M.Arakelian, and Y.R.Shen, Opt. Lett. **6**, 411 (1981);
[40] E.Santamato, Y.R.Shen, Opt. Lett. **9**, 564 (1985); F.Bloisi, L.Vicari, F.Simoni, G.Cipparrone, and C.Umeton, J. Opt. Soc. Am. B **5**, 2462 (1988);
[41] I.C.Khoo, P.Y.Yan, and T.H.Liu, J. Opt. Soc. Am. B **4**, 115 (1987);
[42] For the $\chi^{(3)}$-effects see the following classical book on nonlinear optics by Y.R.Shen, 'The Principles od Nonlinear Optics', J.Wiley & Sons, 1984; A.Yariv, 'Quantum Electronics', J.Wiley & Sons, 1967;
[43] S.D.Durbin, S.M.Arakelian, and T.R.Shen, Opt. Lett **7**, 145 (1982); I.C.Khoo, Phys. Rev. A **27**, 2747 (1983); Thermal index grating was also produced recently in PDLC: see, F.Simoni, G.Cipparrone, D.Duca, I.C.Khoo, Opt. Lett. **16**, 360 (1991);
[44] I.C.Khoo, IEEE J. Quantum Electr. **QE-22**, 1268 (1986); I.C.Khoo, Opt. Eng. **25**, 198 (1986);
[45] I.C.Khoo and T.H.Liu IEEE J. Quantum Electr. **QE-23**, 171 (1986); T.H.Liu and I.C.Khoo, IEEE J. Quantum Electr. **QE-23**, 2020 (1986);
[46] B.Ya.Zel'dovich, A.V.Sukhov, and N.V.Tabiryan, Mol. Cryst. Liq. Cryst. **143**, 7 (1987); I.C.Khoo, R.R.Michael, P.Y.Yan, IEEE J. Quantum Electr. **QE-23**, 1344 (1986);

[47] For a review on Optical Bistability see, for example, E.Abraham and S.D.Smith, Rep. Prog. Phys. **45**, 815 (1982);
[48] M.M.Cheung, S.D.Durbin, and Y.R.Shen, Opt. Lett. **8**, 39 (1983)
[49] M.Kreutzer, W.Balzer, and T.Tschudi, Appl. Opt., **29**, 579 (1990);
[50] E.Santamato, G.Abbate, R.Calaselice, P.Maddalena, A.Sasso, Phys. Rev. A **37**, 1375 (1988).
[51] E.Santamato, A.Sasso, R.Bruzzese, Y.R.Shen, Opt. Lett. **11**, 452 (1986).
[52] F.T.Arecchi, G.Giacomelli, P.L.Ramazza, and S.Residori, Phys. Rev. Lett. **67**, 3749 (1991);
[53] G.A.Swartzlander,Jr., D.R.Andersen, J.J.Regan, H.Yin, and A.E.Kaplan, Phys. Rev. Lett. **66**, 1583 (1991);
[54] G.D'Alessandro and W.J.Firth, Phys. Rev. Lett. **66**, 2597 (1991);
[55] R.MacDonald and H.J.Eichler, Opt. Comm. **89**, 289 (1992);
[56] M.Tamburrini, M.Bonavita, S.Wabnitz, and E.Santamato, Opt. Lett. **18**, 855 (1993);
[57] K.C.Chu, C.K.Chen, and Y.R.Shen, Mol. Cryst. Liq. Cryst. **59**, 97 (1980);
[58] S.D.Durbin, S.M.Arakelian. and Y.R.Shen, Opt. Lett. **7**, 145 (1982);

PHOTO-INDUCED REFRACTIVE INDEX CHANGES IN BULK SEMICONDUCTORS

H. M. VAN DRIEL
Department of Physics, University of Toronto
and Ontario Laser and Lightwave Research Centre
Toronto, M5S-1A7
Canada

ABSTRACT. This article reviews the physical processes underlying photo-induced refractive index changes in bulk semiconductors. The mechanisms considered here include: 1) virtual electronic processes such as Kerr and Stark effects caused by below-band gap radiation; 2) free carrier effects which occur during and following single photon or multi-photon absorption processes and 3) thermally induced contribution. Bound carrier effects are generally weak but occur essentially instantaneously while free carrier effects are generally larger and disappear through carrier diffusion or recombination on a picosecond or longer time scale. Thermally induced effects, which decay with heat diffusion, can also be significant especially for long pulses and can remain for milliseconds. Various experimental techniques which have been used to determine refractive index dynamics in semiconductors are considered. Results of recent experiments probing the evolution of the refractive index in II-VI semiconductors on a femtosecond/picosecond time scale are used to illustrate some of the above concepts.

1. Introduction

For decades information technology has been dominated by electronics. However optics is replacing electronics in various transmission and storage applications and continues to make in-roads in other areas as well. One key application that remains stubbornly in the domain of electronics is switching and routing. Indeed, an all-optical switching device has become the "holy grail" of optical communications technologies. The underlying physical mechanism which forms the basis for much research in all-optical switching involves photo-induced refractive index changes [Miller, 1981; Shen, 1984; Gibbs, 1985; Miller, 1993]. It is envisioned that a gate optical pulse can be used to alter the refractive index of a material and thus modify either the direction of propagation, phase, or transmission of another optical pulse passing through the device. For large scale information processing, suitable candidate materials for such a switch must be able to offer *a large* and *fast* response to the inducing pulse with little light absorption. In addition there may be requirements on photo-chemical stability, cost, etc. Many different types of materials have been and continue to be investigated including bulk, superlattice and quantum-well semiconductors, pure glasses, semiconductor-doped glasses, and polymers [Miller, 1981; Stegeman, 1988; Haug, 1988; Gibbs, 1990]. Also, several different configurations have been researched for switching applications based on Fabry-Perot interferometers, etalons, waveguides, diffraction, and scattering [Stegeman, 1988]. Many of the materials and geometries are covered elsewhere

A. Miller et al. (eds.), Nonlinear Optical Materials and Devices for Applications in Information Technology, 141–181.

in these proceedings. Here, the emphasis is on some fundamental concepts of photo-induced refractive index changes in bulk semiconductors. Although bulk materials may not provide any or all of the desirable properties for optical switching in applications, many of the concepts related to this field have first been investigated in such materials.

There are four general mechanisms which have been considered in connection with photo-induced refractive index changes in semiconductors. These are related to: a) bound electrons, b) free-carriers, c) thermal effects and d) carrier transport. The bound electron mechanism involves virtual electronic excitation while free-carrier effects are related to excitation of actual electrons and holes with removal of photons from the beam. Free carrier induced changes to the refractive index can be further divided into interband and intraband contributions, which have different strengths depending on the ratio of photon and band-gap energies. Thermal effects are most often associated with changes in the refractive index brought about by a change in the band gap with lattice temperature while transport related processes have one of their most dramatic manifestations in photo-refractive processes associated with the separation of photoexcited electrons and holes. The four different mechanisms can also be discussed with respect to strength and time evolution of the refractive index change. In general, virtual effects are weak (unless the photon energy is close to the band-gap), have a fast response time but low associated beam attenuation while free carrier effects generally have a larger response per inducing photon but a longer decay time since carrier recombination or diffusion must occur. Turn-on times are simply related to the optical pulse width. In what follows we will refer to the virtual effect as instantaneous although strictly speaking the intrinsic response time is dictated by the uncertainty principle. Photo-refractive mechanisms can also yield a large response through a modification of the dielectric properties of the material by the electric field associated with separated charge carriers. The response time is again governed by carrier recombination (usually out of traps) which in some cases can be as short as picoseconds. Lattice heating or phonon induced changes can similarly be large but these disappear on a time-scale governed by relatively slow heat diffusion, typically on a time scale of milliseconds [Wherrett, 1990]. The strength and speed of the refractive index change to a certain extent can be controlled through beam parameters (e.g., spot size, pulse width, wavelength) or by tailoring the sample geometry through size or mounting.

The various intrinsic and extrinsic properties of material obviously define its possible applications. One can define various figures of merit for an optical switch depending on which combination of parameters one wishes to optimize. For example, a common figure of merit is the ratio of the the beam attenuation distance to the distance required to achieve a phase change of π. Another figure of merit is this ratio divided by the recovery time of the refractive index change. A high figure of merit then occurs with a strong bound electron effect which offers little absorption of energy and has a fast response time. Although specific applications, such as telecommunications, use specific wavelengths and put stringent requirements on materials, optical sources, and geometries, in a broader sense all-optical switching has been pursued with many different sources and materials, even if one only considers bulk semiconductors. Fig. 1 illustrates in part the range of semiconductor band-gaps and photon energies of optical sources available for such studies.

In what follows the salient features of how bound and free carriers along with lattice thermal mechanisms can alter the refractive indices in bulk crystalline semiconductors will be explored. Elsewhere in these proceedings D.A.B Miller describes effects in semiconductor quantum wells and superlattices while A. Glass covers photorefractive effects. This contribution begins with a general overview of the relevant nonlinear optics and illustrates some of the features of optically induced refractive index changes by considering a two level atom picture and applying it to excitonic effects. A two-band model of a semiconductor is then used to consider bound electron and free carrier induced changes to the refractive index, and this is followed by a discussion of thermal effects. Finally some of the above features are illustrated through presentation of some of our recent femtosecond time-resolved experiments which monitor optically induced refractive index changes in II-VI semiconductors.

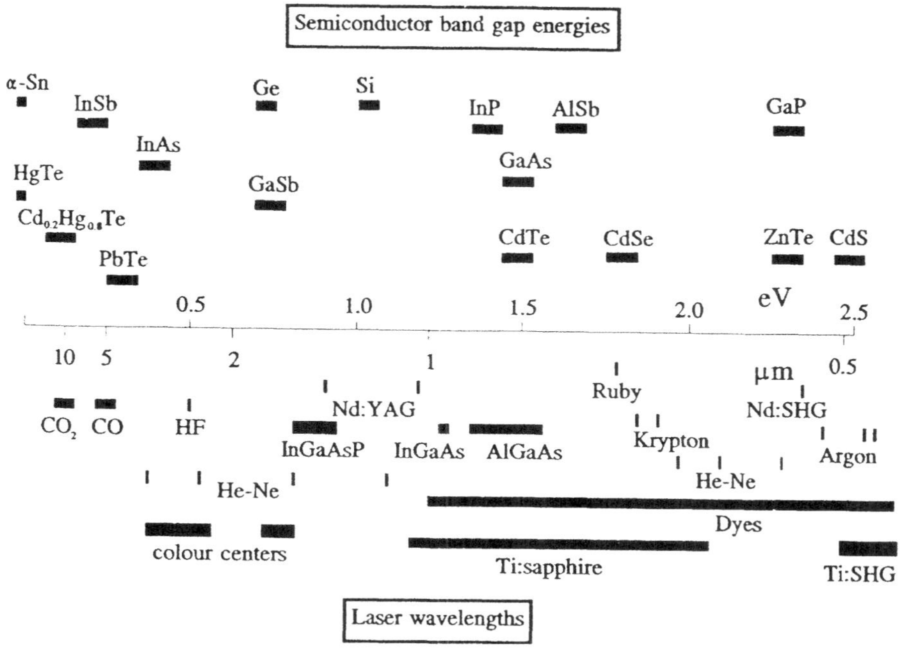

Figure 1. Band-gap energies of bulk semiconductors ($0 < T < 300$ K) and photon energies of typical laser sources [Miller; 1993].

2. Refractive Index Changes; A Nonlinear Optics Perspective

Changes in the refractive index of a semiconductor can be understood from both a macroscopic and microscopic picture. We begin first by offering the macroscopic perspective in which the refractive index can be considered as one aspect of how a medium responds to electromagnetic fields. When optical radiation interacts with a solid the optical response can be described by the superposition of the responses to the individual monochromatic components which make up the beam. Although the pulses we consider later are as short as 100 fs, this duration is still long compared to the sub-femtosecond period of typical laser pulses. To a good approximation then we are justified in taking the interaction to be quasi-monochromatic at a single, centre frequency ω and describe the incident electric field by $E(\omega)$. The response to the field can be written in terms of an induced polarization density which can be expanded in terms of powers of the monochromatic field components to yield a variety of nonlinear source terms for the wave equation. These terms lead to linear absorption, second harmonic generation, stimulated Raman scattering, third harmonic generation, two-photon absorption, etc. [Shen, 1984]. Because we are only interested in refractive index changes induced by the beam itself, we need only consider contributions to the polarization density at the original frequency, *viz.*, $\vec{P}(\omega)$. The two lowest order non-zero (linear and nonlinear) induced responses are given by

$$\vec{P}(\omega) = \vec{\vec{\chi}}^{(1)}(-\omega,\omega)\cdot\vec{E}(\omega) + \vec{\vec{\chi}}^{(3)}(-\omega,\omega,-\omega,\omega)\cdot\vec{E}(\omega)\vec{E}(-\omega)\vec{E}(\omega) \ . \tag{1}$$

For simplicity the anisotropy in the linear and nonlinear susceptibilities will be ignored allowing the tensor character of the responses to be dropped. The product $E(\omega)E(-\omega)$ is related to the instantaneous irradiance of the beam by

$$I(\omega) = \frac{1}{2\pi} c\, n_0(\omega)E(\omega)E(-\omega) \tag{2}$$

where $n_0(\omega)$ is the refractive index at low intensities and c is the speed of light. The scalar polarization density then becomes

$$P(\omega) = \left[\chi^{(1)}(-\omega;\omega) + \frac{2}{cn_0\varepsilon_0}\chi^{(3)}(-\omega;\omega,-\omega,\omega)I(\omega)\right]E(\omega) \tag{3}$$

$$= \chi^{(1)}_{\text{eff}}(-\omega,\omega)E(\omega) \tag{4}$$

where the effective susceptibility, $\chi^{(1)}_{\text{eff}}(-\omega;\omega)$, is, in general, a complex number. From Eq. 3, and a little algebra, the effective susceptibility leads to a complex refractive index, $\hat{n}(\omega) = n_0(\omega) + i\kappa(\omega)$, whose real part can be written as

$$n(\omega) = n_0(\omega) + n_2(\omega)I(\omega) \ . \tag{5}$$

The quantity $n_2(\omega)$ is often referred to as the nonlinear refractive index or the Kerr coefficient.

From the relation between κ and the absorption coefficient α through $\alpha = 4\pi\kappa\lambda^{-1}$, one also has

$$\alpha(\omega) = \alpha_0(\omega) + \beta(\omega)I(\omega) \tag{6}$$

where $\alpha_0(\omega)$ (hereafter referred to as α_0) is the low irradiance, single photon absorption coefficient while $\beta(\omega)$ can be identified with two-photon absorption (2PA).

The Kramers-Krönig analysis of the linear optical response of a material relates the refractive index, $n_0(\omega)$, and the absorption coefficient, $\alpha_0(\omega)$ by

$$n_0(\omega) = 1 + \frac{c}{\pi} P\int_0^{\infty} d\omega' \, \frac{\alpha_0(\omega')}{\omega'^2 - \omega^2}. \tag{7}$$

Several authors have shown [Miller, 1981; Hutchings, 1992] that one can extend this relation to take into account various perturbations which alter the absorption coefficient. If the perturbation can be characterized by a parameter ζ then the associated change in refractive index is found from

$$\Delta n(\omega,\zeta) = \frac{c}{\pi} P\int_0^\infty d\omega' \, \frac{\Delta\alpha(\omega',\zeta)}{\omega'^2 - \omega^2} . \tag{8}$$

The parameter ζ may represent strain, free carrier density, electric field, lattice temperature, light irradiance, etc. The importance of this result is that a knowledge of the changes in frequency dependent absorption characteristics under a perturbation (over a large frequency range!) will determine the refractive index change at a particular frequency. Since absorption strengths are much easier to calculate or measure than Δn, determinations of the refractive index change are easily obtained. For example, from Eq. 6 and 8 above it follows that the Kerr coefficient and the general two-photon absorption coefficient are related. This will be discussed in more detail below.

The generation of real carriers in a semiconductor through single or multi-photon absorption can also lead to changes in the refractive index, through alteration of $\chi^{(1)}$. If one knows how the absorption characteristics change because of the redistribution of real carriers among energy levels, one can use Eq. 8 to determine the associated refractive index change. Similarly, if there is renormalization of the band-gap and a change in the energy level spectrum and absorption coefficient due to lattice heating or carrier effects, refractive index changes can also occur. The change in the refractive index can therefore occur directly through $n_2(\omega)$ from virtual excitations while a second contribution can occur through real carriers or an increased phonon population (e.g., through lattice heating) which can then modify n_0. The total change in refractive index, Δn, can be written as the sum of real (r) and virtual (v) excitation contributions so that

$$\Delta n = (\Delta n)_v + (\Delta n)_r \tag{9}$$

where

$$(\Delta n)_v = n_2 I . \tag{10}$$

The contributions to Δn from real excitations can be related to free carriers of density N and nonequilibrium phonons of density N_p through

$$(\Delta n)_r = \gamma N + \upsilon N_p . \tag{11}$$

From Eqs. 10 and 11 one sees that Δn will have an instantaneous contribution plus one that may turn on as quickly as the optical pulse but which disappears only when the carriers or excess phonons do. In cases where sufficiently long pulses are used so that all the relevant time constants that govern the evolution of the real excitations are short compared to the pulse width, and if these densities scale linearly with the pulse irradiance, the carrier and phonon densities will follow the pulse profile. In this case, Eqs. 9-11 can be combined to yield

$$\Delta n = (n_2)_{eff} I \tag{12}$$

whereby one can speak of an effective Kerr coefficient, and by inference from Eq. 4, an effective $\chi^{(3)}$ which incorporates virtual and real excitations. This leads some authors to use the term refractive nonlinearities and describe all refractive index changes through a Kerr coefficient. A caveat is needed, however. For example, in the case of femtosecond or picosecond optical pulse excitation and probing, as will be seen later, this language is not appropriate and the separation of real and virtual carrier effects is quite apparent.

The simultaneous occurrence of different nonlinear optical processes can also sometimes be viewed as effective higher order processes. For example, generation of carriers by two photon absorption occurs through a $\chi^{(3)}$ process, while the carriers themselves alter $\chi^{(1)}$. It follows that Δn is related to the product of a fourth rank and a 2nd rank tensor which

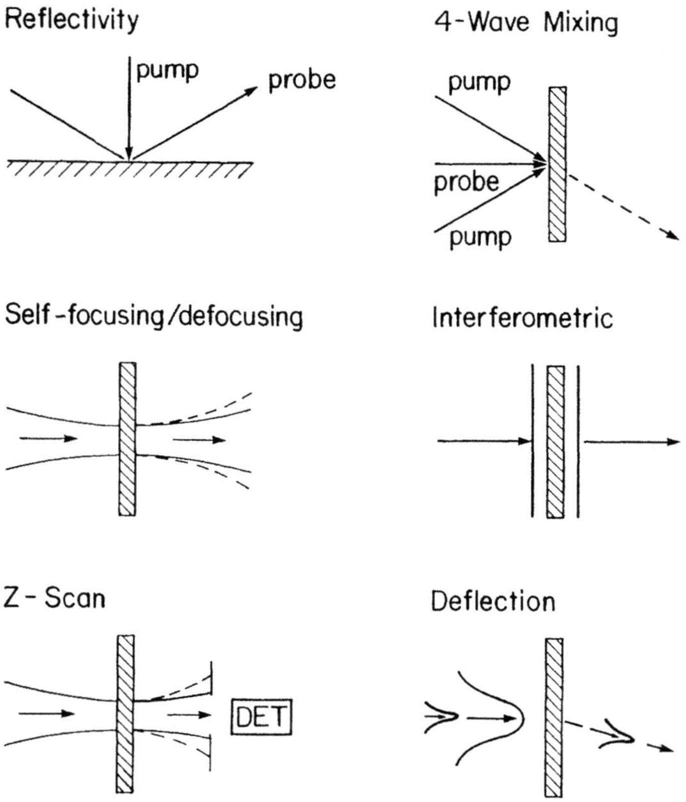

Figure 2. Techniques for studying optically induced refractive index changes in bulk semiconductors.

behaves effectively as a 6th rank tensor which can be associated with an effective $\chi^{(5)}$ process, viz.,

$$\Delta n \propto \chi^{(1)}\chi^{(3)} I^2 \propto [\chi^{(5)}]_{eff} I^2 \quad (13)$$

The cascading of nonlinearities can lead to substantial changes in the refractive index [Said, 1992], although the usual tradeoffs concerning strength and time of response usually also occur. It should also be emphasized again, that the description in Eq.13 in terms of effective nonlinearities is subject to the same considerations as those governing Eq. 12.

3. Measurement Techniques

Over the years several techniques have been used to study the magnitude and time dependence of Δn in semiconductors. Since the refractive index defines the propagation vector of a light wave, one could anticipate that the various techniques would be based on phase measurements, beam deflection or some form of beam distortion. Figure 2 illustrates schematically some of the techniques that have been used. The simplest relates to time-resolved reflection changes. An exciting pulse generates changes in the absorption coefficient or refractive index which, in turn, alter the reflectivity of a solid. Monitoring the reflectivity with a time-delayed probe pulse can yield information about carrier diffusion, recombination, energy relaxation, etc. For visible pump and probe pulses these effects are quite small and this technique is not preferred, but in the infrared real-carrier induced

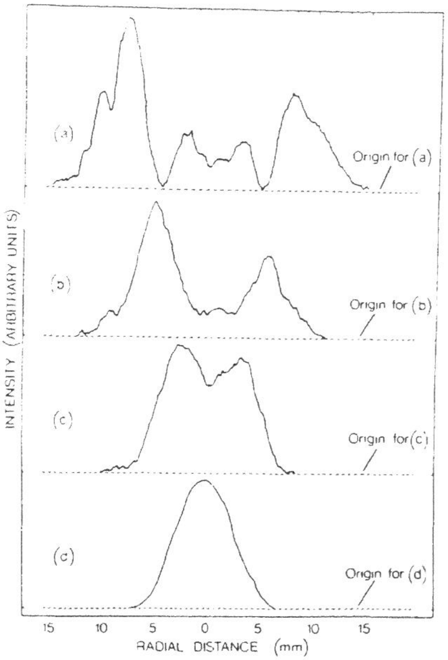

Figure 3a. Beam profiles of a cw CO laser (decreasing irradiance; a-d) after transmission through InSb (Miller ; 1978).

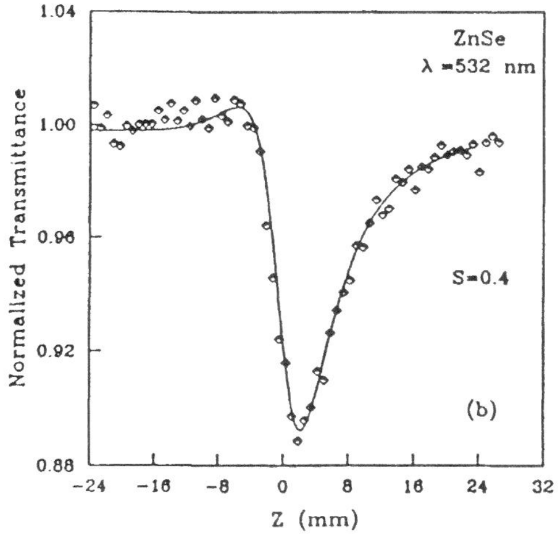

Figure 3b. Normalized z-scan data for a 2.7 mm sample of ZnSe probed by 532 nm, 27 ps pulses(Said;1978).

changes to the refractive index can be substantial [Auston, 1978; Gallant, 1981]. Another technique exploits the fact that a spatially Gaussian beam can induce spatially dependent refractive index changes, thereby generating an effective lens which causes the beam to focus or defocus [Miller, 1978]. Fig. 3a shows how the beam profile of a CO laser alters as a function of laser irradiance after passing through a thin piece of InSb. This effect was refined into a "z-scan" technique [Hill, 1982] to obtain quantitative information on nonlinear refraction effects and has been widely used to characterize a variety of materials[Sheik-Bahae, 1989, 1991]. An aperture is placed behind the sample which is moved longitudinally through the Rayleigh range of the beam. The irradiance dependent lens causes more or less light to pass through an aperture and the effective transmission of the system can be used to evaluate the refractive index change. Fig. 3b shows the results of a z-scan experiment for ZnSe probed by 532 nm, 27 ps pulses. Of course the technique spatially and temporally averages over effects in the beam and some modelling is required to extract information about the index change or two-photon absorption coefficient. Nonetheless useful estimates of the magnitude and sign of the refractive index change can be obtained quickly. A recent improvement in the technique recognizes that the wings of the beam suffer the largest relative irradiance change in an induced lens. By blocking the central portion of the beam and only measuring the "leakage" around the edges of the blocking element, high sensitivity measurements can be performed [Xia; 1993] equivalent to measuring phase delays below $10^{-3}\lambda$. A fourth technique is based on four-wave mixing. Here, a pair of beams are incident together on a sample to generate a refractive or absorptive index grating. A third incident pulse can be deflected from the grating into a direction specified by the grating wavevector and the time-resolved amplitude of this pulse gives a measure of the induced refractive index change, the magnitude of Δn and its time evolution. Note, however, that one cannot obtain information about the sign of the nonlinear refractive index change from this method. A fifth technique makes use of a Fabry-Perot interferometer and the sensitivity of the transmission to small changes in the refractive index of a sandwiched material. The final technique considered here, and one which we have used recently, is a refinement of the beam-induced lens effect but with time-resolution capabilities. Here, a pump beam induces a lens in the sample and a probe pulse is incident on the sample off-axis of the lens. The probe beam suffers a deflection the magnitude and sign of which depend on the characteristics of the refractive index change. As with the other techniques some modelling must be employed to account for spatial averaging.

4. Two Level Atom Picture of Induced Refractive Index Changes

A semiconductor has many electronic and atomic vibrational degrees of freedom which do not allow one to obtain a complete description of the energy levels or the dynamics of the system. Before considering the microscopic origins of the refractive index in a bulk semiconductor it may be helpful to consider a much more elementary system such as an atom. Even here a very useful model to obtain the essential features of the optical response is the two level atom in which there is near resonance of incident light with a pair of radiatively coupled levels. A large collection of such atoms ultimately affords a macroscopic description using density matrices [Allen and Eberly; 1975]. The application of perturbation theory shows that three parameters control the response, namely, p_{12}, the dipole matrix element of the transition, T_1, the population or energy relaxation time, and T_2, the dephasing time of the macroscopic polarization density. In the absence of local field and many-body effects, for a homogeneously broadened system one finds that the macroscopic response is governed by a susceptibility whose real and imaginary parts are given by

$$\chi'(\omega) = \pi \frac{|p_{12}|^2}{\varepsilon_0 \hbar}(N_1 - N_2)(\omega_0 - \omega)T_2 g(\omega) \tag{14}$$

and $$\chi''(\omega) = \pi \frac{|p_{12}|^2}{\varepsilon_0 \hbar}(N_1 - N_2) g(\omega) \tag{15}$$

where |1> and |2> are the two states whose frequency separation is ω_0, N_i is the population of state |i>, and $g(\omega)$ is the homogeneously broadened line shape function given by

$$g(\omega) = \frac{1}{2\pi}\left[\frac{2/T_2}{(\omega-\omega_0)^2 + (1/T_2)^2}\right]. \tag{16}$$

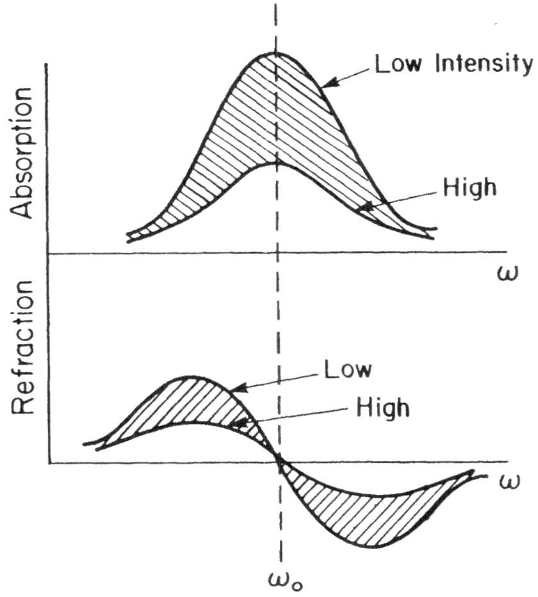

Figure 4. Refractive index and absorption coefficient for a two level system interacting with a strong and a weak beam.

The population difference, in steady state with a monochromatic field of field amplitude E_0, is given by:

$$(N_1 - N_2) = (N_1 - N_2)_{eq}\left[\frac{1+(\omega-\omega_0)^2 T_2^{\,2}}{1+(\omega-\omega_0)^2 T_2^{\,2} + \frac{|p_{12}|^2 E_0^2 T_1 T_2}{\hbar^2}}\right]. \tag{17}$$

where "eq" refers to the equilibrium population in the absence of the optical beam. From the susceptibility one obtains the refractive index and absorption coefficient as

$$n = 1 + \frac{\chi'(\omega)}{2} \qquad \alpha = \frac{2\pi\chi''(\omega)}{\lambda} \,. \tag{18}$$

Figure 4 shows the refractive index and absorption coefficient in the presence of a weak and a strong beam.

Let us consider only the case where the population of the lower level is larger than that of the upper level.) For $\omega < \omega_0$ the refractive index change is negative in the presence of a strong beam, but for $\omega > \omega_0$ the change is positive. The effect in either case is largest at frequencies half the linewidth above or below the resonance frequency, and is zero on resonance or far from resonance. From equations 14, 17 and 18, in the case of small E_0 one can perform a power series expansion of the refractive index in E_0 to find that the lowest order correction to n is proportional to $(E_0)^2$ or I and therefore the saturation characteristics are associated with a $\chi^{(3)}$ effect. Note that a transfer of population from the lower to the upper level via incoherent pumping has a similar effect on the refractive index to that of field-induced saturation. One can also use the two-level atom approach to consider other effects such as resonance frequency renormalization. If, for example, ω_0 decreases due to some interaction process, then for $\omega << \omega_0$ the refractive index will increase, while an increase will occur for $\omega >> \omega_0$.

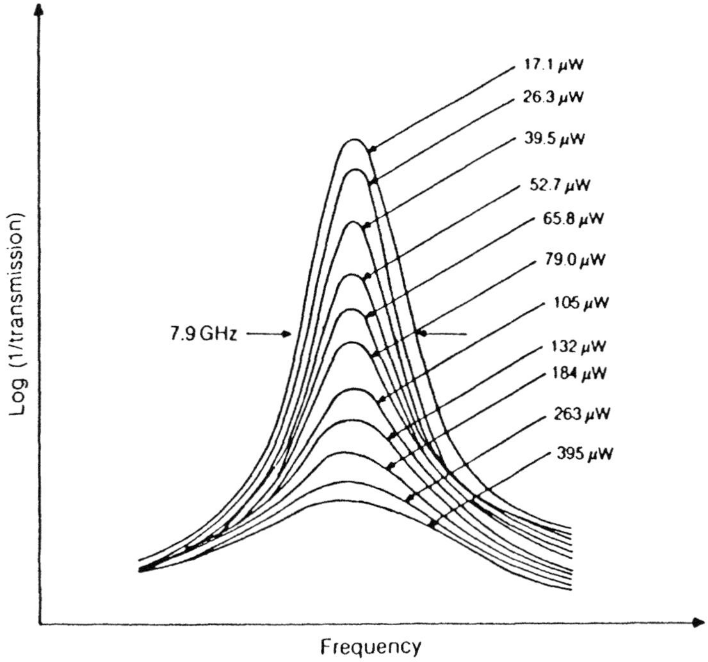

Figure 5. Saturation of the absorption of the bound exciton in CdS at 4K (Dagenais;1983).

As will be seen later, some of the basic ideas here are transferrable to interband effects in semiconductors, but they can already be applied to excitonic effects. Excitons are bound electron-hole pairs which are *formed* by absorption of below-gap photons (typically a few meV less than the band gap energy). Because of this low binding energy, excitons are stable only at cryogenic temperatures. Excitons, like atoms, have highly resonant features with large oscillator strength (i.e., large dipole matrix elements) and therefore are easily saturated. Figure 5 shows the absorption saturation characteristics associated with the creation of bound excitons in CdS. Only modest power levels are required to induce substantial changes in the absorption coefficient. The change in the absorption characteristics of course leads to alteration in the refractive index in a manner discussed above. However, caution has to be used when considering the saturation of excitonic resonances since their "Bohr radius" is of the order of 10 nm, so that significant exciton interaction begins to occur for densities near 10^{16} cm^{-3} whereby absorption may broaden and be reduced at high optical excitation. This effect may also be discussed as exhaustion of available states for excitons. The low power required to induce substantial effects make excitons attractive for optical switching phenomena. Indeed for short pulses one can use fluence levels as low as pJ/cm^2 to obtain saturation. However, one of the chief drawbacks for switching applications, besides the need for low temperatures, relate to their long recombination time, typically tens of ns. The need for low temperatures can be overcome by using quantum-well geometries which lead to higher excitonic binding energies; the long lifetimes typically remain however.

5. Contributions To Refractive Index Changes in Semiconductors

5.1. ELECTRONIC ENERGY LEVELS

We now turn to bound and free electronic effects in bulk semiconductors. For crystalline semiconductors, which possess translational symmetry, momentum is a good quantum number and a simplified picture of the electronic levels emerges in terms of energy bands as a function of wavevector. The physically unique set of wavevectors for which this can be

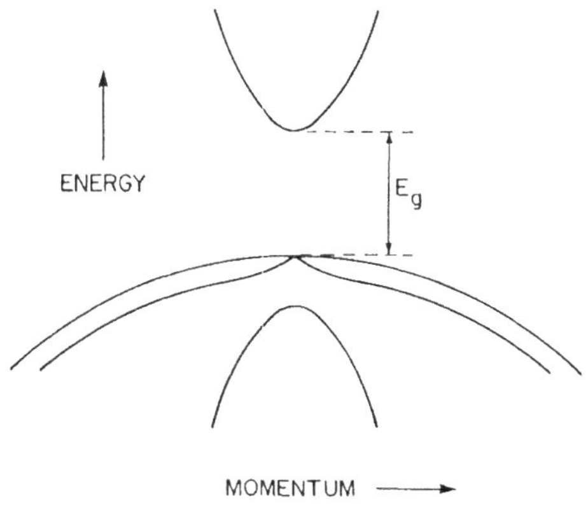

Figure 6. Band structure of a direct gap semiconductor.

done defines the Brillouin zone. This set can also be used to label the vibrational degrees of freedom leading to the concept of phonon dispersion curves. Characteristic of the electronic energy bands in a semiconductor, there exists a band gap of value E_g, typical values of which are given in Fig. 1. For photon energies, $\hbar\omega$, below the band gap, the (ideal) semiconductor is transparent at low intensities, while photon energies larger than the band gap lead to linear absorption in which electrons are created in the upper (conduction) band leaving positively charged holes in the lower (valence) band. Because the joint density of optically coupled states increases with increasing energy, the absorption coefficient does likewise. For effects of interest here we are concerned only with photon energies in the vicinity of the band-gap energy or below. As a result most effects can be understood in terms of the bands immediately adjacent to the band gap. Fig. 6 shows such bands for a typical direct-gap semiconductor in which the conduction and valence band edges occur at the same wavevector (usually zero) in the Brillouin zone. Near the conduction and valence band edges, the electrons and holes have dispersion relations with characteristic masses smaller than that of the free electron mass. The electrons, for example, have a dispersion relation given by $E = \hbar^2 k^2 / 2m_e$ in which m_e is as small as a few percent of the free electron mass. The hole bands can be considered similarly, the upper band being referred to as the heavy hole band and the other band (which becomes degenerate with it at the zone center) being called the light hole band. In addition there is a lower third band, the spin-orbit split off band. Interband absorption between the various hole bands plays an important role in defining the infrared optical properties of semiconductors where free holes exist, but for now we concentrate on valence-conduction band transitions. From the perspective of the two-level atom picture above and for a pair of bands we see that semiconductor can be considered to be a two-level "atom" for which the energy levels are inhomogeneously broadened in momentum space. For a direct gap semiconductor the absorption coefficient corresponding to k-conserving interband transitions for a single pair of conduction and valence bands can be obtained within the effective mass approximation [Johnson, 1967] as

$$\alpha(\omega) = \frac{2^{5/2} e^2 m_r^{3/2} \langle p \rangle^2}{m_0^2 n_0 c \hbar^3 \omega} (\hbar\omega - E_g)^{1/2} \tag{19}$$

where $\langle p \rangle$ is the momentum matrix element averaged over all interband transitions, and $m_r = (m_e^{-1} + m_h^{-1})^{-1}$ is the reduced electron-hole mass. For a system with multiple hole bands the total absorption coefficient can be obtained by adding the contributions from each pair of conduction and valence bands.

In indirect gap semiconductors such as Ge and Si, the band edges occur at different points of the Brillouin zone and absorption or corresponding emission processes require phonon participation to provide momentum. Absorption is weak as is any resulting change in the refractive index and these types of semiconductors will not be considered further here. The various absorption events define the possible changes in the refractive index that can occur. For convenience we will begin by considering a semiconductor in which the valence band is completely filled with electrons and the conduction band is empty. We consider first excitation with photons below the band-gap and then later consider effects related to above-gap excitation. We will distinguish these as non-resonant and resonant excitation effects although, as seen above, they are sometimes referred to as virtual (bound) and free carrier effects.

5.2. NONRESONANT EFFECTS

The modified Kramers-Krönig relation (Eq.8) can be used to calculate Δn for below-band-gap radiation. Absorption in this case can only occur if the bound electron distribution is itself modified by the presence of a light field. If a material is illuminated by monochromatic radiation at a frequency ω' then the induced absorption change at a frequency ω due to the presence of a light field can be expressed to lowest order as

$$\Delta\alpha(\omega) = \theta(\omega, \omega')\, I(\omega') \tag{20}$$

where the coefficient $\theta(\omega,\omega')$ depends on the nature of the absorption change. When Eqs. 8 and 20 are combined, the change in refractive index brought about by virtual electronic effects is of the form of Eq. 10.

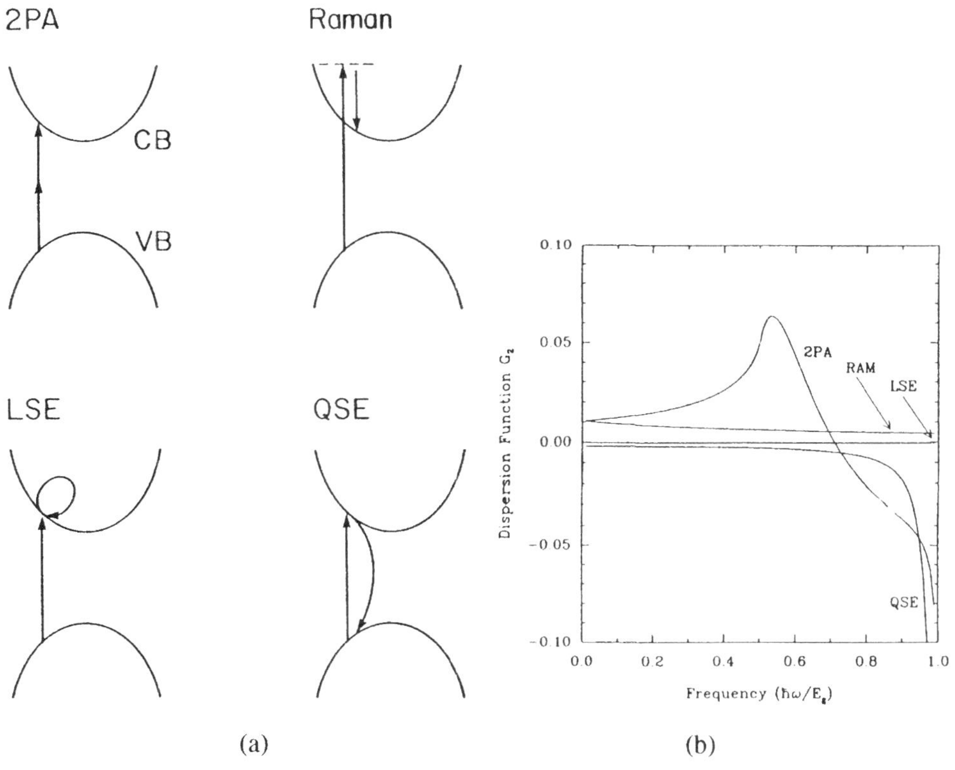

Figure 7. a) Illustration of two-photon absorption, Raman, linear Stark, and quadratic Stark effects in a semiconductor. b) Contribution of each of these effects to n_2 as a function of frequency.

Sheik-Bahae et al. [1991] have shown that when $\hbar\omega < E_g$ the two most important contributions to $\Delta\alpha$ are from two-photon absorption (2PA) and the quadratic (or optical) Stark effect (QSE), while effects due to the linear Stark effect (LSE) and the electronic Raman effect (RAM) are of a lesser significance. Each of these different processes is illustrated in Fig. 7 and will now be discussed in turn.

For nondegenerate 2PA an optical transition involves two photons of unequal energy. Phenomenologically, nondegenerate 2PA can be described as a change in the absorption coefficient at one frequency, ω, due to the presence of a light field at a second frequency, ω', so that Eq. 20 becomes

$$\Delta\alpha(\omega) = \beta(\omega,\omega')I(\omega') \tag{21}$$

where $\beta(\omega,\omega')$ is the nondegenerate 2PA coefficient which contributes to the imaginary part of the third order optical susceptibility, Im $\chi^{(3)}$ (-ω; ω,ω',-ω'). Expressions for $\beta(\omega,\omega')$ have been derived by Sheik-Bahae et al. [1991] for direct band gap semiconductors using second order perturbation theory within an effective mass approximation. General expressions[Wherret, 1984] for the degenerate 2PA show that β (in cmGW $^{-1}$) can be written

$$\beta = (3.1 \pm 0.5)\mathrm{x}10^3 \frac{E_p^{1/2} F_2(2\hbar\omega / E_g)}{n_0^2 E_g^3}, \tag{22}$$

Figure 8 shows the degenerate 2PA coefficient for a selection of semiconductors at 1.06 or 0.53 μm as measured by Van Stryland et al. [1985].

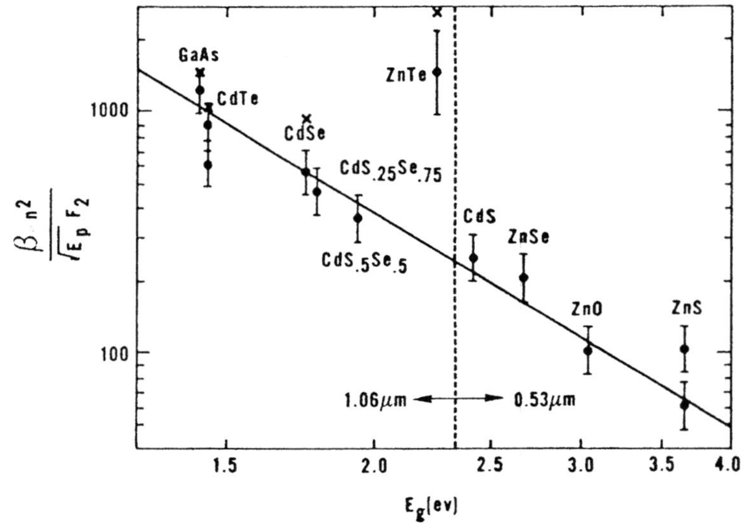

Figure 8. Experimental 2PA coefficients in various semiconductors compared to theory [Van Stryland;1988].

where E_p is the Kane energy parameter in eV (typically about 20 eV) and F_2 is a function

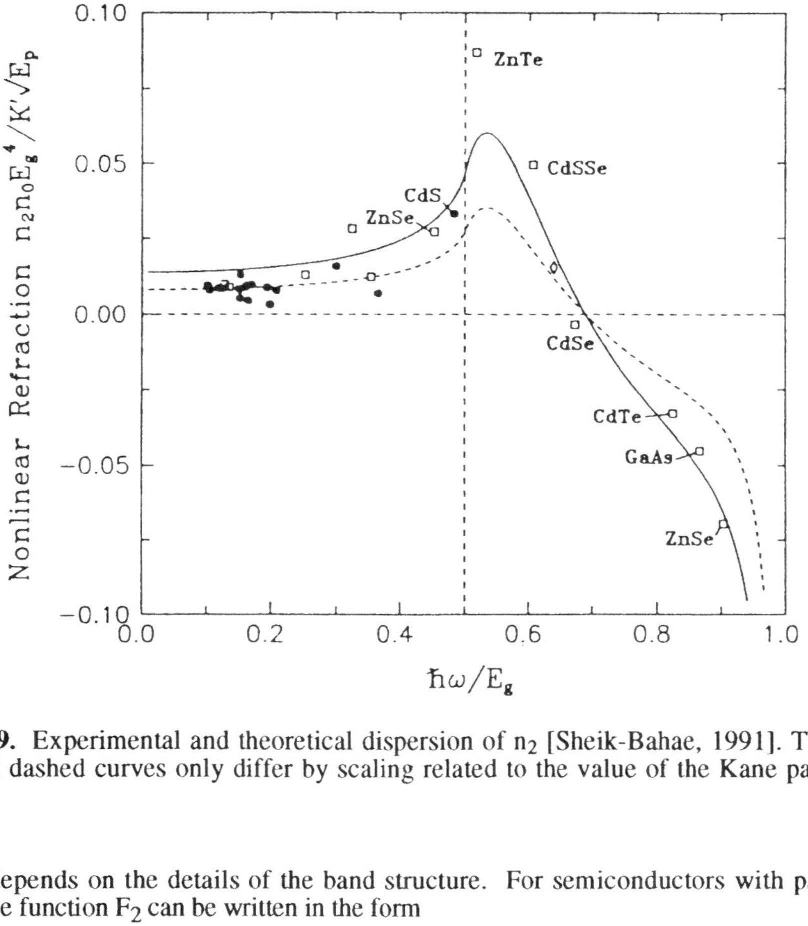

Figure 9. Experimental and theoretical dispersion of n_2 [Sheik-Bahae, 1991]. The solid line and dashed curves only differ by scaling related to the value of the Kane parameter chosen.

which depends on the details of the band structure. For semiconductors with parabolic bands the function F_2 can be written in the form

$$F_2(x) = \frac{(x-1)^{3/2}}{x^5}. \tag{23}$$

The general expression for the non-degenerate two photon absorption coefficient can be used to deduce a general scaling behavior for n_2 as a function of $\hbar\omega / E_g$. It was found that the value of n_2 scales with $E_g^{-4} n_0^{-2}$. It is possible to obtain a dimensionless dispersion function, $G_2(\hbar\omega / E_g)$ for n_2 if the latter is divided by $6\pi x 10^8 c^{-1} E_g^{-4} n_0^{-2} E_p^{1/2}$. The dispersion function G_2 is plotted in Fig. 9. One sees that the value of n_2 is positive when $\hbar\omega / E_g <$ 0.7 and negative otherwise. Near the band-edge the magnitude of n_2 increases dramatically because of the quadratic Stark effect (virtual band blocking). The details of this behavior have been observed by Fujimoto and co-workers in GaAs [Lagasse, 1989; Anderson, 1990] using femtosecond pulse interferometry. Note that the dispersion of n_2 also extends to frequencies below which 2PA can occur. This is similar to what is observed for linear optical effects. The results here are important in that they offer a simple model for two photon absorption which can yield results in good agreement with experiment and can offer accurate predictions for n_2 as well. Furthermore they indicate that it is possible to obtain significant values of n_2 well below the threshold for two-photon absorption, a process which can only contribute to beam attenuation and perhaps undesirable thermal effects. This has been exploited by Villeneuve et al. [1992] to obtain good figures of merit in AlGaAs based on a fundamental beam at 1.5 μm, a wavelength of prime importance in optical communications.

The effect of the quadratic (or optical) Stark effect is to blue shift the electronic band gap in the presence of intense optical radiation. The shift can be calculated by considering the coupling of the valence and conduction band states by the non-secular part of the radiation Hamiltonian. To first order it is found that the induced absorption changes for photon energies less than E_g scale linearly with the irradiance. This is identical to the scaling displayed by 2PA-induced changes. It follows that the quadratic Stark effect contribution to n_2 can be calculated through the Kramers-Krönig transformation. Since the absorption changes induced by the quadratic Stark effect are greatest in the immediate vicinity of the band edge, the greatest changes in the refractive index occur for photons with energy just below the band gap. Similar to the 2PA contribution to n_2 in the near band edge regime, the quadratic Stark effect contribution to n_2 is negative and increases sharply in magnitude as the photon energy approaches the band gap energy. Its effect is indicated in Fig. 7.

The two other processes which contribute to n_2, the linear Stark effect and the electronic Raman effect, can largely be ignored. Linear Stark effect contributions arise from a reduction in the oscillator strength due to the coupling of the conduction band to itself (or of the valence band to itself) by the radiation field. Electronic Raman contributions arise from two photon transitions where the difference in the energies of the two photons is equal to the energy of the electronic transition (a photon at one energy is absorbed while a photon at a second energy is created). The contributions of these two processes to n_2 are also plotted in Fig. 7.

5.3. RESONANT EFFECTS

The generation of free carriers has long been known to profoundly effect the refractive index of semiconductors through a variety of mechanisms. Indeed, historically these effects were appreciated before the bound electron effects discussed in the previous section, which are usually only observed at very high intensities and, for long optical pulses, are obscured bythe much larger free carrier effects. A major breakthrough for optical switching came when Miller et al. [1978] observed substantial changes in the refractive index of InSb excited close to the band gap energy. The change in the refractive index here occurs simply with a change in the linear optical properties brought about by carrier excitation, or in the language of Eq. 8, $\zeta = N$. Because the parameter γ of Eq.11 depends on details of the plasma, but also because the plasma itself depends on nonlinear optical effects for its existence, it is worth reviewing some elements of optical absorption of radiation and the evolution of photoexcited plasmas before we delve further into the theoretical aspects of contributions to refractive index changes from real carriers.

Linear and two-photon absorption absorption lead to a reduction of the beam irradiance with propagation distance, z, in the sample. Free carrier absorption also contributes to attenutation, but not to carrier generation. The energy lost from the radiation field due to

FCA leads only to carrier heating. In many cases FCA is described in terms of a cross section, σ_{FCA}, so that its contribution to the absorption coefficient is $\sigma_{FCA}N$. Typical values

Table 1 Optical properties for select II-VI semiconductors

Parameter	CdS	CdSe	$CdS_{.75}Se_{.25}$	ZnSe
E_g (eV) (300K)	2.53	1.74	2.33	2.58
n_0 (λ=620 nm) (a)	2.5	2.13	2.41	2.56
dn_0/dT_L (K^{-1}) (λ=620 nm) (a)	$1.1x10^{-4}$	-	$2.4x10^{-4}$	$1.6x10^{-4}$
n_2 (cm^2GW^{-1})	$-5.2x10^{-5}$	-	$-2.6x10^{-4}$	$-1.8x10^{-5}$
γ_{FCA} (cm^3)	$-4.0x10^{-22}$	-	$-4.5x10^{-22}$	$-5.1x10^{-22}$
γ_{BF} (cm^3) (T=300 K)(b)	$-8.5x10^{-22}$	-	$-3.6x10^{-21}$	$-9.4x10^{-22}$
γ_{BF} (cm^3) (T=300 K)(c)	$-1.3x10^{-21}$	-	$-4.4x10^{-21}$	$-7.4x10^{-22}$
γ_{BF} (cm^3) (high T)	$-3.1x10^{-22}$	-	$-5.2x10^{-22}$	-
γ_{BF} (cm^3) (nonthermal)	$-7.8x10^{-23}$	-	$-8.7x10^{-23}$	-
γ_T (cm^3)	$+1.6x10^{-23}$	-	$+4.2x10^{-23}$	$+1.8x10^{-23}$
β (2PA) ($cmGW^{-1}$) (620 nm)	6.8	-	9.0	5.7

a) from Landolt -Börnstein[1982]
b) from Auston[1978]
c) full band-filling result for 10^{17} cm^{-3}

of σ_{FCA} are of the order of 10^{-18} cm^{-2} for visible and near infrared radiation. The equation governing the irradiance at depth z, I(z) is given by :

$$\frac{dI(z)}{dz} = -(\alpha_0 + \beta I(z) + \sigma_{FCA} N(z)) I(z). \tag{24}$$

For sub-bandgap radiation only two photon absorption leads to carrier generation in intrinsic semiconductors and one electron-hole pair is generated for two absorbed photons while for single photon absorption, assuming unity quantum efficiency, one electron-hole pair is generated per photon absorbed. The local carrier generation rate can then be written as:

$$\frac{\partial N}{dt}\Big|_{gen} = \frac{1}{2\hbar\omega}\beta I^2(z) + \frac{1}{\hbar\omega}\alpha_0 I(z). \tag{25}$$

The local irradiance and carrier density are calculated by numerically solving Eqs. 24 and 25. Of particular interest is the ratio of the energy absorption leading to carrier production to the total energy absorption, the difference often leading to undesirable thermal effects in all-optical switching.

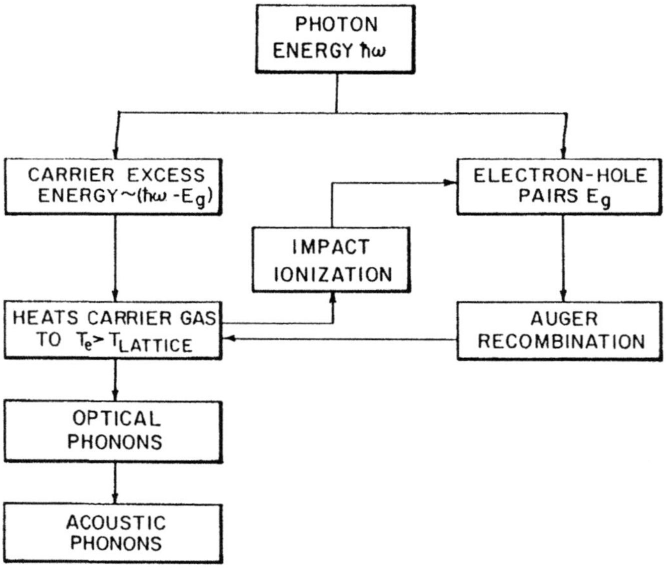

Figure 10. Flow of Energy in carrier and phonon system following a single above-band-gap-photon absorption.

The electron-hole pairs are generated with an ionization energy of E_g and total kinetic energy of $\hbar\omega - E_g$ $(2\hbar\omega - E_g)$ for linear (two photon) absorption, respectively. For scalar effective masses and linear absorption only, simple considerations of energy and momentum conservation in absorption processes establish that the initial excess electron (hole) energy is $(\hbar\omega - E_g)m_r / m_{e,h}$. Figure 10 [van Driel;1987] illustrates how the carrier kinetic energy and carrier density evolve as the semiconductor returns to equilibrium following optical pulse excitation. The carriers rapidly thermalizes on a time scale of 100's of fs or faster for carrier densities above 10^{17} cm^{-3} so electrons(holes) form Fermi-Dirac distributions characterized by temperatures T_e (T_h) and Fermi level F_e (F_h) such that

$$f_e(E_e) = [1 + \exp\{\beta_e(E_e - F_e)\}]^{-1} \tag{26}$$

$$f_h(E_h) = [1 + \exp\{\beta_h(E_h - F_h)\}]^{-1} \tag{27}$$

with $\beta_e = (k_B T_e)^{-1}$ and $\beta_h = (k_B T_h)^{-1}$. The carrier density can then be written as

$$N = N_c F_{1/2}(\beta_e \eta_e) = N_v F_{1/2}(\beta_h \eta_h) \tag{28}$$

where η_e (η_h) is the Fermi level of the electrons (holes) relative to their band edges, N_c and N_v are the effective densities of states in the conduction and valence bands respectively and F_j is the Fermi-Dirac integral of order j. The temperatures and Fermi levels for the electrons and holes are determined by self-consistently solving the equations which govern the carrier density and kinetic energy content of the plasma, *viz.*:

$$\frac{m_r}{m_e}\left[\hbar\omega - E_g\right] = \frac{3}{2} k_B T_e N_c F_{3/2}(\beta_e \eta_e), \tag{29a}$$

$$\frac{m_r}{m_h}\left[\hbar\omega - E_g\right] = \frac{3}{2} k_B T_h N_v F_{3/2}(\beta_h \eta_h). \tag{29b}$$

It should be noted that, to be exact, each hole with a distinct effective mass must be treated as a separate carrier species. It can be shown that for low density plasmas this calculational scheme leads to the well known result for Boltzmann statistics that the initial temperature of each carrier species is equal to $(2/3)k_B^{-1}$ multiplied by the initial energy of that species. It is not uncommon that this temperature is 1000's of degrees.

Following internal thermalization the electrons and holes evolve so as to reach a common carrier temperature through collisions with each other. Depending on the carrier density and band structure, this process can occur simultaneously with the internal thermalization process. If, as usually is the case, the effective mass of the electrons is much smaller than that of the holes, the electrons initially have a much higher temperature than that of the holes but lose much of their energy to the holes on a picosecond time scale. However, on a time scale of a few picoseconds the carriers in polar semiconductors also cool by emission of longitudinal optic (LO) phonons through the Fröhlich interaction [Seeger, 1982]. At high excitation it is possible that the optical phonon distribution is driven far from equilibrium (hot phonon effect) so that the carrier cooling rate is inhibited [van Driel, 1979; Vasconcelos, 1980; Pötz, 1983]. If this occurs then both carrier temperature and optical phonons decay at a rate determined by the optical phonon lifetime, typically 5 ps. The optic phonon modes and subsequently excited acoustic modes come to equilibrium with the remaining modes in the lattice through anharmonic lattice interactions which occur on a time scale of several ps to several ns depending on the particular modes.

Finally, the carrier density is reduced through various recombination processes. Recombination through traps and defects is proportional to the carrier density and is generally most effective at low densities. Radiative recombination, a two body (electron-hole) process dominates in direct gap semiconductors for densities of $10^{16}\,cm^{-3}$ - $10^{18}\,cm^{-3}$ and can lead to recombination times as short as nanoseconds. At higher densities Auger recombination is most effective. This three body process yields recombination times as short as 100 ps for densities of $10^{20}\,cm^{-3}$. The overall recombination rate can then be written as

$$\frac{\partial N}{\partial t}\Big|_{rec} = -AN - BN^2 - CN^3 \quad . \tag{30}$$

Finally, diffusion redistributes carriers spatially. The electrons and holes do not diffuse independently since an electric field develops along with a space charge as they atttempt to separate. As a results, although they have different diffusion coefficients in majority species situations, in equal numbers they diffuse with an ambipolar diffusion coefficient which has a value of the order of 10's of cm^2sec^{-1}. The particular value depends on the temperature of carriers as well as the lattice. The overall evolution of the electron-hole plasma is governed by the equation of continuity

$$\frac{dN}{dt} = \frac{\partial N}{\partial t}\Big|_{gen} + \frac{\partial N}{\partial t}\Big|_{rec} + \vec{\nabla}\cdot(D\vec{\nabla}N) \ . \tag{31}$$

Note that for radiative recombination the emitted photons have an energy higher than that of the band-gap. They can therefore be reabsorbed leading to carrier generation in other regions of the sample, in some cases not more than a fraction of a micron from where the initial recombination event occurred. This photon recycling process [Chandresekhar, 1950; Dumke, 1957; Epifanov, 1976; Tsarenkov, 1979; Fox, 1993] leads to an effective transport of carriers. It is also possible that emitted photons can be reabsorbed by free carriers leading to a heating of the plasma. This would have the effect of altering Fermi levels and carrier temperatures and changing the recombination kinetics. The point is that all of these processes are inter-related and the evolution of a spatially inhomogeneous plasma is very complicated, even without having to consider normal ambipolar diffusion processes. Nonetheless, for an initial laser absorption depth of 1 μm, it typically requires 10-100ps for the plasma to double its depth.

We are now able to consider how free carriers contribute to the refractive index change through alteration of the absorption coefficient. Such absorption changes include those due to band filling (BF), free carrier absorption (Drude effect), band gap renormalization (BGR) and plasma screening of Coulomb interactions . Band filling effects are associated with the bleaching of optical interband transitions due to the filling of electron states in the conduction band and hole states in the valence band. Free carrier absorption, as mentioned briefly above, is a process by which electrons and holes make intraband transitions in the conduction and valence bands respectively with the participation of both a photon and a phonon. BGR is a many-body effect whereby the energy band gap is reduced by the presence of free carriers and thus causes changes in optical absorption as a result of shifting energy levels. Plasma screening effects are mainly associated with excitonic contributions to the refractive index; as the plasma density increases the Coulomb interaction which binds the electron-hole pair is reduced, which results in a reduction in the oscillator strength of the excitonic transition and an enhancement of oscillator strength for the continuum states. We consider first those effects which can be considered to result from interband absorption effects.

Typically, the most important free carrier contribution to the refractive index change is from band filling in which originally vacant conduction (or valence) band states become occupied by photoexcited electrons (or holes). Since occupied states cannot participate in interband optical transitions it follows that band filling is accompanied by a reduction in the magnitude of the absorption coefficient, and hence also by a change in the refractive index. As we saw above, at a qualitative level one can treat the semiconductor as a collection of simple Lorentz oscillators in which one oscillator exists for each allowed interband transition. If the refractive index is to be evaluated for optical frequencies which lie below the minimum oscillator frequency, since each oscillator contributes positively to the net refractive index in this regime, and since the contributions of oscillators corresponding to blocked (bleached) transitions vanish, it follows that band filling must contribute negatively to the net refractive index (i.e. $\Delta n_{BF} < 0$). Also, since the magnitude of the refractive index at a particular optical frequency is most affected by oscillators with resonances at nearby frequencies, it follows that the magnitude of Δn_{BF} for photon energies less than the band gap energy will be most affected by the occupancy of the band edge states. The magnitude of Δn_{BF} should also increase as the optical frequency approaches the band edge frequency.

In order to quantitatively predict the magnitude of Δn_{BF} a more sophisticated treatment than that offered in the previous paragraph is required. The approach which is usually taken is to derive an expression for the change in the absorption coefficient due to band filling, and then to integrate this expression using Eq. 8. The analysis begins by considering the expression for the absorption coefficient in the case where electrons and holes exist [Johnson, 1967]. In this case

$$\alpha(\omega) = \frac{2^{5/2}e^2 m_r^{3/2}\langle p\rangle^2}{m_0^2 n_0 c\hbar^3\omega}(\hbar\omega - E_g)^{1/2}[1 - f_e(\hbar\omega) - f_h(\hbar\omega)] \tag{32}$$

where $f_e(\hbar\omega)$ and $f_h(\hbar\omega)$ are the electron and hole occupancy factors for the states coupled by photon energy $\hbar\omega$. The effects of bandfilling are expressed through the occupancy factors. From Eq. 8 it follows that the change in the absorption coefficient due to the presence of photoexcited carriers can be written:

$$\Delta\alpha(\omega) = -\frac{2^{5/2}e^2 m_r^{3/2}\langle p\rangle^2}{m_0^2 n_0 c\hbar^3\omega}(\hbar\omega - E_g)^{1/2}[f_e(\hbar\omega) + f_h(\hbar\omega)]. \tag{33}$$

The calculation of $\Delta\alpha(\omega)$, and hence also $\Delta n(\omega)$, is thus reduced to a calculation of the occupancy factors for specific situations. The simplest case to deal with is that in which the carriers are in quasi-equilibrium at the band edge (i.e., in the limit of a two-level atom). This case has been considered by Auston et al.[1978] who have derived an analytic expression for Δn_{BF} in which the magnitude of $\Delta n(\omega)$ is predicted to scale linearly with the density of photoexcited carriers:

$$\Delta n(\omega) = -\frac{2\pi N e^2}{n_0\omega^2 m_r}\frac{\omega_g^2}{\omega^2 - \omega_g^2}. \tag{34}$$

In the derivation of this expression the average momentum matrix element in Eq. 33 has been replaced by an approximate value $\hbar\omega_g m_0^2 / 4m_r$ derived from $k \cdot p$ perturbation theory.

Eq. 34 was obtained only for a single pair of conduction and valence bands. In general one must consider contributions from each combination of hole and conduction bands. Also, it should be noted that Eq. 34 cannot be used to predict the magnitude Δn_{BF} as the ratio ω / ω_g approaches unity since damping effects have been neglected. For bulk semiconductors damping effects are very difficult to incorporate into what is essentially a model of an inhomogeneously broadened transition. Table 1 offers *as examples* values of Δn per carrier (or γ_{BF}) for CdS, $CdS_{0.75}Se_{0.25}$, and ZnSe and for $\hbar\omega = 2$ eV. As expected since $\hbar\omega < E_g$ for these materials, the values of γ_{BF} are all negative and increase in magnitude as $\hbar\omega$ approaches $\hbar\omega_g$.

For more realistic carrier densities and/or for carrier distributions characterized by a non-zero temperature, the Fermi-Dirac distribution functions must be used in deriving the refractive index change. For the II-VI semiconductors considered above, since the split off bands are located between 0.45 eV and 0.55 eV below the valence band edges, only the heavy and light hole bands need be considered. The only remaining unknown is the average momentum matrix element $\langle p \rangle$. Following Auston et al., we consider an analysis based on $k \cdot p$ perturbation theory with $\langle p \rangle$ replaced by the Kane momentum parameter, $P = h\langle p \rangle / m_o$, which can in turn be related to the nearly material-independent Kane energy parameter, E_p. While Auston et al. used an analytical expression for E_p, often a value is derived from fits to band structure measurements making the approach somewhat more empirical in nature. Because of this discrepancy the calculations of Δn_{BF} considered here have been made using the experimentally measured absorption coefficient. The predicted values of γ_{BF} are listed in Table 1 for a carrier density of 10^{17} cm^{-3} and a plasma temperature equal to 300 K. For this density γ_{BF} is constant, i.e. the refractive index change scales linearly with carrier density. Fig. 11 shows γ_{BF} as a function of carrier density for densities between 10^{17} cm^{-3} and 10^{21} cm^{-3} for all three materials. The value of γ_{BF} is constant for low plasma densities, but falls rapidly for carrier densities in excess of 5×10^{19} cm^{-3} after degeneracy effects set in. It follows that in the low to moderate density regime the magnitude of the band filling contribution to Δn is a good measure of plasma density.

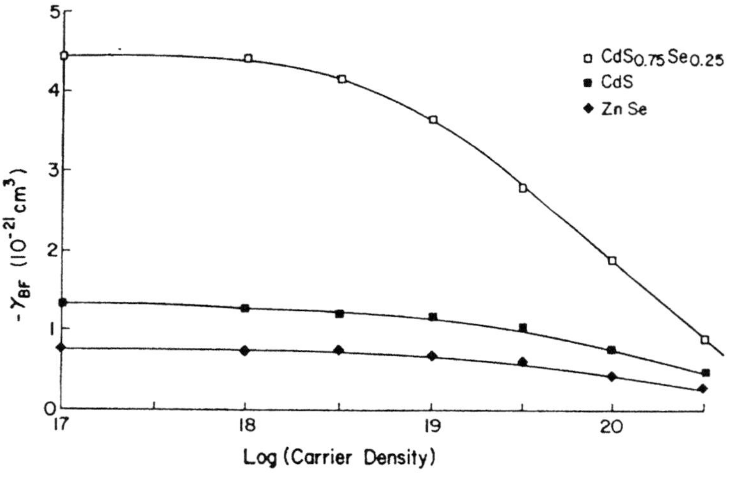

Figure 11. Value of γ_{BF} as a function of carrier density in the case of near-band-edge filling of various II-VI semiconductors.

It is also apparent that the contribution to Δn_{BF} depends on the details of the carrier distribution and, for thermal distributions, the carrier temperature. In an idealized picture, prior to thermalization the electrons and holes occupy states which are optically coupled by the exciting radiation and which can be far removed from the band edges. These states cover an energy range equal to that of the exciting radiation which for 100 fs pulses can be as large as 20 meV. The refractive index changes per carrier can be calculated analytically using the formalism above, and the results are listed for CdS and $CdS_{0.75}Se_{0.25}$ in Table 1 for a carrier density of 10^{17} cm^{-3} excited by 100 fs, 2 eV photons. One sees that γ_{BF} is much smaller for nonthermal distributions than for thermalized distributions, a result consistent with the fact that the number of occupied states near the band edge is significantly reduced for the nonthermal case.

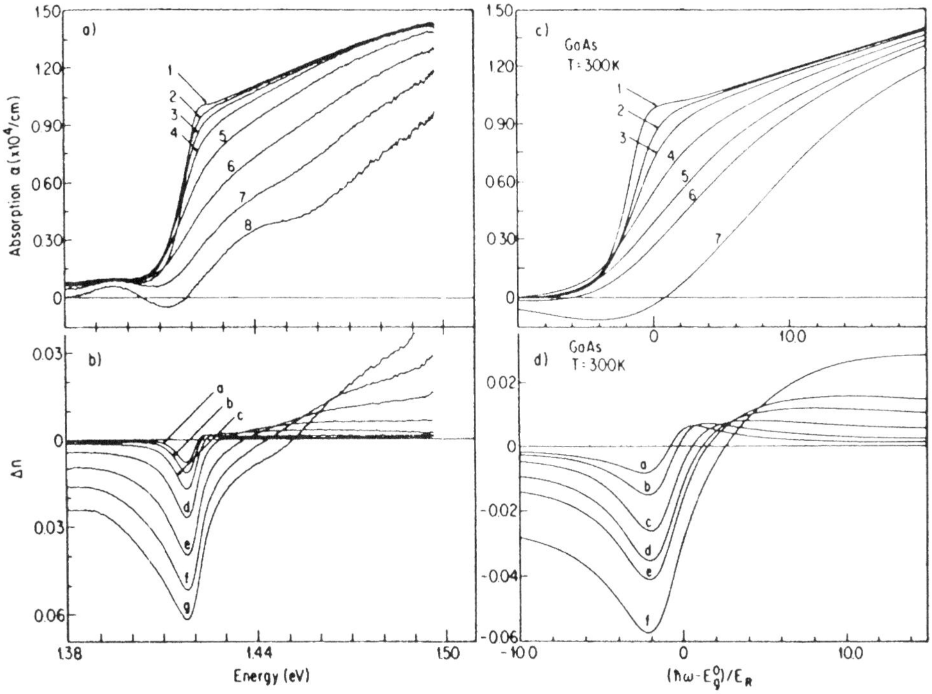

Figure 12. Experimental and theoretical absorption and refractive index spectra for GaAs at room temperature for different excitation powers at 821 nm [Lee; 1986]. The numbers 1-8 in a) indicate average power levels from 0- 50 mW, while the theoretical densitites corresponding to 1-7 in c) are from 10^{15} to $1.5x10^{18}$ cm^{-3}.

The next stage in the relaxation process for which the carrier distribution functions can be expressed analytically is that following thermalization but prior to significant carrier cooling. One can easily calculate the initial temperatures (approx 10,000 K for electrons, 2000 K for holes) following two photon absorption of 620 nm radiation for the materials listed in the Table 1; the calculated values of γ_{BF} for these temperatures are listed as well. The band filling contribution to Δn has also been calculated for combinations of electron and hole temperatures between the initial values and 300 K. It is found that the results are primarily sensitive to the electron temperature. For example, in CdS with a fixed electron temperature T_e=10,400 K (corresponding to the initial energy expected with two photon excitation at λ temperature T_e=10,400 K (corresponding to the initial energy expected with two photon excitation at λ = 620 nm), Δn increases by only 6% when a hole temperature T_h=300 K is used rather than the initial hole temperature T_h=2,220 K, whereas Δn increases by a factor of almost 4 when the electron and hole temperatures decrease from their initial values to room temperature. It follows that to a good approximation the temperature dependence of the refractive index can be treated by fixing the hole temperature at 300 K and only considering the changes in the electron temperature. The effect of bandgap renormalization on Δn has been investigated [Fox, 1993] using the same formalism as was outlined for the band filling problem. BGR occurs when carrier densities become large enough that exchange and correlation self-energies are significant relative to the energy band gap. The electron and hole energies are reduced (at a qualitative level one can understand this in terms of the energy that the carriers gain by avoiding each other spatially) which in turn leads to a reduction in the band gap energy and hence to an increase in the refractive index for $\omega < \omega_g$. Simple expressions for the BGR in terms of carrier density are not possible for this many-body effect. Most model calculations are developed for zero temperature and predict a rigid shift in the band edge which scales as $N^{1/3}$. A more sophisticated analysis performed by Zimmermann [1988] predicts a band edge shift given by the following expression:

$$\Delta E_g(N,T_c) = -\frac{4.64(Na_B^3)^{1/2}R_H}{\left[Na_B^3 + (0.107k_BT_c / R_H)^2\right]^{1/4}}, \tag{35}$$

where a_B is the Bohr excitonic radius, R_H is the exciton Rydberg energy, and T_c is the carrier temperature. This expression predicts band gap reductions of 45 meV for 10^{18} cm^{-3}, 90 meV for 10^{19} cm^{-3}, and 160 meV for 10^{20} cm^{-3} in CdS and $CdS_{0.75}Se_{0.25}$, and approximately 5% lower values for ZnSe, all at 300 K. These predicted shifts are greater than the blue shifts of the band edge due to band filling. The predictions of this model are in sharp contrast to experimental results which indicate that in materials like CdS the effects of BGR dominate over band filling only for low plasma temperatures [Said, 1992]. In particular, for N= 2x10^{18} cm^{-3} it was found that the absorption edge was blue shifted by 20 meV which is roughly what one would expect for this carrier density if only band filling were operative. Nonetheless, calculations in which BGR is considered show that positive values of γ_{BGR} are predicted (see Table 1) for all but the highest plasma densities.

The band-gap renormalization effect may be somewhat offset by another many body effect associated with the presence of free carriers, namely the plasma screening of the Coulomb interaction between electrons and holes. Even where bound excitons do not exist (e.g. at high temperatures), or where photon energies are not in resonance with exciton levels, electron-hole correlations remain which contribute to the oscillator strength, i.e. through Coulomb enhancement of the continuum states. One way in which the strength of exciton creation resonance can be modified by the presence of a plasma (and hence contribute to Δn) is through screening which reduces the strength of the Coulomb interaction and thereby reduces the strength of the interband absorption. The above-edge Coulomb absorption enhancement and its screening must be incorporated in any model of free carrier induced refractive index effects, especially in large gap semiconductors. Banyai and Koch [1986] and Lee et al. [1986] have considered this along with band-gap renormalization and band filling in GaAs. The influence of these many body effects on Δn is shown in Fig. 12.

5.4. INTRABAND CARRIER EFFECTS

Contributions related to changes in the intraband absorption play a significant role particularly at high plasma temperatures where the interband contributions decrease in magnitude or in the infrared region of the spectrum. Here the dominant contribution to $\Delta\alpha$ is due to FCA which, like band filling, scales linearly (to a first approximation) with carrier density as indicated earlier. The calculation of the magnitude of γ_{FCA} can be treated in a similar manner to the calculation of γ_{BF} (i.e. one derives an expression for FCA from first principles and performs the integration in Eq. 21), however detailed expressions for FCA with the lattice temperature and carrier temperature dependences included are in general difficult to develop. Instead, the problem can be treated approximately within the context of a simple Drude model whereby the contribution of a specific carrier species to Δn can be expressed as follows:

$$\Delta\chi'(\omega) = \frac{-Ne^2}{m_c}\frac{\tau_c^2}{1+\omega^2\tau_c^2} \tag{36}$$

$$\Delta\chi''(\omega) = \frac{Ne^2}{\omega m_c}\frac{\tau_c}{1+\omega^2\tau_c^2} \tag{37}$$

where τ_c is the momentum relaxation time and m_c is the effective mass of the carrier. From these expressions, if one neglects carrier damping effects one finds that a (small) refractive index change is given by

$$\Delta n = -\frac{2\pi Ne^2}{\varepsilon\omega^2 m_c n_0^2} = -\frac{1}{2}\frac{\omega_p^2}{\omega^2} \tag{38}$$

where ε is the background dielectric constant of the semiconductor, i.e. due to bound electrons, and ω_p is the plasma frequency. When both electrons and holes are present, both contribute to the change in susceptibility and refractive index and the plasma frequency is now as indicated in Eq. 38 except that the carrier effective mass is replaced by the reduced effective mass of electrons and holes. It follows that γ_{FCA} can be written

$$\gamma_{FCA} = -\frac{2\pi e^2}{\varepsilon\omega^2 m_c n_0^2}. \tag{39}$$

This effective Drude contribution for each carrier species has been calculated for the three II-VI semiconductors using the above expression with the electron and two hole bands considered separately. The total Drude contribution to the refractive index is listed in Table 1.

In the visible region of the spectrum the refractive index change due to intraband effects is at best comparable to that of interband effects. From Eq. 39 one can see however that in the infrared intraband effects can dominate. Spitzer and Fan [1957] have carried out considerable work over many years using infrared reflectivity spectroscopy beyond approximately 5 μm to characterize the properties of extrinsically doped semiconductors. Gallant et al.[1981] illustrated how infrared reflection spectroscopy can be used to study kinetic plasma properties and the changes in the refractive index of Ge and Si using time resolved techniques. For example, the properties of electron-hole plasmas with densities up to a few times 10^{19} cm^{-3} excited in Ge by a 80 ns, 1.06 μm or 0.53 μm pulses have been studied. The reflectivity change of Ge was monitored using a 10.6 μm, cw $C0_2$ laser beam and a Ge:Au detector with a response time of 1 ns. Typical time dependent traces of the reflectivity are shown in Fig. 13. During the excitation pulse the carrier density at the surface increases due to pumping and then decreases due to a combination of recombination and diffusion. When the increasing density reaches $2x10^{19}$ cm^{-3}, corresponding to the plasma frequency for 10.6 μm radiation, the reflectivity reaches a minimum as the refractive index drops to a value near unity. As the peak plasma density increases with increasing pump irradiance, the real part of the dielectric constant becomes negative, indicative of a metallic state, and the reflectivity rises substantially. After the peak of the pulse the surface carrier density drops and the reflectivity retraces its earlier behavior. The fact that the

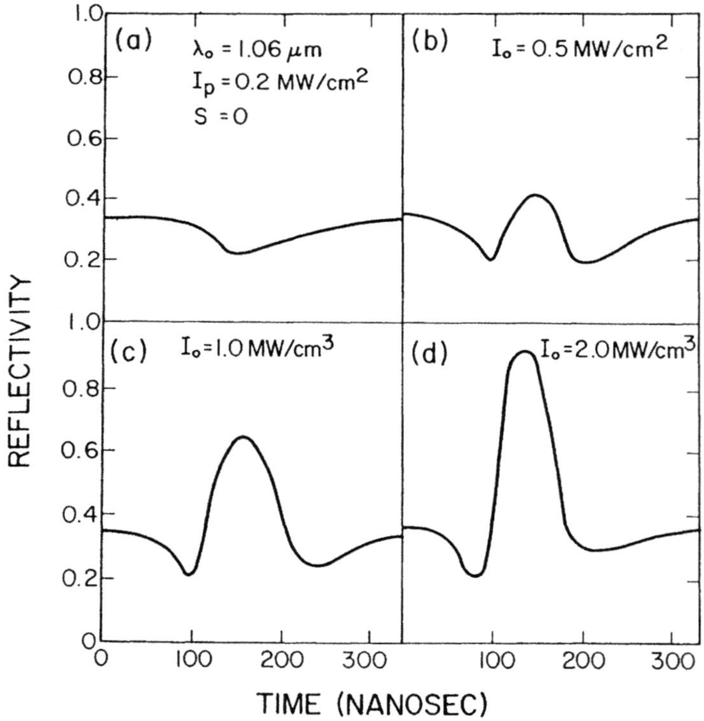

Figure 13. Time-resolved reflectivity in Ge at 10.6 μm after excitation by 1.06 μm, 80 ns pulses.

reflectivity minimum value is not zero on either side of the maximum is due to damping of the plasma by momentum relaxation of carriers. The observation that the second minimum is typically higher than the first is consistent with heating of the lattice by the pulse, leading to an increase in the damping rate by carrier-phonon interactions. The collision times measured are of the order of 100's of fs, consistent with mobility values in Ge. Similar time-resolved experiments have been carried out using 25 ps, 1.06 and 0.53 μm excitation of Ge and Si to monitor plasmas with densities higher than $3x10^{20} cm^{-3}$. The infrared probe beams in this case had wavelengths of 1.5, 1.9 or 2.8 μm and were derived by stimulated Raman scattering from the 1.06 μm pulses.

The large refractive index changes induced in semiconductors in the infrared can be used for all optical switching phenomena and generation of short pulses as demonstrated simultanesously by Alcock and Corkum[1975] and Samson et al.[1976]. Alcock and Corkum [1978] later used a Brewster geometry and picosecond 1.06 μm excitation of Ge to produce subpicosecond pulses at 10.6 μm. In a Ge switch , the quasi-cw CO_2 pulse develops a sharp rising edge as the semiconductor is excited by the 1.06 μm radiation. When this pulse is sent into a second Ge slab in a transmission geometry, with an appropriately synchronized excitation pulse, the induced plasma reflects the tail of the incident pulse leaving an ultrashort transmitted pulse with subpicosecond duration.

5.5 LATTICE THERMAL EFFECTS

The final process which will be considered to yield a contribution Δn is lattice heating. Any absorption of laser light leads to some lattice heating. For pulses short compared to carrier recombination times, the consideration of lattice heating separately from free carrier effects in optically excited semiconductors is somewhat artificial since the increase in lattice energy is due to energy transfer from the carrier system to the phonon system as the carriers cool. Of course, the *time evolution* of carrier and lattice phonon effects will be different. On a time scale of picoseconds, the phonons generated by cooling or recombining carriers may not have established an equilibrium distribution characterized by an elevated lattice temperature (T_L) although after nanoseconds the concept of a temperature to describe the plasma is perhaps appropriate. In what follows we will nonetheless use an effective lattice temperature to discuss lattice heating effects, since a more detailed model is not justified at this time. As the lattice temperature rises the band gap decreases due to many body effects involving the interaction of carriers and the phonons, and due to changes in the lattice constant as the solid expands. In addition the expansion of the lattice itself can change the optical thickness (nL) the beam sees. It follows that the optical thickness is governed by

$$\frac{d(nL)}{dT_L} = nL\left(\frac{1}{n}\frac{dn}{dT_L} + \frac{1}{L}\frac{dL}{dT_L}\right). \tag{40}$$

Normally the thermo-optic effect makes a much larger contribution than thermal expansion with the value of dn/dT_L being about 10^{-4} while the expansion coefficient is typically about an order of magnitude smaller. The change of the refractive index can be directly related to the variation of the band gap

$$\frac{dn}{dT_L} = \frac{dn}{dE_g}\frac{dE_g}{dT_L}. \tag{41}$$

To a first approximation the band gap decreases linearly with temperature; Fig. 14 shows typical values of the band-gap thermal coefficient.

The values of dn/dT_L (for $\hbar\omega = 2$ eV) for the three II-VI materials we have consdiered throughout are listed in Table 1. Two values of $\Delta n/N$ are calculated following two photon excitation by 620 nm sub-picosecond pulses; the first assumes that the carriers cool to the bandedge and hence transfer $(2\hbar\omega - E_g)$ worth of energy per carrier to the lattice. The second value assumes that the carriers recombine nonradiatively so that each carrier contributes $2\hbar\omega$ to the lattice. In both cases the lattice heating contribution to ΔN is at least two orders of magnitude smaller than the bandfilling contribution . For much *longer* pulses where recombination reduces the carrier density due to diffusion or recombination, the surface carrier density effects will reach steady-state while heating effects will increase with laser pulse width. Under these circumstances, thermal effects can dominate the free carrier effects. The extent of the refractive index change will also depend on the proximity of the photon energy to the band-gap. A Kramers-Krönig analysis based on Eq. 8 can be used here with $\zeta = T_L$. This has been considered by Wherret et al. [1988] who have shown that the thermo-optic coefficient varies as:

$$\frac{dn}{dT_L} \propto E_g \left[\frac{E_g}{E_g - \hbar\omega} \right]^{1/2} . \tag{42}$$

As with free carrier contributions the largest effects occur when the photon energy approaches the band-gap.

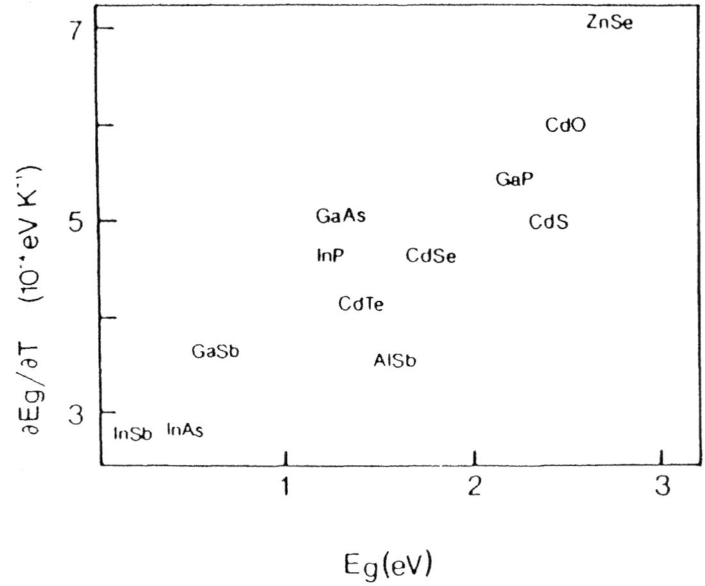

Figure 14. Temperature dependence of band gaps for certain semiconductors [Wherret; 1988].

6. Time-Resolved Effects

To illustrate some of the different effects contributing to changes in the refractive index of semiconductors we present here some of our recent results in which time-resolved techniques have been used to separate bound electronic and free carrier effects and to observe how these different contributions scale with incident irradiance. The results have

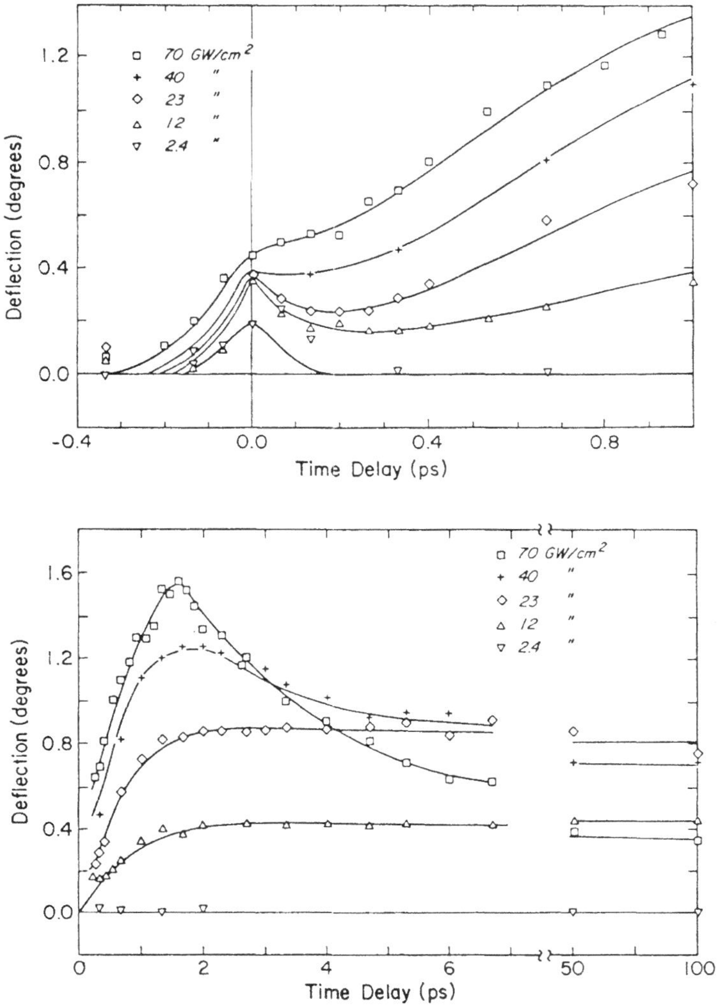

a)

b)

Figure 15. Pump beam induced probe deflection in $CdS_{0.75}Se_{0.25}$ a) near zero delay time and b) at longer times.

been obtained using beam deflection and induced diffraction grating techniques. The beam deflection measurements were performed in 100 μm thick slabs of CdS and $CdS_{0.75}Se_{0.25}$. The laser induced grating experiments were carried out using these same samples and also in

20 μm thick ZnSe. Experiments were carried out with a 620 nm (2 eV) 120 fs pulses for peak pump irradiances up to 700 $GWcm^{-2}$ and for lattice temperatures between 77 and 300K. Qualitatively the same behavior was observed in the different samples, so most of thediscussion will focus here on one of the samples, $CdS_{0.75}Se_{0.25}$.

6.1. TIME RESOLVED BEAM DEFLECTION AND INDUCED DIFFRACTION RESULTS

The beam deflection results are treated first. A typical result is displayed in Fig. 15 where the probe pulse deflection angle in $CdS_{0.75}Se_{0.25}$ is plotted versus the pump-probe time delay, t, for a series of different pump pulse (peak surface) irradiance, I_0. The radial position of the probe pulse, r_p, and the e^{-2} radius of the pump pulse, r_0, are both equal to 40 μm (the quoted irradiance is that evaluated at r_p and for these beam deflection experiments will be referred to as I_{BD}). The positive deflection angles measured correspond to a (depth averaged) $\Delta n < 0$ in all cases consistent with what is expected theoretically for negative changes in the refractive index which occur when the photon energy is greater than $0.7E_g$. For the lowest excitation conditions the observed deflection follows the temporal profile of the pump pulse. For larger values of I_{BD} the peak at t = 0 is still observed but is followed by a subsequent deflection which peaks at a delay time of 1.5 ps and then decreases on a time scale larger than 200 ps. With increasing excitation the non-instantaneous contribution to Δn is seen to completely dominate the instantaneous ones. For $I_{BD} > 25$ $GWcm^{-2}$ the beam deflection still peaks at 1.5 ps, decreases sharply on a time scale as short as 3 ps, and then decays more slowly for times > 100 ps. The data in this high excitation regime is distinctive in that the smooth curves drawn through data points corresponding to high values of I_{BD} cross through those corresponding to lower values of I_{BD} so that the maximum deflection at long time delays corresponds to an intermediate excitation level. This behavior will be referred to as "curve crossing" and the excitation level which produces the largest

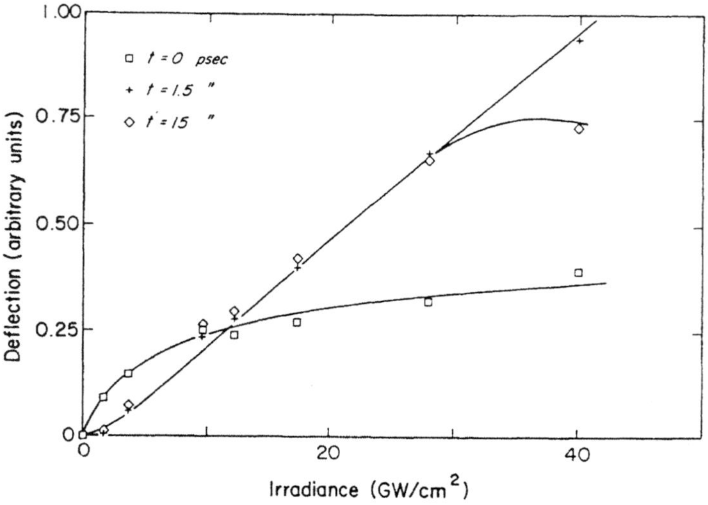

Figure 16. Scaling behavior of deflection with irradiance for $CdS_{0.75}Se_{0.25}$ ($r_0 = 60$ μm, $r_p = 40$ μm).

beam deflection for long time delays, and which marks the onset of rapid signal decay, will be referred to as the threshold irradiance. This irradiance is dependent on the location of theprobe pulse relative to the centre of the pump pulse. For $r_0 = r_p = 40$ mm a threshold value for $I_{BD} = 25$ GWcm^{-2} is measured. For the same pump spot size but with r_p=15 μm the threshold corresponds to $I_{BD} = 110$ GWcm^{-2}, and for $r_p = 0$ the measured threshold is equal to 120 GWcm^{-2}. Qualitatively similar behavior is seen in CdS with the exception that the deflections are slightly smaller for the same excitation strength and the delayed peak in the deflection occurs at t = 2.0 ps.

The main features of the data include the instantaneous deflection at t=0, the peak in the non-instantaneous deflection at t =1.5 ps, and the partial recovery in the deflection angle for large values of I_{BD} and time delay greater than 10 ps. Measurements of these deflection signals at t = 0, 1.5 and 10 ps were made as a function of excitation level and are presented in Fig. 16 for $CdS_{0.75}Se_{0.25}$. It can be seen that the instantaneous deflection grows linearly with I_{BD} for small values of I_{BD} and saturates for larger values. The magnitude of the deflection at t =1.5 ps grows superlinearly with peak irradiance for small values of irradiance and afterwards grows linearly with irradiance. The deflection magnitude for large time delays is observed to follow that at t = 1.5 ps for values of I_{BD} up to the threshold value of 25 GWcm^{-2}, but decreases for large values (curve crossing). These scaling laws are discussed in more detail below but are generally consistent with the bound and free carrier contributions to the refractive index change discussed above. The deflection varies as $\overline{\Delta n}d$ where the bar indicates an average over the sample and d is the thickness of the sample or the two-photon absorption depth, which ever is greater.

To ensure that the deflection results are not due to artefacts of the technique, the 4-wave mixing technique was used. The diffraction grating data displays features similar to those observed using the beam deflection technique. Typical results are shown in Fig. 17 for $CdS_{0.75}Se_{0.25}$. The values of I_G refer to the peak value of the irradiance at the sample surface and evaluated at the peaks in the diffraction grating at the centre of the pump spot; since the two pump spots which produce the gratings have the same irradiance, I_G is four times as large as the peak irradiance from a single pump pulse. Because the measured signal consists of a magnitude only, the sign of the refractive index change can not be deduced from the data. No peak in the diffraction efficiency is observed at t = 0 for any excitation strength. Instead the diffraction efficiency grows with increasing time delay beginning with the arrival of the pump pulse. The growth continues until a time delay of 2 ps is reached. For the lower values of I_G the diffraction efficiency decreases slowly beyond t = 2 ps with a time constant greater than 100 ps. For values of I_G greater than 200 to 250 GWcm^{-2} the diffraction efficiency peaks and then displays a rapid decay on a time scale as short as 3 ps. The diffraction data displays a curve crossing behavior similar to that observed in the beam deflection results. Again the onset of the fast decay behavior occurs at an irradiance for which maximum signal amplitudes are measured at long time delays. The threshold irradiances in these experiments were not the same as those obtained from the deflection experiments for reasons to be explained below. Similar results are obtained in CdS and ZnSe although for the thin ZnSe sample the signal decay time for large values of I_G can be as short as 4 ps and have a greater depth of recovery than in the thicker samples.

Scaling measurements similar to those made using the beam deflection technique were also performed using the diffraction technique. For low excitation the peak height at t = 2ps scales as $[I_G]^{3.5}$, while for high excitation the peak height saturates. This scaling behavior is consistent with detailed theoretical modeling based on free carrier effects. Measurements of the absolute diffraction efficiency indicate that in the saturation regime the diffraction efficiency is equal to15% for each of the two diffracted beams.

We will now consider the data in the three main time regimes relative to the theoretical results presented earlier.

6.1.1 *Instantaneous Behavior.* Instantaneous contributions to Δn are associated with virtual excitation of the electronic system and are manifested by a response which is proportional to the instantaneous excitation strength. Although no such response is observed in the diffraction data, the peak at $t = 0$ in the beam deflection data (Fig. 15) is consistent with such a nonlinearity. From the scaling measurements presented in Figs.15 and 16 the irradiance dependence of the refractive index change can be deduced. For small excitation strengths the height of the peak at $t = 0$ grows linearly with I_0, while for large excitation the peak height saturates and becomes constant. In the discussion presented earlier it was shown that this type of scaling behavior follows if Δn is linearly proportional to the instantaneous irradiance (and hence linearly proportional to I_0). This is consistent with the theoretical expectation that the optical Stark effect and 2PA should dominate the instantaneous nonlinear refractive index, n_2, for the combinations of materials and excitation wavelength used. Moreover, both of these effects contribute negatively to Δn, consistent with the measured changes. In order to extract absolute magnitudes for n_2 a full treatment of the beam propagation problem would be required. Nonetheless the magnitudes of the measured beam deflection angles for $t = 0$ are consistent at the order of magnitude level with the theoretical values of n_2 listed in Table 1.

The lack of a peak at $t = 0$ in the diffraction data is expected. For low irradiances the diffraction efficiency at $t = 0$ should scale as I_0^2, while for larger irradiances the diffraction efficiency is predicted to scale sublinearly with I_0. This is due to the 2PA dominated attenuation of the exciting radiation and due to 2PA attenuation of the diffracted light. Using the theoretically predicted refractive nonlinearity values listed in Table 1 the scaling calculations predict that for the range of I_0 values used in the diffraction experiments the instantaneous diffraction efficiency should be at least 2 orders of magnitude less than the diffraction efficiency which follows for nonzero time delays.

6.1.2 *Intermediate Time Behavior.* For delay times between 0 and 3 ps both the deflection and diffraction results display changes in the (depth-averaged) refractive index which remain after the pump irradiance has vanished. For time delays between $t = 0$ and 1.5 to 3 ps these refractive index changes increase in magnitude before either saturating or exhibiting the onset of a decay back to zero. The scaling data shows that the height of the peak beam deflection at $t = 1.5$ to 3 ps scales superlinearly with I_0 for small values of I_0 and linearly with I_0 for large values of I_0. The magnitude of the peak diffraction efficiency scales as $I_0^{3.5}$. The models presented and discussed in the theory section predict these types of scaling behavior if the local value of Δn is proportional to the local carrier density where carrier generation is dominated by 2PA. Moreover the beam deflection data shows that Dn is negative. These results suggest that the refractive index change measured between $t = 0$ and 1.5 to 3 ps is due to free carriers with Δn being dominated by bandfilling and Drude contributions.

The fact that Δn is not reached until a few ps following photoexcitation is due to the thermalization and cooling of the carriers. The photoexcited carriers are initially located in a narrow band of high energy states coupled by 2PA. Carrier-carrier and carrier-phonon scattering produce thermalized electron and hole distributions which subsequently cool by phonon emission. The thermalization and cooling processes both lead to a redistribution of the carriers in which states close to the band edge become preferentially occupied. Although these processes have no effect on the plasma contribution, the refractive index change due to band filling is highly sensitive to the carrier distribution and increases with thermalization and carrier cooling. Since the pulse width used is 120 fs, and since the carrier densities are larger than 10^{18} cm^{-3}, it is unlikely that much of the increase in beam deflection strength or diffraction efficiency can be attributed to thermalization (or even that any evidence of

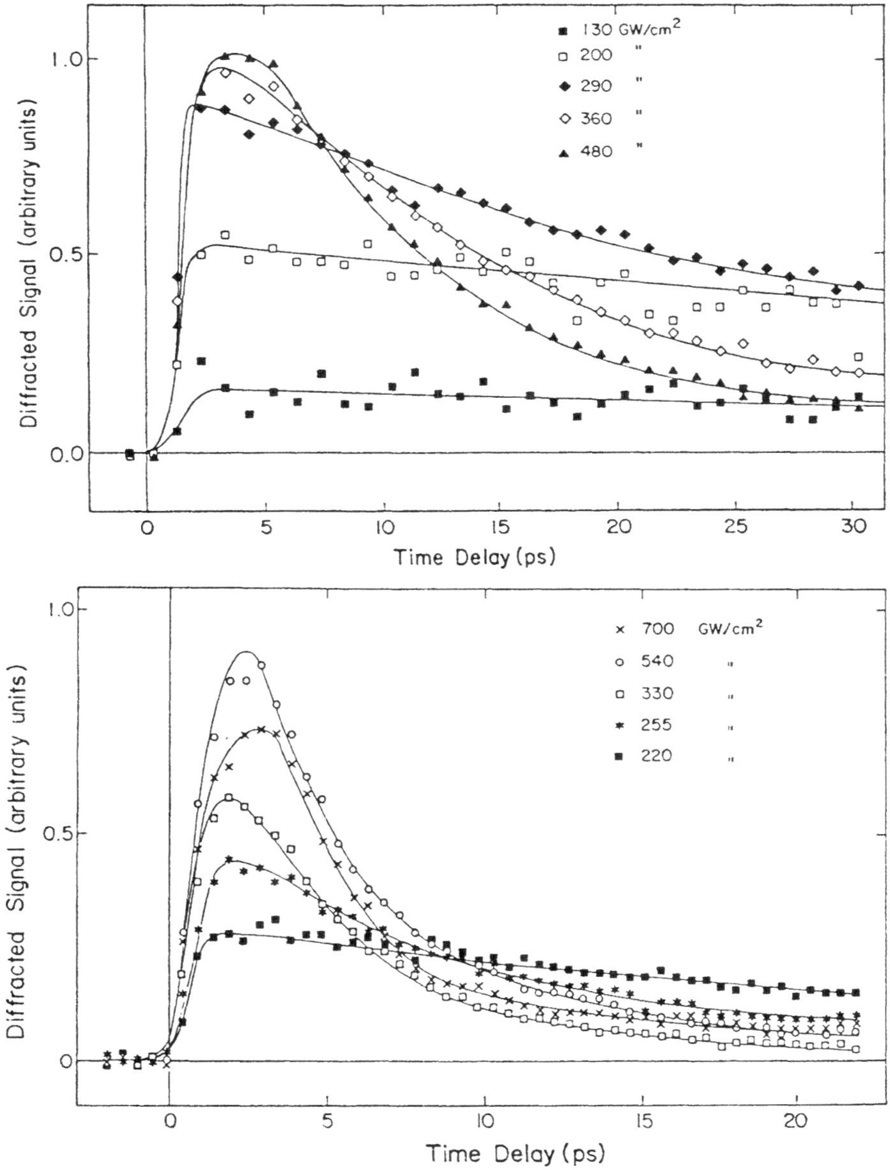

a)

b)

Figure 17. Time-resolved diffraction in CdS for a) low values of I_G and b) high values of I_G.

nonthermalized distributions exists in the data). Since carrier cooling is expected to occur on a time scale of a few ps, however, most of the signal changes can likely be attributed to this mechanism.

The results of calculations of the refractive index change per carrier for non-thermalized, hot and lattice temperature carrier distributions were presented earlier. For low to moderate carrier densities ($< 10^{19}$ cm^{-3}) the refractive index change per carrier for $CdS_{0.75}Se_{0.25}$ (CdS) should increase by a factor of 1.8 (1.5) as the carriers thermalize and by a further factor of 5 (2.5) as they cool from their initial temperature to the lattice temperature. The ratios are largest for $CdS_{0.75}Se_{0.25}$ and smallest for ZnSe since the Drude contribution is a smaller fraction of the total refractive index change per carrier for $CdS_{0.75}Se_{0.25}$ than for ZnSe. Experimentally the changes in signal magnitudes are as follows. For $CdS_{0.75}Se_{0.25}$ (CdS) the diffraction efficiency changes by a factor of 4 (4) between t = 0 and 0.1 ps, and by a factor of 20 (8) between t = 0.1 ps and t = 2 ps. For $CdS_{0.75}Se_{0.25}$ the beam deflection magnitude changes by a factor of 4 between t = 0.1 ps (extrapolated) and t = 1.5 ps (the free carrier contribution to beam deflection at t = 0 is masked by the instantaneous contribution). No results are quoted for ZnSe since the lower signal sizes recorded on this sample make estimates of the signal sizes in the region about t = 0 difficult. For the purposes of comparison between experiment and theory it is assumed that half of the total carrier generation has occurred by t = 0, and that carrier generation is complete by t = 0.1 ps. If it is assumed that the beam deflection magnitude scales linearly with Δn and that the diffraction efficiency scales quadratically with Δn then the experimental results indicate that Δn increases by a factor of 4.2 (2.8) between t = 0.1 ps and t = 2 ps. These increases in Δn are in close agreement with the theoretically predicted increases which accompany the cooling of the carriers from their initial temperatures to room temperature. It follows that the changes in the beam deflection angles and diffraction efficiencies between t = 0.1 ps and t = 2 ps can be attributed to carrier cooling, and that the cooling time is of the order of 1 ps. For the theoretically predicted increases in Δn which accompany carrier thermalization the diffraction efficiency should increase by a factor of 13 (9). Since the measured changes in diffraction efficiency between t = 0 and t = 0.1 ps are smaller, it can be concluded that carrier thermalization must take place within 120 fs and that significant carrier cooling does not occur during excitation.

The measured cooling times of 1.5 to 3 ps are consistent with those reported for these materials in the literature [Auston, 1978]. Moreover the calculations described in the theory section predict the electron temperature to decrease at a rate of roughly 10,000 Kps^{-1} until temperatures close to the lattice temperature are reached. This too is consistent with cooling times of between 1 and 3 ps. To investigate the cooling behavior in more detail, diffraction efficiencies were measured in CdS for time delays between t = 0 and t = 5 ps. A range of different excitation levels were used producing peak carrier densities between $6x10^{18}$ cm^{-3} and $1.5x10^{20}$ cm^{-3}. The data however display little dependence on the excitation level indicating that the cooling rate is not highly affected by hot phonons or screening of the Fröhlich interaction[Kocevar, 1985]. This is in marked contrast to materials like GaAs where such processes become important for carrier densities of the order of 10^{18} cm^{-3}. It is interesting to note however that the Fröhlich interaction predicts that the carrier temperature should decrease linearly with time. A detailed analysis of our data shows that the functional dependence is close to exponential with time, and in any event, that there is a period of rapid cooling followed by a reduced cooling rate over the 2-3 ps time. We believe that this may be due to the electrons initially cooling by transferring their energy to holes. When the two distributions reach approximately the same temperature they cool by the phonon-mediated processes.

6.1.3 *Long time Behavior.* The data plotted in Figs. 15 and 17 show that for small values of I_0 the refractive index change saturates and decays slowly with time. For larger values, the measured signal magnitudes display a partial recovery which can be described by a decay time as short as 4 ps. For the deflection data the threshold is dependent on the location of the probe pulse and varies from I_{BD} = 25 GWcm^{-2} for $r_p = r_0$ = 40 μm to I_{BD} = 120 GWcm^{-2} for r_p = 0 and r_0 = 40 μm (both for CdS and $CdS_{0.75}Se_{0.25}$). For the diffraction data the threshold value of I_G is of the order of 240 GWcm^{-2} for CdS and $CdS_{0.75}Se_{0.25}$, and is closer to 200 GWcm^{-2} for ZnSe. We will now consider possible reasons for the ultrafast recovery observed and why it occurs at a threshold that varies with the technique used.

Possible reasons for the decrease in the beam deflection magnitude or diffracted signal strength are first considered. The beam deflection magnitude and diffraction efficiency are both proportional to the strength of the local radial gradient in Δn. Decreases in either the beam deflection magnitude or diffraction efficiency may be due to decreases in the (spatially averaged) value of Δn or to decreases in the local radial gradient in Δn (e.g. with no accompanying change in the average local value of Δn). Since the signal magnitudes measured between t = 0 and t = 2 to 3 ps have been interpreted in terms of a free carrier effect it would be consistent to assume that the measured signal decays may in part be due to carrier recombination (decrease in average magnitude of Δn) and/or carrier diffusion (decreases in the magnitudes of the transverse gradients). Finally, it is possible that Δn is not due to free carriers only and that the signal decreases are due to a process which contributes positively to Δn and which continues to grow in strength after carrier cooling has ceased.

Lattice heating and band gap renormalization are two processes with positive contributions to Δn. As the carriers cool and transfer their energy to the phonon bath, phonon relaxation occurs more slowly than carrier cooling. If the lattice contribution to Δn increases in magnitude as the phonons thermalize internally then their contribution to Δn might not be maximized until many ps after carrier cooling has finished (i.e. at t > 3 ps). The calculations which were outlined above, however, indicate that the magnitude of the lattice heating contribution to Δn should be at least 2 orders of magnitude less than the contribution of the free carrier contribution and hence is negligible. Moreover, lattice heating effects should scale linearly with carrier density and are not predicted to display a threshold behavior. Band gap renormalization is unlikely to be responsible for the observed signal decays, however, since band gap renormalization occurs nearly instantaneously with carrier generation, and since it is predicted to scale sublinearly with carrier density.

Carrier diffusion has been used to explain observations of rapid plasma expansion in II-VI semiconductors upon high excitation above some threshold [Cornet, 1981, Junnarkar, 1986; Rudolph, 1990; van Lap, 1991]. In these cases diffusion coefficients up to 10^5 times larger than the low density value (5 cm^{2}s^{-1}) have been invoked. The signal decays displayed in our beam deflection and diffraction measurements have been considered in terms of this enhanced diffusion process. The beam deflection results are considered first. The deflection strength at a time t is proportional to the radial gradient of the carrier density evaluated at the location of the probe pulse. If the time dependence of the carrier density is governed by diffusion with ambipolar diffusion coefficient D, it is simple to show that the local density varies according to

$$N(r,z,t) = N(0,z,0)\frac{r_0^3}{(16Dt + r_0^2)^{3/2}}\exp[-\frac{4r^2}{(16Dt + r_0^2)}]. \tag{43}$$

For the combination of spot size and probe spot location used in the measurement of the data in Fig. 15 that D, must exceed 5×10^5 cm^{2}s^{-1} to explain the factor of 2 reduction in the deflection strength within the first 5 ps. This value of D is consistent with those which have been reported in the literature, but our plasma density threshold for the onset of the rapid signal decays is considerably lower than the reported threshold values.

A similar analysis can be applied to the diffraction grating results. In this case it is assumed that the diffusive decay of the grating structure can be described as follows

$$N(r,z,t) = \frac{N(0,z,0)}{2}\{1+\exp(-\frac{4D\pi^2 t}{\Lambda^2})\cos(2\pi x / \Lambda)\} \tag{44}$$

where N(0,z,0) is the carrier density at peaks in the grating structure at a depth z below the surface, and Λ is the grating period. Since the data displayed in Fig. 17 corresponds to a grating period of 1 μm it follows that a 4 ps decay time can be described by a diffusion coefficient of the order of 60 cm^2s^{-1}. This value is closer to the accepted value of D for low density plasmas, but is 4 orders of magnitude less than the value of D which describes the decay of the beam deflection data.

In an attempt to determine the extent to which diffusion plays a role in the signal decays, time resolved measurements of the diffraction efficiency were performed using three different grating spacings between Λ = 1.2 and 2.9 μm and a (constant) value of I_G larger than the threshold for the onset of rapid decays. If diffusion is responsible for the fast decays observed in Fig.17 then the time constants which describe the decays should vary by a factor of 6 according to Eq. 44. Since there was no measurable difference in the decay times it appears that diffusion does not play a significant role. This is consistent with the fact that similar decay times are observed using both the beam deflection and grating diffraction techniques despite the fact that the carrier density gradients differ by a factor of 10^2.

Carrier recombination should be dominated by radiative and trapping processes for most of the carrier densities investigated in the beam deflection experiments. Auger processes may contribute for the very highest excitation strengths. Using a value of 10^{-10} cm^3sec^{-1} [Dneprovskii, 1988] for the radiative recombination coefficient it follows that for I_0=25 $GWcm^{-2}$ (i.e., threshold for the onset of decay in the beam deflection data; peak carrier density N=5x10^{17} cm^{-3}) the decay time should be of the order of 10 ns. Decay times for recombination via traps are highly dependent on sample quality, but usually fall in the ns regime. No reliable measurements of the Auger coefficient have been reported for the CdSSe family, however most estimates fall in the range of 10^{-29} cm^6sec^{-1} [Dneprovskii, 1990]. Using this value a decay time of the order of a few hundred ns is calculated for I_o = 25 $GWcm^{-2}$. Since the measured decay time from Fig. 7 is of the order of a few ps it follows that none of these recombination processes can credibly account for the observed decay. Stimulated emission recombination has been used to explain rapid decays of carrier density in samples where the carrier distribution is degenerate. However, the peak carrier density at the location of the probe spot at the threshold irradiance for the onset of decay is an order of magnitude below the critical density for efficient local stimulated emission recombination. The threshold irradiance for the diffraction data are higher (> 200 $GWcm^{-2}$; peak carrier density between 2x10^{19} and 4x10^{19} cm^{-3}) and so the predicted recombination times are closer to the measured values. Using the same recombination coefficients as above, both the radiative and Auger recombination times are predicted to be of the order of 100 to 500 ps, still considerably longer than the measured 4 to 20 ps times. The carrier densities, however, are sufficiently larger than the critical density for stimulated emission that stimulated emission induced recombination may play a role [Fox, 1989; Dubard, 1987] can adequately account for the short time constants observed for above threshold excitation at t > 3 ps. By itself, however, stimulated emission cannot explain the signal decays (and curve crossing) observed in both the beam deflection and diffraction experiments. Carrier recombination, then, might explain the rapid signal decays observed using the diffraction technique, but *cannot by itself* explain the signal decays observed using the beam deflection technique. Rapid carrier transport is also involved. It is easily shown that if the local rate of carrier loss through recombination is proportional to the local carrier density alone then curve crossing cannot occur. However, a process such as stimulated emission recombination coupled with some nonlocal process for carrier transfer might explain the rapid signal decays, the threshold behavior, and the curve crossing.

One process which we have proposed and which is consistent with the data is a photon assisted carrier transfer process [Fox, 1992b; Fox, 1993a]. This process is the stimulated emission analog of spontaneous emission processes which have been used to deal with such disparate phenomena as transport of radiation in stellar atmospheres [Chandreskkhar, 1950] to carrier transport in semiconductors themselves [Dumke, 1957, Epifanov, 1976, Tsarenkov, 1979]. Through this mechanism carriers are moved from regions of high to low density when the conditions for stimulated emission are met. The light generated by stimulated emission recombination is only weakly reabsorbed (e.g. through free carrier absorption) within the region excited above the gain threshold. The light instead propagates until it is reabsorbed in a surrounding region characterized by a nondegenerate carrier distribution thereby increasing the local carrier density. The transport occurs mainly parallel to and in the vicinity of the front surface since the lateral extent of the gain region is about 60 μm while the two photon absorption depth, or the depth of the degenrate region is of the order of 10 μm. By this process a carrier may be lost from a highly excited part of the distribution and redeposited many microns away nearly instantaneously. This *nonlocal* process which reduces the lateral carrier density gradients leads to a reduction in the deflection or diffraction signal on a time scale consistent with that measured. Curve crossing can then occur since the stimulated emission process in a plasma of a given density can remove the density gradient from a certain region, leading to smaller deflection, while a slightly less dense plasma does not undergo such a rapid transformation of its density profile. The different threhsolds observed for the rapid recovery in the deflection and diffraction experiments are related to the different geometries (2-D versus 1-D) in which the stimulated emission process takes place.

7. Conclusions

This article has attempted to review the physical processes underlying photo-induced refractive index changes in bulk semiconductors. The mechanisms discussed have included virtual electronic processes such as Kerr and Stark effects caused by below-band gap radiation, free carrier effects which occur during and following single photon or multi-photon absorption processes and thermally induced changes in the index related to band-gap changes. Bound carrier effects are generally weak but occur essentially instantaneously while free carrier effects are generally larger and disappear through carrier diffusion or recombination on a picosecond or longer time scale. Thermally induced effects, which decay with heat diffusion, can also be significant and can remain for milliseconds. Although many of the features have been illustrated here for II-VI semiconductors, the results carry over to other semiconductors as well. As Wherret [1988] has pointed out many of the contributions to the refractive index change can be formulated in terms of expressions which only depend on the ratio of photon and band-gap energies. As the femtosecond data has shown, however, particularly on a picosecond time scale,where much of the interest in all optical switching exists, the dynamics of an induced plasma through temperature and density effects can play a large role in determining the evolution of the change of the change in refractive index.

In the introduction various figures of merit were addressed to assess the potential of materials for all-optical switching. These involve trade-offs in optical nonlinearity, beam attentuation and time of response. Elsewhere in this volume one can see how the properties of bulk materials can enable all-optical switching to come closer to fruition in certain devices.

8. Acknowledgments

I gratefully acknowledge financial support for work conducted in my laboratory from the Natural Sciences and Engineering Research Council of Canada and Technology Fund Ontario (Canada). Drs. E. Fox and E. Canto-Said have been responsible for taking much of the femtosecond-resolved data reviewed here and contributing to the interpretation.

REFERENCES

Alcock, A.J., Corkum, P.B, and James, D. J., 1975, A fast scalable switching technique for high-power CO_2 laser radiation, *Appl. Phys. Lett.*, **27**, 680.

Alcock, A.J. and Corkum, P.B, 1979, Ultrafast switching of infrared radiation by laser produced carriers in semiconductors, *Can. J. Phys.* **57**, 1280.

Allen and Eberly, 1975, Optical Resonance and Two-level atoms, John-Wiley, New York.

Anderson, K.K., Lagasse, M.J., Wang, C.A, Fujimoto, J.G., Haus, H.A., 1990, Femtosecond dynamics of the nonlinear index near the band edge in AlGaAs waveguides, *Appl. Phys. Lett.* **56**,1834.

Auston, D.H., McAfee, S., Shank, C.V., Ippen, E.P., Teschke, O., 1978, Picosecond spectroscopy of semiconductors, *Solid State Electron.* 21, 147.

Banyai, L., and Koch,S.W., 1986, A simple theory for the effects of plasma screening on the optical spectra of highly excited semiconductors, *Z. Phys.B*, **63**, 283.

Chandresekhar, S., 1950, "Radiative Transfer," Dover, New York.

Combescot, M., 1979, Hydrodynamics of an electron-hole plasma created by a pulse, *Solid State Commun.* 30, 81.

Cornet, A., Pugnet, M.,. Collet, J,. Amand, T, Brousseau, M., 1981, Spatial expansion of hot electron-hole plasma at high density in CdSe, *J. de Phys.* **C7**, 471.

Cotter, D., Ironside, C.N., Ainslie, B.J., Girdlestone, H.P., 1989, Picosecond pump-probe interferometric measurement of optical nonlinearity in semiconductor-doped fibers, *Opt. Lett.* **14**,317.

Dagenais, M., 1983, Low power optical saturation of bound excitons with giant oscillator strength, Appl. Phys. Lett. **43**, 742.

Ding, Y.J., Guo, C.L. Swartzlander, G.A., Jr., Khurghin, J.B., Kaplan, A.E., 1990, Spectral measurement of the nonlinear refractive index in ZnSe using self-bending of a pulsed laser beam, *Opt. Lett.* **15**, 1431.

Dneprovskii, V.S., Klimov, V.I. , Novikov, M.G., 1988, Recombination dynamics of an electron-hole plasma in cadmium sulfide, *Sov. Phys. Solid State* **30**, 1694.

Dneprovskii, V.S., Efros, A.L., Ekimov, A.I., Klimov, V.I., Kudriavstev, I.A., Novikov, M.G., 1990, Spontaneous and stimulated collapse of high density electron-hole system in CdSe, *Solid State Commun.* **74**, 555.

Downer, M.C. and Shank, C.V., 1986, Ultrafast heating of silicon sapphire by femtosecond optical pulses, *Phys. Rev. Lett.* **56**, 761.

Dubard, J., Oudar, J.L., Alexandre, F., Hulin, D., Orszag, A., 1987, Ultrafast absorption recovery due to stimulated emission in GaAs/AlGaAs Multiple quantum wells, *Appl. Phys. Lett.* **50**, 821.

Dumke, W.P., 1957, Spontaneous radiative recombination in semiconductors, *Phys. Rev.* **105**, 139.

Eichler, H.J., Gunter, P., Pohl, E.W., 1986, "Laser Induced Dynamic Gratings," Springer-Verlag.

Epifanov, M.S., Galkin, G.N., Bobrova, E.A., Vavilov, V.S., Sabanova, L.D., 1976, Photon transfer of excitation of nonequilibrium carriers in gallium arsenide, *Fiz. & Tekh. Poluprovodn.* **10,** 889 (*Sov. Phys. Semicond.* **10**,526).

Fork, R. L., Greene , B.I., Shank, C.V., 1981, Generation of optical pulses shorter than 0.1 psec by colliding pulse mode locking, *Appl. Phys. Lett.* **38**, 671.

Fox, A.M., Manning, R.J., Miller, A., 1989, Picosecond relaxation mechanisms in highly excited GaInAsP, *J. Appl. Phys.* **65**, 4287.

Fox, E.C., Canto-Said, E.J., van Driel, H.M., 1991, Femtosecond time-resolved refractive index changes in $CdS_{0.75}Se_{0.25}$ and CdS, *Appl. Phys. Lett.* **59**, 1878.

Fox, E.C., Canto-Said, E.J., van Driel, H.M., 1992a, Separation of bound and free carrier contributions to the refractive index change induced in II-VI semiconductors by femtosecond pulses, *Semicond. Sci. Technol.* **7B,** 183.

Fox, E.C., Canto-Said, E.J., van Driel, H.M., 1992b, Femtosecond time-resolved refractive index changes in CdSSe, SPIE meeting on "Ultrafast Phenomena in Semiconductors and Superconductors," Sommerset, N.J., U.S.A.

Fox, E.C. and van Driel, H.M., 1993a, Ultrafast carrier recombination and plasma expansion via stimulated emission in II-VI semiconductors, *Phys. Rev. B* **47**, 1663.

Fox, E.C., 1993s, Ph.D. thesis (University of Toronto) (unpublished).

Friberg, S.W. and Smith, P.W., 1987, Nonlinear optical glasses for ultrafast optical switches, *I.E.E.E. J. Quantum Electron.* **QE-23**:2089.

Gallant, M.I.A., and van Driel, H.M., 1981, Infrared reflectivity probing of thermal and spatial properties of laser-generated carriers in germanium, *Phys. Rev. B*, **26**, 2133.

Gibbs, H.M., 1985, "Optical Bistability: Controlling Light with Light," Academic Press, New York.

Gibbs, H.M., Khitrova, G., Peyghambarian, N., 1990, "Nonlinear Photonics," Springer Verlag, Berlin.

Goebel, E.O., Hildebrand, O., Lohnert, K., 1977, Wavelength dependence of gain saturation in GaAs lasers, *I.E.E.E. J. Quantum Electron.* **QE-13**,848.

Haug, H., 1988, "Optical Nonlinearities and Instabilities in Semiconductors," Academic Press, San Diego.

Hill, J., Parry, G. and Miller A., 1982, Nonlinear refractive index changes in CdHgTe at 175K with 10.6 µm radiation, *Opt. Commun.*, **43**, 151.

Hutchings,D., Sheik-Bahae,M., Hagan, D.J., and Van Stryland, E.W., 1992, Opt. Quant. Elec., **24**, 1.

Johnson, E.J., 1967, in "Semiconductors and Semimetals," vol. 3, ed. Willardson, R.K. and Beer, A.C., Academic Press, London.

Junnarkar, M.R. and Alfano, R.R., 1986, Photogenerated high-density electron-hole plasma energy relaxation and experimental evidence for rapid expansion of the electron-hole plasma in CdSe, *Phys. Rev. B* **34**, 7045.

Kalafati, Y. D. and Kokin, V.A., 1991, Picosecond relaxation processes in a semiconductor laser excited by a powerful ultrashort light pulse, *Sov. Phys. J.E.T.P.* **72**,1003.

Kobayashi, A., Sankey, O.F., Volz , S.M., Dow, J.M., 1983, Semiempirical tight-binding band structures of wurtzite semiconductors: AlN, CdS, CdSe, ZnS, and ZnO, *Phys. Rev. B* 28, 935.

Kocevar, P., 1985, Hot phonon dynamics, *Physica* **134** B+C,155.

Kressel, H. and Butler, J.K., 1977, "Semiconductor Lasers and Heterojunctions," Academic Press, New York.

LaGasse, M.J., Anderson, K.K., Haus, H.A., Fujimoto, J.G., 1989, Femtosecond all-optical switching in AlGaAs waveguides using a time division interferometer, *Appl Phys. Lett.* **54**, 2068.

Landot and Börnstein, 1982, "Numerical Data and Functional Relationships in Science and Technology, New Series," vol. 17 & 22, Springer-Verlag

Lee,Y.H., Chavez-Pirson, A., Koch, S.W., Gibbs, H.M. Park, S.H., Morhange, J., Jeffrey, A., Peyghambarian, N., Banyai,L., Gossard,A.C., and Wiegman, W., 1986, Room temperature Optical Nonlinearities in GaAs, *Phys. Rev. Lett.*, **57**,2446.

Majumder, F.A., Swoboda, H.-E., Kempf, K., Klingshirn, C., 1985, Electron-hole plasma expansion in the direct-band-gap semiconductors CdS and CdSe, *Phys. Rev. B* **32**, 2407.

Miller, A., Miller, D.A.B., and Smith, S. D., 1981, Dynamic non-linear optical processes in semiconductors, *Adv. in Physics* **30**, 697.

Miller, A. in Nonlinear Optics in Signal Processing, 1993, (Eds., R.W. Eason and A. Miller) Chapman and Hall (London).

Miller, D.A.B., Mozolowski,M., Miller, A., and Smith, S.D., 1978, *Optics Commun.* **27**,133.

Miller, D.A.B., Seaton,C.T., Prize,M.E. and Smith,S.D., 1981,Band-gap-resonant nonlinear refraction in III-V Semiconductors, *Phys. Rev. Lett.*, 47,197.

Nurmikko, A.V. and Gerritsen, H.J., 1976, Fast transient spectroscopy of the free-carrier plasma edge in Ge, *App. Phys. Lett.* **29**, 640.

Pötz, W. and Kocevar, P., 1983, Cooling of highly photoexcited electron-hole plasma in polar semiconductors and semiconductor quantum wells: a balance-equation approach, *Phys. Rev. B* **82**, 7040.

Pugnet, M., Collet, J., Cornet, A., 1981, Cooling of hot electron-hole plasmas in the presence screened electron-phonon interactions, *Solid State Commun*. **38**, 531.

Puls, J., Rudolph, W., Henneberger, F., Lap, D., 1988, Femtosecond studies of room temperature optical nonlinearities in wide-gap II-VI semiconductors, *Phys. Stat. Sol. (b)* **150**, 419.

Rinker, M., Swoboda, H.-E, Majumder, F.A., Klingshirn, C., 1989, Diffusive and thermal properities of the electron-hole plasma in CdS and CdSe, *Solid State Commun.* **69**, 887.

Rudolph, W., Puls, J., Henneberger, F., Lap, D., 1990, Femtosecond studies of transient nonlinearities in wide-gap II-VI semiconductor compounds, *Phys. Stat. Sol. (b)* **159**, 49.

Said, A.A., Sheik-Bahae, M., Hagan, D.J., Wei, T.H., Wang, J., Young, J., Van Stryland, E.W., 1992, Determination of bound and free-carrier nonlinearities in ZnSe, GaAs, CdTe, and ZnTe, *J.Opt. Soc. Am. B* **9**, 405.

Saito, H. and Göbel, E.O., 1985, Picosecond spectroscopy of highly excited Cds, *Phys. Rev. B* **31**, 2360.

Seeger, K., 1982, "Semiconductor Physics, an Introduction," Springer Verlag, Berlin.

Shah, J., 1989, Photoexcited hot carriers: from CW to 6 fs in 20 years, *Solid State Electron.* **32**, 1051.

Shank, C.V., Auston, D.H., Ippen, E.P., Teschke, O., 1978, Picosecond time resolved reflectivity of direct gap semiconductors, *Solid State Commun.* **26**, 567.

Sheik-Bahae, M., Said, A.A., Van Stryland, E.W., 1989, High-sensitivity, single-beam n_2 measurements, *Opt. Lett.* **14**, 955.

Sheik-Bahae, M., Hutchings, D.C., Hagan, D.J., Van Stryland, E.W., 1991, Dispersion of bound electronic nonlinear refraction in solids, *I.E.E.E. J. Quantum Electron.* **QE-27**,1296.

Shen, Y.R., 1984, "Principles of Nonlinear Optics," John Wiley & Sons, Toronto.

Solymar, L. and Cooke, D.J., 1981, "Volume Holography and Volume Gratings," Academic Press, New York.

Spitzer, W.G., and Fan,H.Y., 1957, Determination of Optical Constants and Carrier Effective Mass of Semiconductors, Phys. Rev. **106**, 883.

Stegeman, G.I. and Stolen, R.H., 1988, "Nonlinear Guided Wave Phenomena," special issue of *J. Opt. Soc. Am. B* **5**, 264.

Tsarenkov, G.V., 1979, Drift of recombination in a variable gap semiconductor, *Sov. Phys. Semicond.* **13**, 641.

Valdmanis, J.A., Fork, R.L., Gordon, J.P., 1985, Generation of optical pulses as short as 27 femtoseconds directly from a laser balancing self-phase modulation group velocity dispersion, saturable absorption, and saturable gain, *Opt. Lett.* 10, 131.

van Driel, H.M., 1979, Influence of hot phonons on energy relaxation of high-density carriers in germanium, *Phys. Rev. B* **19**, 5928.

van Driel, H.M., Lompré, L-A. and Bloembergen, N., 1983, *Appl. Phys. Lett.* **44**, 285

van Driel, H.M., 1987, Kinetics of high-density plasma generated in Si by 1.06- and 0.53-mm picosecond laser pulses, *Phys. Rev. B* **35**, 8166.

van Lap, D., Peschel, U., Ponath, H.E., Rudolph, W., 1991, Investigation of carrier temperature relaxation with femtosecond transient grating experiments in CdS_xSe_{1-x} semiconductors, Inst. Phys. Conf. Ser. No. 126: Section V, 357, presented at Int. Symp. on Ultrafast Processes in Spectroscopy, Bayreuth.

Van Stryland, E.W., Vanherzeele, H., Woodall, M.A., Soileau, M.J., Smirl, A.L., Guha, S., Boggess, T.F., 1985, Two photon absorption, nonlinear refraction, and optical limiting in semiconductors, *Opt. Eng* **24**, 613.

Van Stryland, E.W., Wu,Y.Y., and Hagan,D.J., 1988, Optical Limiting with semiconductors, *J. Opt. Soc. Am.B* **5**, 1980.

Vasconcellos, A. and Luzzi, R., 1980, Coupled electron-hole plasma-phonon system in far-from-equilibrium semiconductors, *Phys. Rev. B.* **22**, 6355.

Villeneuve, A., C.C. Yang, Wigley, P.G.J. , Stegeman, G.I., Aitchison and Ironside, C.N., 1992, Ultrafast all-optical switching in semiconductor nonlinear directional couplers at half the band gap, *App. Phys. Lett.* **61**, 147.

Wherrett, B.S., 1984, Scaling Rules for Multiphoton interband absorption in semicondcutors, *J. Opt. Soc. Am.B* **1**, 67.

Wherrett, B.S., 1988, Nonlinear Refraction for CW Optical Bistability in "Optical Nonlinearities and Instabilities in Semiconductors," ed. Haug, H., Academic Press, San Diego.

Wherrett, B.S., Darzi, A.K., Chow, Y.T., McGuckin, B.T., Van Stryland, E.W., 1990, Ultrafast thermal refractive nonlinearities in bistable interference filters, *J. Opt. Soc. B* **7**, 215.

Xia, T., Walmsley, C., Wang, Z., Said, A.A., Sheik-Bahae, M., Hagan, D.J., and Van Stryland, E.W. CLEO, 1993, paper CFI1

Zimmermann, R., 1988, Nonlinear optics and the Mott transition in semiconductors, *Phys. Stat. Sol.* (b) **146**, 371.

NONLINEAR OPTICAL EFFECTS IN ACTIVE SEMICONDUCTOR DEVICES

PAOLO SPANO
Fondazione Ugo Bordoni
Via B. Castiglione, 59
00142 Rome
ITALY

ABSTRACT. In recent years nonlinear effects in active semiconductor devices, lasers and amplifiers, gained a larger and larger interest for two reasons. First, for the influence of these nonlinear effects on the ultimate performances of the devices, in particular of quantum well and strained quantum well devices where the optical nonlinearities are very high. Second, for the attempt to use the fast semiconductor nonlinearities to design new photonic devices like, for instance, wavelength converters.

1. Introduction to Optical Nonlinearities in Active Semiconductor Devices

The nonlinear optical interactions in active semiconductor devices [1-3] can be traced back to three different processes occurring when an optical field interacts with an electron-hole gas. The first one is due do carrier density variations, so it is an inter-band effect. The second and the third are due to a redistribution of the carrier density inside the bands, hence they are intra-band effects. Let us examine these processes in some details.

- The first effect is connected with the variation of the carrier density induced by the increase of the stimulated emission rate in the presence of the injected optical field. The variation of the carrier density induces a change in both gain and refractive index, the latter through the linewidth enhancement factor α. The characteristic time for the field-induced carrier density variation is the spontaneous lifetime, which is ~0.1-1 ns depending upon the values of the carrier density and of the field intensity inside the cavity.
- The second effect is due to a variation of the carrier distribution inside the semiconductor bands for a fixed total carrier number. It is due to the burning of a hole in the spectral distribution of the carrier energy caused by the stimulated emission. The process which tends to restore the original carrier distribution is the fast carrier-carrier scattering which occurs on a time scale ~100 fs.
- The last effect is the variation of the carrier temperature induced by the stimulated transitions and the free carrier absorption of the electromagnetic field while it propagates through the active device. After the above-mentioned fast thermalization of the electron gas, in fact, the Fermi distribution of the carriers will assume, in general, a higher temperature than in the absence of the

A. Miller et al. (eds.), Nonlinear Optical Materials and Devices for Applications in Information Technology, 183–205.

electromagnetic field. This is mainly due to the excess of high energy carriers caused by free carrier absorption [4]. The time for the carrier gas to recover to the original lattice temperature is the electron-phonon scattering time, typically $\lesssim$1 ps.

The above mentioned nonlinear interactions affect both gain and refractive index of the semiconductor material. The resulting modulation effects have been recently exploited in different photonic devices [5-8].

To evaluate the characteristic times of the nonlinearities and their strength in active semiconductor devices, in particular in semiconductor amplifiers, both time [9] and frequency domain [10] techniques have been reported in the literature.

In the first class of experiments, a strong and short optical pulse (pump) is injected in the device under test. The induced variation of the material gain is probed by a weak field which can be delayed with respect to the pump field, as shown in Fig. 1a). This technique can be also used, with some modifications, for the measurement of the nonlinear refractive index [11, 12]. In this case, a third pulse (reference) is sent through the amplifier before the pump pulse, and the refractive index variation is detected by means of the interference between the probe and the reference beam as shown in Fig. 1b).

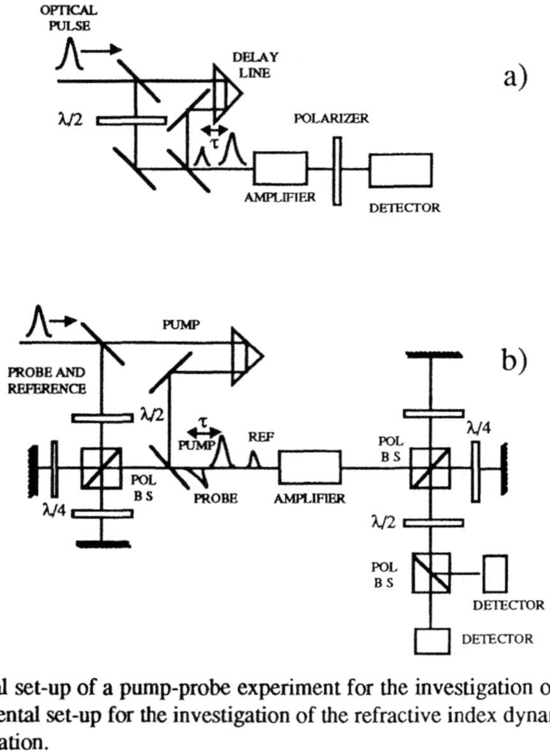

Fig. 1 - a) typical set-up of a pump-probe experiment for the investigation of the gain dynamics. b) experimental set-up for the investigation of the refractive index dynamics in the pump-probe configuration.

The second class of experiments is based on a frequency domain approach. Two CW pumps at frequencies ω_2 and $\omega_3=\omega_2+\Omega$ are injected into the active device, producing two side-bands at frequency $\omega_1=\omega_2-\Omega$ and $\omega_4=\omega_3+\Omega$ (see Fig. 2). The nonlinear dynamics of gain and refractive index is investigated by measuring the intensity of the side-bands as a function of the frequency detuning Ω.

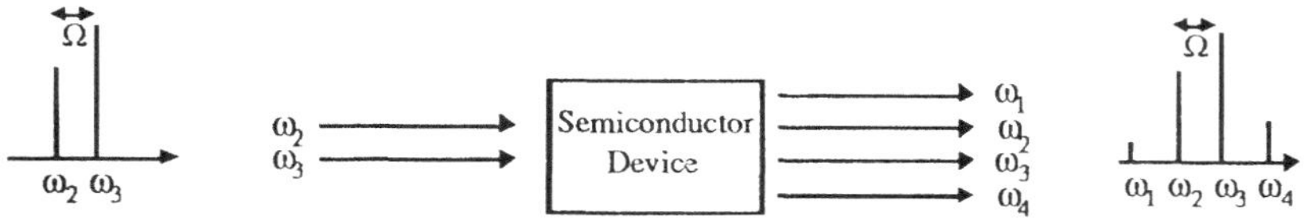

Fig. 2 - Schematic of a four-wave-mixing experiment used to investigate semiconductor nonlinearities of active devices in the frequency domain.

2. Theoretical Approach to Nonlinearities in Active Semiconductor Media

An elegant treatment of the nonlinear optical interactions in semiconductor active waveguides, based on the density matrix formalism, has been recently developed by A. Uskov and co-workers [13]. In this section we will summarise their approach.

Let us consider first a simple two level system. The energy of the upper level be E_2 the energy of the lower level E_1 . Let us define the wave-function of the system,

$$\Psi(\mathbf{r},t) = \sum_{n=1}^{2} c_n \exp(-\frac{iE_n t}{\hbar})\psi_n(\mathbf{r}) \tag{1}$$

The operator $H=H_o+H_I$ is the Hamiltonian of the system, H_o is the unperturbed term and $H_I = e\mathbf{r}\cdot\mathbf{E}$ the interaction term, which represents the potential energy of the atom electric dipole in the electric field $\mathbf{E}$, and $\omega_o = (E_2 - E_1)/\hbar$ is the Bohr frequency.

The standard treatment of the time dependent interactions [14] gives a set of equations connecting the time derivatives of the coefficients $\rho_{11}=|c_1|^2$, $\rho_{22}=|c_2|^2$, $\rho_{12}=c_1c_2^*\exp(i\omega_o t)$, $\rho_{12}^*=\rho_{21}=c_1^*c_2\exp(-i\omega_o t)$

$$\dot{\rho}_{11} = -\frac{iE}{\hbar}\left[d_{12}\rho_{21} - d_{21}\rho_{12}\right] \tag{2a}$$

$$\dot{\rho}_{22} = -\dot{\rho}_{11} \tag{2b}$$

$$\dot{\rho}_{12} = (i\omega_o - \frac{1}{\tau_2})\rho_{12} - \frac{iE}{\hbar}d_{12}\left[\rho_{22} - \rho_{11}\right] \tag{2c}$$

$d_{12}=e\int\psi_1^* r\psi_2 d^3r$ being the component of the dipole operator along the direction of the field and τ_2, the dephasing time for the polarization. Eqs. (2a) and (2b) describe the rate of change of the electron population in the upper and lower state of the atom, eq. (2c) describes the rate of change of the dipole momentum of the atom.

A semiconductor is far from being a two level system. Rather, it can be considered an ensemble of interacting two level systems. The extension of the density matrix equations to this system including the interactions among the different levels of the valence and conduction bands, reads

$$\dot{\rho}_{c,k} = -\frac{\rho_{c,k}(t) - f_{c,k}(t)}{\tau_{1c}} - \frac{\rho_{c,k}(t) - f^L_{c,k}(t)}{\tau_{hc}} - \frac{\rho_{c,k}(t) - f^{eq}_{c,k}}{\tau_s} - \frac{iE}{\hbar}\left[d^*_k \rho_{cv,k} - d_k \rho_{vc,k}\right] + \Lambda_{c,k} \tag{3a}$$

$$\dot{\rho}_{v,k} = -\frac{\rho_{v,k}(t) - f_{v,k}(t)}{\tau_{1v}} - \frac{\rho_{v,k}(t) - f^L_{v,k}(t)}{\tau_{hv}} - \frac{\rho_{v,k}(t) - f^{eq}_{v,k}}{\tau_s} - \frac{iE}{\hbar}\left[d^*_k \rho_{cv,k} - d_k \rho_{vc,k}\right] + \Lambda_{v,k} \tag{3b}$$

$$\dot{\rho}_{cv,k} = -(i\omega_k - \frac{1}{\tau_2})\rho_{cv,k}(t) - \frac{iE}{\hbar} d_k\left[\rho_{c,k}(t) - \rho_{v,k}(t) - 1\right] \tag{3c}$$

k is the electron wave-vector, $\hbar\omega_k$ is the energy difference between electron and holes having the same k. The subscript c labels quantities of the conduction band, v of the valence band, so $\rho_{v,k}$ and $\rho_{c,k}$ are the occupation probability of the k-th state in the valence and conduction band respectively, and $\rho_{cv,k}$ is the off-diagonal matrix element of the density matrix between the k-th conduction and valence state. In eqs. (3), the meaning of ρ_v is different from that of ρ_{22} in eq. (2), as ρ_v represents the hole (not the electron) occupation probability in the valence band. Hence the electron and hole density is defined by

$$n(t) = \frac{1}{V} \sum_k \rho_{x,k}(t) \tag{4}$$

and Λ_k is the pumping term related to the total current flowing through the device by

$$C = e \sum_k \Lambda_{c,k} = e \sum_k \Lambda_{v,k} \tag{5}$$

The first three terms on the right of eqs. (3a) and (3b) are phenomenological terms accounting for the intraband and the interband interactions.

The first term is the action of the electron-electron scattering. This process is responsible for the relaxation of the perturbed carrier distribution function to a Fermi distribution f(t). This is the fastest relaxation process of the electron gas, its characteristic time τ_1 is ~50-100 fs [3, 15]. The fast relaxation of the carrier distribution washes out the spectral hole dug by the electric field in the carrier distribution. This process is also responsible for the thermalization of the high energy carriers generated by free carrier and two photon absorption.

The second term is the action of electron-phonon scattering. This process is responsible for the balance of the carrier and lattice temperature, $f^L(t)$ being the Fermi distribution corresponding to the lattice temperature. It is characterized by the time constant τ_h~0.5-1 ps, longer than the previous one. After thermalization of the electron and hole gas through the e-e scattering, the carrier temperature is in general higher than the lattice temperature because the generation of high energy carriers, through free carrier and two photon absorption, overcomes the removal of low energy carriers due to stimulated emission. It is worth noting that both the electron-electron scattering considered above, and the electron-phonon scattering do not alter the total number of carriers in the bands. Moreover, electron-electron scattering process keeps the total energy of the carrier gas constant.

Finally, the third term takes into account the electron hole recombination. This process tends to restore the carrier distribution f^{eq} corresponding to the actual current injection. It is characterized by a time constant τ_s that ranges from near 0.1 ns for high carrier density to near 1 ns for low carrier density.

It is worth to note, however, that the set of equations (3) does not describe exhaustively the interactions between the electric field and the carriers in the band. It properly includes the relaxation processes through the phenomenological terms containing the Fermi distributions, but the nonlinear excitation processes are neglected. These processes, namely the two photon absorption and the free carrier absorption, are the very processes responsible for the deviation of the electron-hole distributions from the Fermi distribution, as depicted in Fig. 3 a). Furthermore, the set of eq. 3 holds in the case of a bulk device in which the carriers are injected directly in the active region. In QW devices, a two-step injection process is effective. The carriers are injected in the optical confinement region and they are captured in the wells after a delay, which is the sum of the transport time in the confinement region and the quantum capture time in the wells. A sketch of the evolution of the carrier distribution after the interaction of the semiconductor with a short optical pulse is shown in Fig. 3 b), and the above-mentioned processes are pictorially illustrated.

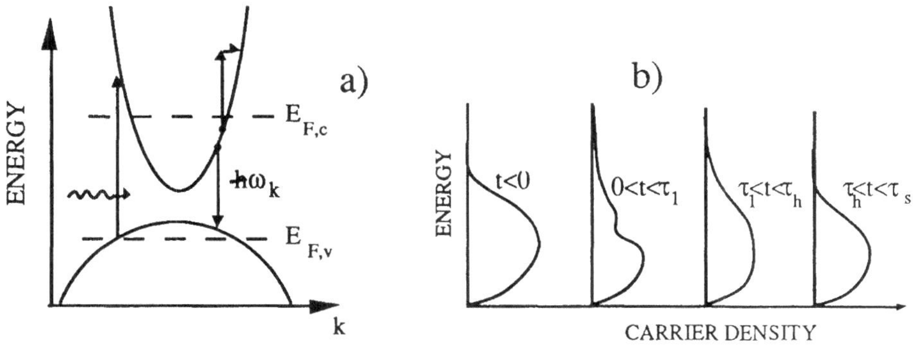

Fig. 3 - a) Schematic representation of the interactions between the semiconductor and the electromagnetic field. b) Carrier density evolution in the semiconductor after the interaction with a short optical pulse.

The solution of eqs. (3) allows one to obtain the macroscopical polarization induced by the optical field E(t), which can be expressed as

$$P(t) = \frac{1}{V} \sum_{k} d_k(\rho_{cv,k}+\rho_{vc,k}) \tag{6}$$

where d_k is the dipole moment of the transition [16].

The polarization P(t) is related to the susceptibility by the general relation

$$P(t) = \varepsilon_o \Big[\int_{-\infty}^{t} d\tau_1 \chi^{(1)}(t-\tau_1)\, E(\tau_1) + \int_{-\infty}^{t} d\tau_1 \int_{-\infty}^{t} d\tau_2 \int_{-\infty}^{t} d\tau_3 \chi^{(3)}(t-\tau_1;t-\tau_2;t-\tau_3)E(\tau_1)E(\tau_2)E(\tau_3)\Big] \tag{7}$$

which, in the Fourier space, becomes

$$P(\omega) = \varepsilon_o \Big[\chi^{(1)}(\omega)E(\omega) + (2\pi)^{-2} \int_{-\infty}^{\infty} d\omega_1 \int_{-\infty}^{\infty} d\omega_2\, \chi^{(3)}(\omega;\omega_1,\omega_2,\omega_3)E(\omega_1)E(\omega_2)E(\omega_3)\Big] \tag{8}$$

where the frequencies are linked by the relationship $\omega=\omega_1+\omega_2+\omega_3$. In eqs. (7), (8) $\chi^{(1)}$ and $\chi^{(3)}$ are the first and third order susceptibility respectively. Eqs. (7) and (8) are general and can be used to evaluate the value of $\chi^{(3)}$ in both time and frequency domain.

In pump-probe experiments, the susceptibility is directly measured in time domain under simplifying assumptions about the dependence of P(t) on the field. The conditions to be met is that χ is a function of the instantaneous value of the field E(t) and $\chi^{(3)}$ is a (in general non instantaneous) function of the field intensity [17]

$$P(t) = \varepsilon_o \Big[\chi^{(1)}(t) + \int_{-\infty}^{t} d\tau_2\, \chi^{(3)}(t-\tau_2)I(\tau_2)\Big]E(t) \tag{9}$$

From eq. (3c), one can understand that the above-mentioned hypothesis is equivalent to a value of τ_2 much lower than the other times involved in the transition process. This sets a limit on the fastest processes that can be investigated with pump probe experiments. In semiconductor devices, the phenomenon responsible for the loss of the phase of the polarization is the electron-electron scattering, the same process that produces the electron thermalization in the bands. Hence, the two time constants τ_1 and τ_2 must be of the same order, and probably very close to each other. Then, it is not clear if pump-probe experiment can properly investigate processes as fast as the electron-electron scattering.

From eq. (9), the gain and phase variations induced by a pulse of intensity I(t) are

$$\Delta G(t) = -(\omega/\eta c)\ \mathrm{Im}[\ \int_{-\infty}^{t} d\tau_2\, \chi^{(3)}(t-\tau_2) I(\tau_2)]\ L \tag{10a}$$

$$\Delta \Phi(t) = -(\omega/\eta c)\ \mathrm{Re}[\ \int_{-\infty}^{t} d\tau_2\, \chi^{(3)}(t-\tau_2) I(\tau_2)]\ L \tag{10b}$$

where ω is the mean frequency of the pulse, η is the refractive index and L is the total length of the device and Im[...] and Re[...] stand for the imaginary and real part of the argument, respectively.

In pump probe configurations, a strong pump pulse is sent through the active material and a low intensity pulse, delayed by τ, is used to probe the changes induced by the pump through the variation of its amplification or phase. In the following, we will assume that pump and probe pulses are obtained by splitting a pulse delivered by the same source. This will be done for the sake of simplicity, and also as this is the usual experimental arrangement. If the coherence effects due to the field interference are neglected [17], the intensity modulation of the probe pulse $M(\tau)$ can be written as

$$M(\tau) \propto \mathrm{Im}[\ \int_{-\infty}^{\tau} dt_1\, \chi^{(3)}(\tau-t_1)\ \int_{-\infty}^{\infty} dt_2\, I(t_1-t_2)\, I(t_2)\ ] \tag{11}$$

The phase modulation has the same expression with the real part instead of the imaginary one. A measurement of $M(\tau)$ permits the evaluation of the imaginary part of the nonlinear susceptibility $\chi^{(3)}(t)$ after deconvolution with the intensity auto correlation function of the pulse.

In four wave mixing experiments, the superposition of two parallel fields $E=E_2\exp(i\omega_2 t) + E_3\exp(i\omega_3 t)$ is injected into the device. The intensity of the field contains a beating term oscillating at $\Omega = \omega_3-\omega_2$ that produces two side-bands through the saturation of the material gain and refractive index. The dependence of the modulation response on frequency gives us information on the nonlinear processes. Eq. (8), with the input field the sum of two delta functions in frequency domain, describe the four wave mixing process. The amplitudes of the polarization generated at $\omega_1=\omega_2-(\omega_3-\omega_2)$ and $\omega_4=\omega_3+(\omega_3-\omega_2)$ are

$$P(\omega_1) = \varepsilon_o\ [\chi^{(1)}(\omega_1)E(\omega_1) + (2\pi)^{-2}\chi^{(3)}(2\omega_2-\omega_3;\omega_2,\omega_2,\omega_3)E(\omega_2)E(\omega_2)E^*(\omega_3)] \tag{12a}$$

$$P(\omega_4) = \varepsilon_o\ [\chi^{(1)}(\omega_4)E(\omega_4) + (2\pi)^{-2}\chi^{(3)}(2\omega_3-\omega_2;\omega_3,\omega_3,\omega_2)E(\omega_3)E(\omega_3)E^*(\omega_2)] \tag{12b}$$

Entering the above expressions into the propagation equations for the device gives the evolution equations for the amplitude of the fields propagating into the device.

3. Measurements of Optical Nonlinearities in Active Semiconductor Devices

Many measurements of the nonlinearities in semiconductor amplifiers have been reported in the literature for both bulk and Quantum Well (QW) devices grown on GaAs and InP substrates. Both time [18,19] and frequency domain [10, 20] techniques have been used. Typical results of a time domain experiment are shown in Fig. 4 [9].

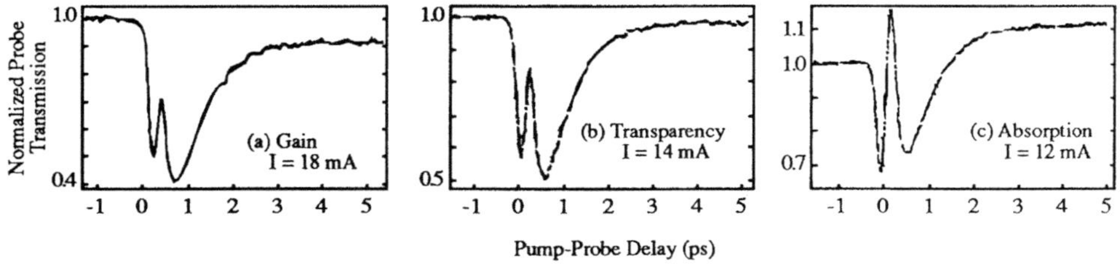

Fig. 4 Probe transmission as a function of pump probe delay in gain, transparency and absorption regions. From Ref. [9].

Here, the transmission of the probe pulse is reported as a function of the pump-probe delay. These results refer to a bulk InGaAsP diode travelling wave (TW) amplifier. Very similar results have been obtained in TW GaAs amplifiers. A tunable laser able to emit short pulses ($\Delta\tau$~100 fs) has been used in the experiment. The different plots refer to pulses with an emission wavelength centred in the absorption, transparency and gain region of the amplifier. Although the outcome is different in the three cases, common trends can be recognized:

- At the very early stage, an instantaneous decrease of probe transmission can be always observed. The fast initial transient, below the temporal resolution of today's experiments, has been attributed to two photon absorption.
- For a little longer times, a decrease or increase of the probe transmission is observed in the gain and absorption region, respectively, which eventually can mask the two photon absorption effect [15]. This effect has been ascribed to either a delay in carrier heating or spectral hole burning.
- For times still longer, a reduction of transmission is observable in all regimes. This reduction, characterized by recovery times of the order of 700 fs in the case of the investigated device, has been attributed to carrier heating.
- The final value of probe transmission, after the fast transient effects have died out, is higher than the initial one in the absorption region and lower in the gain region. This is due to the variation of the total number of carriers which recover their unperturbed value with a time constant (τ_s~100 ps) much longer than the time scale shown in the figure.

These results can be fitted by the simple impulse response [2]

$$h(t) = u(t)\,[a_o - a_2 e^{-t/\tau''}(1 - e^{-t/\tau_{eff}}) + a_1\, e^{-t/\tau'}] + a_3\delta(t) \qquad (13)$$

where τ'=200 fs, τ''=700 fs, $\tau_{eff}=(\tau'\tau'')/(\tau''-\tau')$, $\delta(t)$ is the delta function and u(t) is the step function. It is worth to note that a_o and a_2 do not change sign from the gain to the absorption region, while a_1, the coefficient of the term connected with spectral hole burning, is positive in the absorption region and negative in the gain region as expected. Moreover, eq. (13) shows that carrier heating (accounted for by the term proportional to a_2) is delayed by an amount depending on the characteristic time of the spectral hole burning τ'. This makes good sense, as this is the time the carriers take to recover a (heated) Fermi distribution. The exponential decay of the carrier-density is not included in eq. (13) because the characteristic time of this process is much longer than the time scale of fig 4.

Similar results have been obtained for the fast dynamics of the refractive index. The observed probe phase shift is reported in Fig. 5 as a function of the pump-probe delay.

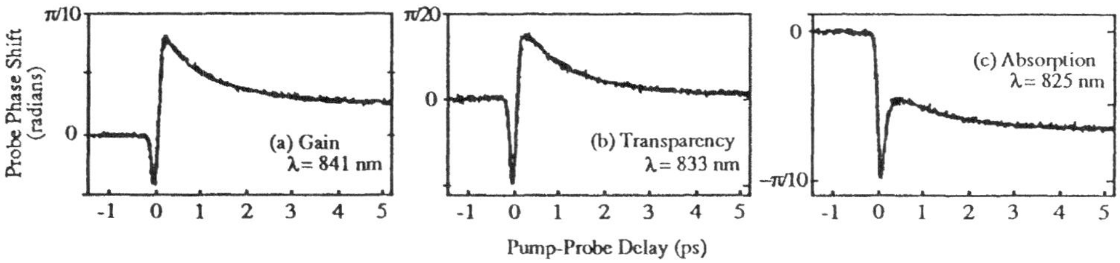

Fig. 5 Measured probe phase shift as a function of pump probe delay in gain, transparency and absorption regions. The data refer to a GaAlAs amplifiers. From Ref. [9].

The response function of the refractive index has the form given by eq. (13). The underlying physical phenomena are the same in both cases.

Measurements on QW devices do not show any basic difference from bulk amplifiers, at least on a time scale of few picoseconds [21].

As far as the FWM technique is concerned, FWM measurements has been recently extended up to frequencies >1 THz [22]. This frequency range corresponds to a time resolution of ~100 fs, the same of the time domain experiments. In most of the experiments, the results have the general trend shown in Fig. 6 [23].

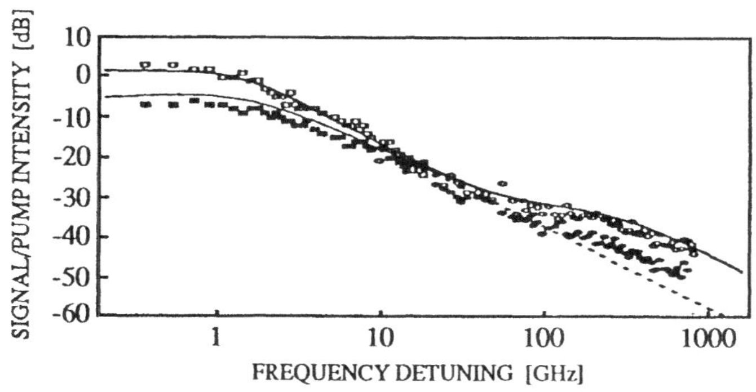

Fig. 6 Ratios of conjugate to pump beam intensities for a 4-well quantum well TW amplifier. From ref. [23].

The amplitude of conjugate beams at frequency ω_1 and ω_4 (see Fig. 2) shows a flat region at low frequency (<2-10 GHz) followed by an amplitude drop of 20 dB/dec. At higher frequencies, a shoulder shows up, followed by a new drop of 20 dB/dec. after a second corner frequency located at ~200-600 GHz. Eventually, a second shoulder can be observed [22]. The first corner frequency is due to carrier density modulation, the second to carrier heating. The origin of the high frequency shoulder is not yet well understood, although it can be ascribed to either spectral hole burning or two photon absorption. Evidence of spectral hole burning can be also found in [20] where the FWM response of a bulk InGaAsP TW amplifier shows a dip in the intensity in one side of the spectrum (corresponding to ω_4 of Fig. 2). The dip shows up at a frequency higher than the corner frequency of the carrier modulation, and lower than the corner frequency of the fast dynamics. After the lower corner frequency, 90 degrees are added to the phase of the modulation on one side of the spectrum, and subtracted on the other. If the effect on the polarization of the carrier modulation and of the fast dynamics are 90 degrees out of phase, the two contributions will be 180 degrees out of phase on one side of the spectrum, and in phase on the other. This is the case, for instance, of amplifiers having both a high value of linewidth enhancement factor, which induces an index modulation overcoming the gain modulation, and a fast dynamics affecting mostly the gain [13]. If the effect on the polarization of the carrier modulation and of the fast dynamics are in phase, after the lower corner frequency the two contributions will be +90 or -90 degrees out of phase in the two sides, and the spectrum will be symmetric.

Other saturation mechanisms can be important in semiconductor devices. Although in most experiments on quantum well devices the effects of carrier transport and capture in the wells cannot be recognized, in ref. [24] this effect shows up as a different drop of conjugate amplification as a function of frequency detuning, which is 40 dB/dec. rather than 20 dB/dec. at intermediate frequencies (10 GHz<f<40 GHz). The reason why this behavior has been observed only in one experiment is still to be understood.

4. Effects of Nonlinearities on Laser Dynamics

The two effects that drove to the investigation of the nonlinear effects in semiconductor lasers and amplifiers were the interaction among the longitudinal modes of Fabry-Perot lasers [25] and the limitations to maximum modulation frequency of both Fabry-Perot and Distributed Feedback (DFB) lasers [26]. In particular, the second effect together with the circuit parasitic is one of the main limitation to the attainment of high bit rate (>10 Gbit/s) optical transmission systems with standard devices.

To understand how the fast nonlinear effects affect the modulation properties of a semiconductor laser, let us resort to the rate equation formalism usually adopted to study the laser behavior. The rate equations can be obtained from eq. (3), assuming the field in the cavity as homogeneous, once the equation for the polarization is adiabatically eliminated. The latter is a good approximation since both τ_2 and τ_1, τ_h are much shorter than the inverse of the modulation bandwidth of laser diodes, hence any nonlinear effect due to the intraband dynamics is practically instantaneous over this time scale. Under this assumption, the intraband dynamics is adiabatically eliminated and one can refer only to the total carrier density n(t). In the simplest case of a single mode laser, the rate equations read [27]

$$\frac{d}{dt} I = \left[\frac{G_n(n-n_t)}{1+\varepsilon I} - \frac{1}{\tau_p} \right] I + \frac{R}{V} \qquad (14\ a)$$

$$\frac{d}{dt} n = -\frac{n}{\tau_s} - \frac{G_n(n-n_t)}{1+\varepsilon I} I + \frac{C}{eV} \qquad (14\ b)$$

where the dependence of the gain on the carrier density is explicitly accounted for by the differential gain $G_n = \partial G / \partial n$. Above, n_t is the carrier density at transparency, R is the spontaneous emission rate into the mode, C is the excitation current, e is the electron charge and, finally, V is the mode volume. In eqs. (14), the gain is assumed to be linearly dependent on carrier density. This approximation fails for QW devices and, in bulk devices, for high values of gain. The term $1+\varepsilon I$ is the phenomenological term which accounts for the fast nonlinear phenomena occurring inside the bands that we mentioned above. Historically, this term has been introduced [28] to match the experimental results, in particular those concerning the modulation response.

In Fig. 7 the cut-off frequency of the small signal modulation, evaluated through eqs. (14), is reported as a function of the output power in the presence and in the absence of ε. In the presence of ε, the value of the cut-off frequency, which increases as the square root of the output power in the absence of any nonlinear effects connected with the intraband dynamics, tends to saturate at high power because of the strong damping coefficient of the relaxation coefficient [27]. In some devices, the saturation bandwidth decreases at even higher powers.

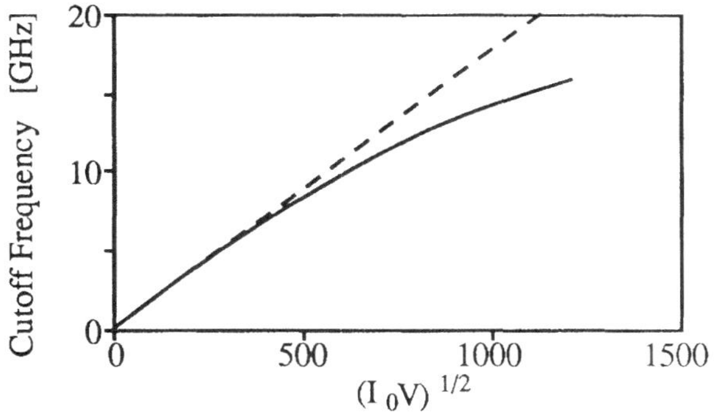

Fig. 7 Cut-off frequency for the modulation of a semiconductor laser as a function of the square root of the photon number inside the cavity. Dashed line refers to $\varepsilon = 0$, solid line to $\varepsilon/V = 0.7\ 10^{-7}$.

5. Examples of New Devices Based on Optical Nonlinearities in Active Semiconductor Devices: Frequency Translators

Although the optical nonlinearities of semiconductor lasers and amplifiers have been initially faced to solve problems connected with fiber transmission systems (the above-mentioned reduction of the modulation bandwidth, crosstalk in amplifiers when two different signals are amplified at the same time, etc...), presently they are exploited to make photonic devices able to perform basic functions in the framework of the advanced systems.

The field where photonic technologies will play a major role is switching. Unfortunately, photonic techniques offer to date modest performances in term of space switching capability, although noticeable progresses have been obtained with InP-based photonic integrated circuits. An alternative approach, driven by the recent advances in the field of tunable lasers [29], is the frequency switching, which can be used in systems based on wavelength division multiplexing technique fully exploiting the large transmission bandwidth of the optical fibres. It has been shown how, in the framework of Asynchronous Transfer Mode (ATM)-type systems (see Fig. 8), the use of network topologies in which the packet information hops among several nodes before reaching the final destination, is particularly convenient. The network complexity, characteristic of this scheme, can be overcome if data packets are encoded in the wavelength domain to determine their optical path through the network [30]. The wavelength switching can be used in these systems as an optical node between stages of a multistage network.

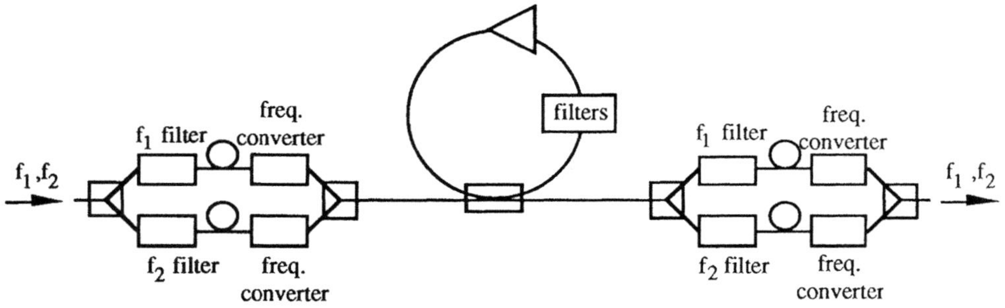

Fig. 8 Basic scheme of a ATM node. From ref. [31]

Although frequency conversion has been reported in devices not based on semiconductor active materials [32,33], it has been obtained so far mainly in semiconductor lasers and amplifiers. In semiconductor lasers and amplifiers frequency conversion is obtained through the nonlinear interaction in the active region. The performances of the device, such as frequency shift amplitude, switching signal amplitude, maximum switch speed, extinction ratio, spectral properties of the switched signal are strongly dependent on the geometry of the frequency converter and on the particular nonlinear effect.

All frequency converters based on nonlinear effects in active semiconductor devices can be divided into two classes. In the first class an optical mode at the converted frequency is already present in the device. This mode might be the resonance frequency of a cavity or the frequency of a radiation injected into the device, for instance. The conversion process transfers, by means of a saturation mechanism, the optical signal to the mode at the new frequency. In the second class the converted frequency is generated by the nonlinear process itself through the FWM process.

6. Frequency Translators Based on Saturation Phenomena

Although these devices are based on quite different mechanisms, the underlying common process is the variation of carrier density of a semiconductor laser or amplifier due to an incoming optical field. One of the main advantages of the method is that, in particular in the case of lasers, the incoming field is allowed to have a polarization perpendicular to the converted field, so that the output fields can be easily separated out of the converter by a simple polarizer.

There is a basic difference between the saturation mechanisms acting in amplifiers and in lasers. In amplifiers, the gain is continuously varied by a saturating field without any sharp change in the operating conditions. In a laser above threshold, on the contrary, the operating conditions of the laser require the gain to be set at the threshold value. All mechanisms proposed to get frequency conversion in lasers (with only one exception [34-36]) operate the switch of the gain from its free running value to a lower value where the lasing (at least at the selected cavity resonance frequency) cannot be maintained.

6.1 SATURATION IN AMPLIFIERS

A promising technique to get frequency conversion is the use of gain dynamics in a semiconductor optical amplifier. Eq. 14 b shows that, in the hypothesis of homogeneous field in the device, the optical gain depends on the power of the input signal

$$G = \frac{G_n(C\tau_s/eV - n_t)}{1+I/I_s}, \tag{15}$$

$I_s = G_n\tau_s/(1+\varepsilon I)$ being the saturation intensity. The dependence of G on the input power is reported in Fig. 9.

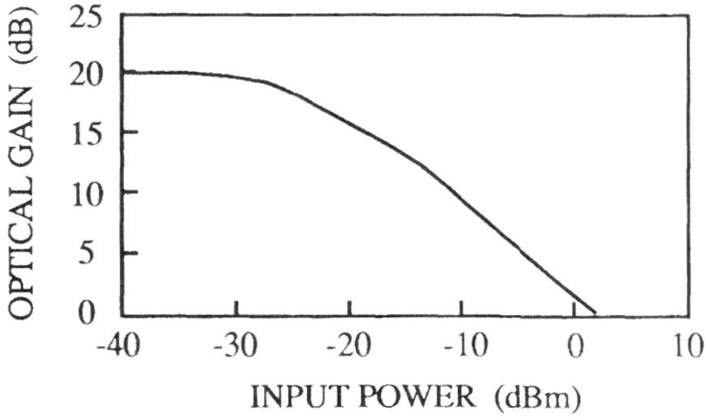

Fig. 9 Gain in a semiconductor optical amplifier vs. input power.

A high intensity modulated signal injected into an amplifier induces, because of gain saturation, a modulation of the amplifier gain if the modulation frequency does not exceed some GHz ($1/2\pi\tau_s$) [37,38]. More precisely, the gain is higher when the input signal is low (zero), and lower when the input signal is high (one). The gain modulation is then (almost) the complement of the temporal behavior of the input signal. Being the saturation common to the whole optical spectrum, the amplification of any other CW signal (having a wavelength inside the gain spectrum) injected into the amplifier is affected by the gain modulation and is consequently modulated. The experimental set-up to attain the wavelength conversion is shown in Fig. 10. By optically filtering the useful wavelength, the transfer of the logical information from one optical carrier to the other is realized with logical inversion.

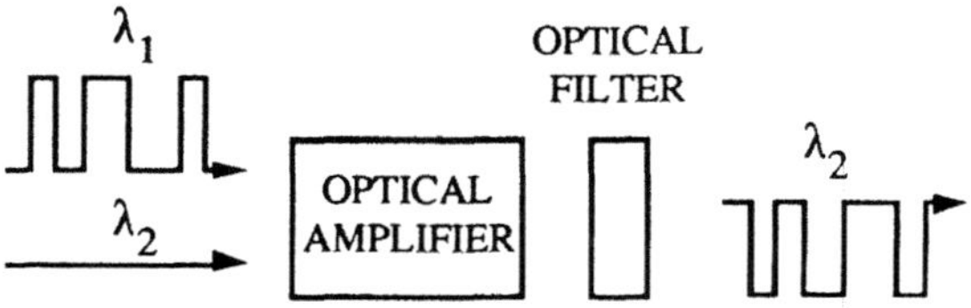

Fig. 10 Experimental set-up to show wavelength conversion by gain saturation in a semiconductor optical amplifier.

The fast attainable switching time (<400 ps) has been successfully exploited in an experiment of wavelength reassignment for fast optical packet switching at 2.5 Gbit/s [37] and in an all-optical de-multiplexing of high speed (4 Gbit/s) time domain multiplexed stream [38]

Key advantages of this technique are the intrinsic broad band and speed of conversion. Their combination leads to the capability of processing signals at high bit rates and to use interference filters to separate the original from the converted wavelengths.

6.2 SATURATION IN LASERS

The simplest device based on saturation is a standard [39] or a multi-electrode distributed feedback (DFB) [35] or a distributed Bragg reflector (DBR) [40] laser (see Fig. 11). The idea is to inject the signal to be converted, at a wavelength λ_1 and intensity modulated (IM), into the cavity of a laser which, in the unperturbed conditions emits a CW field at frequency λ_2. The threshold gain G_{th} does not depend only on the carrier density, but also on the total field intensity in the cavity, through the relation

$$G_{th}=G_n(n-n_t)/(1+\varepsilon I) \qquad (16)$$

Being the value of G_{th} constant above threshold, this equation shows that the presence of the external field, due to the change of I, causes a carrier variation in the laser [41] and a consequent change of the related material refractive index and so of the frequency of the output field. The final result of the process is that a IM field at λ_1 is converted in a frequency modulated (FM) field at λ_2. It has been shown in ref. [35] that an injected power of the order of 200 μW is high enough to get a frequency deviation larger than 300 MHz at λ_2=1.54 μm for λ_1 ranging from 1.50 and 1.58 μm. Moreover, by using a FM/IM Mach-Zehnder converter a 2.5 Gbit/s penalty free transmission has been achieved with a wavelength conversion of 17 nm [36]. It is worth noting, however, that the output field is not purely FM because of the spurious intensity modulation induced by the modulation of the carrier density.

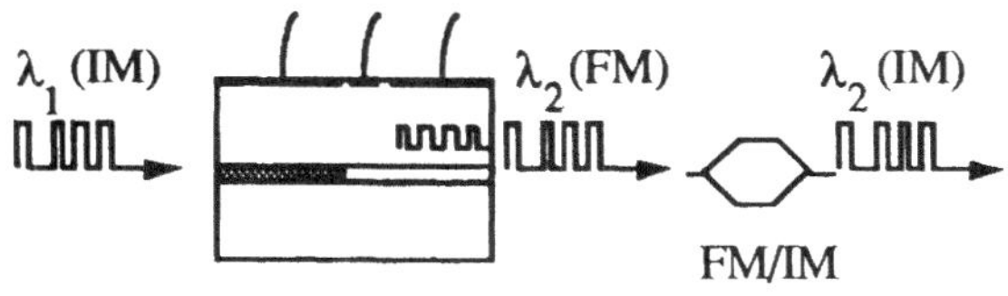

Fig. 11 Basic scheme of the wavelength converter based on gain saturation in DBR lasers.

A similar mechanism is behind the direct wavelength conversion of an IM signal at frequency λ_1 in an IM signal at frequency λ_2 [42]. A DBR laser operates at wavelength λ_2 in CW conditions with an injection current slightly exceeding the threshold. An IM signal at wavelength λ_1 is injected into the cavity. The injected field increases the stimulated emission rate which, if the input power is high enough, drives the laser below threshold. The result is a suppression of the output power at λ_2 when the field at λ_1 is present. The output from the laser is therefore the complement of the injected signal. This method presents three problems. The first one is the high input power needed to switch the laser, at least if the laser operates at appreciable output power. An input power of -4 dBm is needed to switch on and off a laser operating 20 mA above threshold [42]. The second one is the relaxation oscillations typical of lasers operating close to threshold [27], which degrades the quality of the converted signal. Finally, the low obtainable extinction ratio represents also a limitation. A comparison of Bit Error Rate (BER) in a frequency conversion experiment in which a DBR laser, operating close to (IM/IM conversion) or far above (IM/FM conversion) threshold, is used, showed a decrease of the system sensitivity of about 1.5 dB for the former configuration.

It is worth noting that the previous switching mechanisms work as well in standard and tunable lasers and that for tunable lasers the converted frequency can be tuned in a range of some nm [36].

The frequency converters reported above are based on the decrease of the gain in a laser when an external field is injected. Yet the opposite effect, that is a decrease of the losses of a cavity when injected, can be used as well [43]. Devices based on this mechanism take advantage of the insertion of a saturable absorber in a laser cavity. A sketch of the typical device is shown in Fig. 12 a). One part of the active section of a multi-section DFB or DBR laser is biased at a very low level to act as a saturable absorber, while the remaining part of the active region is biased at a positive bias. The power-current characteristic of this laser shows a bistable behavior due to the presence of the saturable absorber [fig. 12 c)]. When the device is biased just below threshold and an optical input at λ_1 is injected into the saturable absorber, the absorption coefficient decreases, light intensity in the cavity increases and lasing at the resonant wavelength of the cavity λ_2 can be obtained if the input intensity is high enough to produce a total gain higher than the cavity loss. When the external input is switched off the process is reversed, the total loss increases and the lasing is inhibited. The main limitations of this method are the high power needed to get the switching (of the order of 300 μW) and the long switching time due to the fact that only spontaneous lifetime is involved in the process, the maximum switching frequency being of the order of 500 MHz. On the other hand a wavelength conversion of more than 4 nm has been obtained with a tunability of the converted wavelength of the order of 4 nm in a three section tunable DBR laser. Recently, the laser structure has been improved by using in a DBR laser an asymmetrical grating [44], a grating with a coupling coefficient much higher on the side of the injection than on the other side. By using this asymmetry and with a careful design of the Bragg reflector, to get a high reflection coefficient only for the laser TE polarized field and not for the injected TM polarized field, the back-reflection from the laser toward the source of the signal at λ_1 can be reduced to negligible values and the output from the other end of the laser increased. With this technique, a ratio of the converted light output from the input end of the device to the output end as high as -30 dB has been obtained.

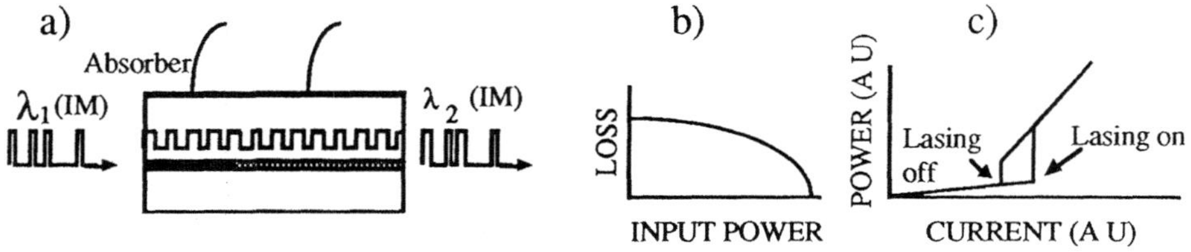

Fig. 12 Wavelength conversion by saturable absorption in a DFB laser. a) basic scheme of the converter. b) cavity loss vs. input power. c) power current characteristics of the laser with the saturable absorber.

A large frequency conversion tunability based upon the injection locking process [45] in a Y-Coupled Cavity Lasers (YCCL) has been obtained in ref. [46]. A similar technique was first suggested to get the small frequency deviations needed to the frequency shift keying (FSK) modulation [47]. In Fig. 13 a schematic top view of the YCCL is reported. The guiding region of the laser is characterized by a Y shape and the contacts are divided into four sections which can be individually accessed.

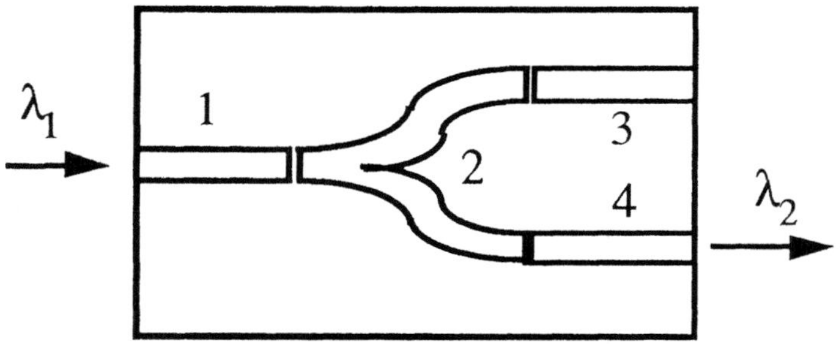

Fig. 13 Top view of a Y-coupled cavity laser.

The presence of a double cavity (1-2-3 and 1-2-4) selects a single longitudinal mode of oscillation of the device which can be electronically tuned over an interval exceeding 20 nm. Wavelength conversion is obtained by injecting in section 1 of the laser, which is emitting at wavelength λ_2, a field at the wavelength λ_1 of a mode below threshold. If the amount of injected power is high enough, the injection at λ_1 decreases the round-trip loss for this mode below that of the mode at λ_2, inducing the switch of the output frequency from λ_2 to λ_1. The output from the laser at λ_2 is therefore the complement of the input as in one of the cases described above. The main advantages of this method are the large frequency conversion range and the intrinsic high speed of the switch which is due to the short characteristic time of the physical process involved, the stimulated emission. A disadvantage is the fine adjustment of λ_1 requested to avoid the typical instability which shows up in injection locking when the injected signal and the cavity resonance are detuned. A wavelength conversion by a 100 μW signal modulated at 2.5 Gbit/s across 45 nm has been obtained with this method [48].

7. Frequency Conversion by Four-Wave Mixing

Referring to the general scheme shown in Fig. 2, the frequency conversion by FWM can be obtained once the signal to be translated is associated to the optical frequency ν_2, while the converting field (pump) is associated to the frequency ν_3. Using the FWM approach, the frequency conversion range is mainly limited by the characteristic time of nonlinear effects, differently from the case of frequency conversion techniques based on saturation effects. The maximum range $2|\nu_3-\nu_2|$ is, in fact, determined by the frequency at which the material nonlinearity can be modulated. The FWM due to the carrier density effects will be referred to as nearly degenerate four wave mixing (NDFWM) from now on. The frequency conversion due to other faster phenomena will be referred to as highly non-degenerate four wave mixing (HNDFWM).

7.1 FREQUENCY CONVERSION IN TRAVELLING WAVE AMPLIFIERS

Frequency conversion has been obtained to date in travelling wave amplifiers mainly using the nonlinearity due to the carrier density modulation. The low value of $\chi^{(3)}$ associated with carrier heating or spectral hole burning effects prevents, in fact, an extensive use of HNDFWM in devices where cavity effects cannot enhance the effective interaction length between probe and conjugate. Only very recently a transmission system based on HNDFWM in amplifiers has been carried out [8]. In this system two Erbium doped fiber amplifiers have been used to raise both the pump power injected into semiconductor amplifier up to +4.3 dBm and the detector sensitivity. With such an experimental set-up a 1.1 dB penalty, at 10^{-9} BER has been demonstrated over 20 nm wavelength conversion.

In the case of NDFWM, two different approaches have been used to get frequency conversion in amplifiers. The first one, the easier, consists in the direct frequency conversion of a signal using a pump at a frequency detuned some hundred MHz from the signal. The power of the conjugate beam coming out the amplifier will depend in general on the power of the pump at frequency ν_3, on the power of the signal at frequency ν_2, and on the strength of $\chi^{(3)}$. The ratio between the conjugate output power and the signal input power can reach values as high as 20 dB [49]. The main limitations of this method are in the limited bandwidth of the incoming signal needed to avoid distortion of the conjugate beam, and in the limited frequency conversion range. Both limitations are directly related to the value of the carrier lifetime, which limits the spectral range over which the nonlinearity is efficient, to a low-pass band with cut-off frequency $B=1/(2\pi\tau_s)$, usually ~1-2 GHz. The practical use of this method is also limited to date by the need of a very sharp optical filter to separate the conjugate beam from the intense pump.

The second approach [50] overcomes the limitations on the frequency conversion range by making positive use of the homogeneous gain saturation. The set-up of the frequency conversion scheme is shown in Fig. 14. Three optical waves, the signal s, the pump p, and the converter c, are coupled into an amplifier. The pump and the signal have a frequency difference $\Omega/2\pi$~1 GHz and, as in the previous case, induce a modulation of the carrier density at frequency difference. The modulation affects the entire gain spectrum because of the homogeneity of the gain saturation. The conversion of the input signal can be achieved at any other wavelength in the gain spectrum by means of the converting wave. Two spectral components at frequency ν_{c1} and ν_{c2}, which are the replica of the input signal, are, in fact, generated at both sides of the converting wavelength because of the carrier modulation. A frequency conversion on an interval Ω^* of 4000 GHz has been obtained using this method [50]. In ref. [51] an extensive study concerning the features of this frequency conversion technique showed that a conversion efficiency (ratio between the emitted power at frequency ν_{c1} and the input signal at frequency ν_s) as high as 16 dB can be obtained. Besides, using this method, the power of the converting wave can be reduced below the power of the converted signal, relaxing the requirements for the optical filter at the output of the amplifier.

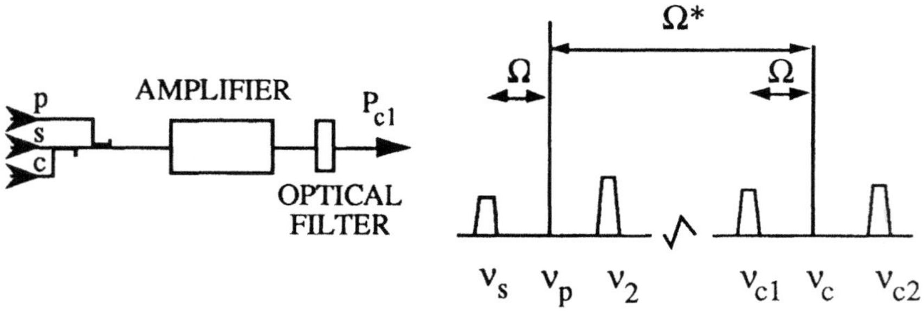

Fig. 14 Basic scheme of frequency conversion in which NDFWM in an optical amplifier is exploited to get high frequency-conversion range.

7.2 FREQUENCY CONVERSION IN LASERS

The main difference between the case of frequency conversion in amplifiers and lasers lies in the presence, in the latter case, of a strong field in the cavity which acts as a pump, thus relaxing the need of the external pump. Other noticeable differences are the presence of cavity resonance in the case of Fabry-Perot configurations, which enhance the conjugate amplification, and the substitution of the carrier lifetime with the much shorter characteristic time of the cavity $(\Gamma G I / \tau_s)^{1/2}$.

Two methods have been proposed to date to get frequency conversion in semiconductor laser devices, namely the use of NDFWM in Fabry-Perot [51] or DFB [52] lasers and the use of HNDFWM in Fabry-Perot devices [53,54].

The first approach is similar to the method used with semiconductor amplifiers. The signal to be converted (with a spectrum centred at ν_2) is fed into the cavity of a laser well above threshold and emitting at frequency ν_3. The amplification of the conjugate beam is almost flat in a band of some GHz (larger than in amplifiers) and the conversion efficiency can be as high as 30 dB [52]. It is worth noting that the band over which the amplification of the conjugate is flat increases as the pump power increases. Unfortunately, it has been proven [55] that this scheme degrades the coherence properties of the conjugate beam. If $\Delta\nu_1$ is the linewidth of the probe and $\Delta\nu_2$ the linewidth of the laser, the linewidth of the conjugate is $\Delta\nu_3=\Delta\nu_2+4\Delta\nu_1$.

The second approach seems to be much more interesting for the high value of the frequency conversion range (>1 THz) [53] which can be achieved.

Experimental results have been obtained in QW InGaAsP lasers [53]. In a first experiment it was shown that the injection of a probe with -11 dBm allows for the generation of a -40 dBm conjugate beam at a frequency detuning as high as 2.2 THz (17.6 nm). In a second step [54] the authors applied this frequency conversion method to a real communication experiment. They performed the experiment using a 1 Gbit/s pattern and measuring a power penalty after frequency conversion on a range of 1 THz of less than 1 dB.

8. Conclusions

The physical processes underlying the nonlinearities in active semiconductor devices have been reported and discussed. The methods that have been proposed to get frequency conversion in a wide range of frequency have been reviewed. A comparison of the different methods have shown that one of the promising methods to get frequency conversion at very large detunings is HNDFWM, although the underlying physical mechanisms are not yet well understood. The reason is that there is no general agreement on the fast saturation mechanisms in semiconductors. In particular, QW devices and SQW devices show a higher nonlinearity of still unknown physical origin.

REFERENCES

[1] H. Nakajima and R. Frey, "Collinear nearly degenerate four-wave mixing in intracavity amplifying media", IEEE J. Quantum Electron., 22, pp. 1319-1354, 1986.

[2] K. L. Hall, J. Mark, E. P. Ippen, G. Eisenstein, "Femtosecond gain dynamics in InGaAsP optical amplifiers", Appl. Phys. Lett., 56, pp. 1740-1743, 1990.

[3] R. Frankenberger, and R. Schimpe "Measurement of the gain saturation spectrum in InGaAsP diode lasers", Appl. Phys. Lett., 57, pp. 2520-2523, 1990.

[4] C. K. Sun, H. K. Choi, C. A. Wang and J. G. Fujimoto " Femtosecond carrier dynamics and stimulated transitions induced carrier temperature change in InGaAs/GaAs strained-layer SQW diode lasers", CLEO '93, Baltimore, paper CThC2.

[5] A. Sapia, P. Spano and B. Daino "Polarization switching in semiconductor lasers driven via injection from an external radiation", Appl. Phys. Lett., 50, pp. 57-59, 1987.

[6] J. E. Midwinter "Photonics in switching: the next 25 years of optical communications", IEE Proc. J, 139, pp. 1-12, 1992.

[7] A. Marrakchi, W. M. Hubbard and S. F. Habiby "Review of Photonic switching device technology", in "Coherent Optical Communications and Photonic Switching" ed. by G. Prati, Elsevier, Amsterdam 1990, pp. 299-312.

[8] M. C. Tatham and G. Sherlock "20 nm wavelength conversion using ultrafast highly nondegenerate four-wave mixing", Integrated Photonics Research, Topical Meeting, Palm Springs, CA, Paper PD1, March 1993.

[9] C. T. Hultgren, K. L. Hall, G. Lenz, D. J. Dougherty and E. P. Ippen "Spectral-hole burning and carrier heating nonlinearities in active waveguides", Ultrafast Electronics and Opto-Electronics, Topical Meeting, San Francisco, 1993.

[10] L. F. Tiemeijer, "Effects of nonlinear gain on four-wave mixing and asymmetric gain saturation in a semiconductor laser amplifier", Appl. Phys. Lett., 59, pp. 499-501, 1991.

[11] C. T. Hultgren, E. P. Ippen, "Ultrafast refractive index dynamics in AlGaAs diode laser amplifiers", Appl. Phys. Lett., 59, pp. 635-637, 1991.

[12] K. L. Hall, A. M. Darwish, E. P. Ippen, U. Koren and G. Raybon "Femtosecond index nonlinearities in InGaAs optical amplifiers", Appl. Phys. Lett., 62, pp. 1320-1322, 1993.

[13] A. Uskov, J. Mørk and J. Mark "Wave mixing in semiconductor laser amplifiers due to carrier heating and spectral-hole burning", to be published on IEEE J. Quantum Electron.

[14] R. Loudon "The quantum theory of light", Oxford University Press, Oxford, 1988.

[15] J. Mark and J. Mørk "Subpicosecond gain dynamics in InGaAs optical amplifiers: experiment and theory", Appl. Phys. Lett., 61, 2281-2283, 1992.

[16] N. Osagawara and R. Ito "Longitudinal mode competition and asymmetric gain saturation in semiconductor injection lasers. II. Theory", Jap. J. Appl. Phys., 27, pp. 615-626, 1988.

[17] E. P. Ippen and C. V. Shank "Techniques for measurement", in "Ultrafast light pulses", ed. by S. L. Shapiro, Springer Verlag, Berlin 1977.

[18] M. P. Kesler, E. P. Ippen, "Subpicosecond gain dynamics in GaAlAs laser diodes", Appl. Phys. Lett., 51, pp. 1765-1768, 1987.

[19] Y. Lai, K. L. Hall, E. P. Ippen and G. Eisenstein "Short pulse gain saturation in InGaAsP diode laser amplifiers", IEEE Photon. Technol. Lett., 2, 711-713, 1990.

[20] K. Kikuchi, M. Kakui, C. Zah, T. P. Lee, "Observation of highly nondegenerate four-wave mixing in 1.5 μm travelling wave semiconductor optical amplifiers and

estimation of nonlinear gain coefficient", IEEE J. Quantum Electron., 28, pp. 151-156, 1992.

[21] J. Mark, J. Mørk and C. P. Seltzer "Sub-picosecond gain dynamics in InGaAsP MQW laser amplifiers" 13th IEEE International Semiconductor Laser Conference, Takamatsu, paper F-2.

[22] K. J. Vahala J. Zhou, N. Park and J. W. Dawson, "TeraHertz four-wave mixing spectroscopy of intraband dynamics in quantum-well amplifiers", CLEO '93, Baltimore, paper CPD2.

[23] M. C. Tatham, G. Sherlock and C. P. Seltzer "Highly non-degenerate (>1 THz) four-wave mixing in multi-quantum well laser structures" Quantum Optoelectronics, Topical Meeting, Palm Springs, CA, Paper QWD4, March 1993.

[24] A. D'Ottavi, E. Iannone, A. Mecozzi, S. Scotti, P. Spano, T. Pietropaoli, J. Landreau, A. Ougazzaden, J. C. Bouley, "Four wave mixing in a strained multiple-quantum-well amplifier", QELS '93, Baltimore, paper QThA5.

[25] A. P. Bogatov, P. G. Eliseev and B. N. Sverdlov "Anomalous interaction of spectral modes in a semiconductor laser", IEEE J. Quantum Electron., 11, pp. 510-515, 1975.

[26] C. B. Su and V. A. Lanzisera "Ultra-high speed modulation of 1.3 μm InGaAsP diode lasers", IEEE J. Quantum Electron., 22, pp. 1568-1578, 1986.

[27] G. P. Agrawal and N. K. Dutta "Long wavelength semiconductor lasers", Van Nostrand Reinhold Company, New York 1986.

[28] J. Manning, R. Olshanski, D. M. Fye and W. Powazinik "Strong influence of nonlinear gain on spectral and dynamic characteristics of InGaAsP lasers", Electron. Lett., 21, 496-497, 1985.

[29] T. L. Koch, U. Koren, "Semiconductor lasers for coherent optical fiber communications", IEEE J. Lightwave Technol., 8, pp. 274-293, 1990.

[30] A. S. Acampora, M. J. Karol, "An overview of lightwave packet networks", IEEE Network, pp. 29-41, 1989.

[31] P. Cinato, B. Costa, A. Daniele, A. de Bosio and E. Vezzoni "Electro-optical switching node based on fast packet and frequency switching techniques", ISS '90, Stockholm, 1990.

[32] T. Morioka, K. Mori, M. Saruwatari, "Ultrafast polarization-independent optical demultiplexer using optical carrier frequency shift through crossphase modulation", Electron. Lett., 28, pp. 1070-1072, 1992.

[33] H.Q. Le, S. Di Cecca, "Ultrafast, room-temperature, resonance-enhanced third-order optical susceptibility tensor of an AlGaAs/GaAs quantum well", Opt. Lett., 16, pp. 901-903, 1991; H.Q. Le, S. Di Cecca, "Ultrafast, multi-THz-detuning, third-order frequency conversion in semiconductor quantum-well waveguides", IEEE Photon. Technol. Lett., 4, pp. 878-880, 1992.

[34] K. Inoue, "Wavelength conversion for frequency-modulated light using optical modulation to oscillation frequency of a DFB laser diode", IEEE J. Lightwave Technol., 8, pp. 906-911, 1990.

[35] P. Pottier, M. J. Chawki, R. Auffret, G. Claveau, A. Tromeur, "1.5 Gbit/s transmission system using all optical wavelength converter based on tunable two-electrode DFB laser", Electron. Lett., 27, pp. 2183-2185, 1991.

[36] B. Mikkelsen, T. Durhuus, R. J. Pedersen, K. E. Stubkjaer, M. Öberg, S. Nilsson, "Penalty free wavelength conversion of 2.5 Gbit/s signals using a tuneable DBR-laser", ECOC '92, Berlin, paper We A10.4.

[37] T. Saitoh, T. Mukai, "Detuning characteristics and conversion efficiency in a 1.5 μm travelling-wave semiconductor laser amplifier" IEEE J. Quantum Electron., 26, pp. 865-874, 1990.

[38] T. Saitoh, T. Mukai, "Recent progress in semiconductor laser amplifiers" IEEE J. Lightwave Technol., 6, pp. 1656-1664, 1988.

[39] A. Cavaciuti, P. Gambini, M. Puleo, E. Vezzoni, F. Delorme, S. Slempkes, B. Roben, "All-optical wavelength reassignment for fast optical packet switching" OFC/IOOC '93 San Jose, paper WD4.

[40] M. Artiglia, A. Cavacuiti, M. Macchi, M. Puleo, E. Vezzoni, "All-optical 8 demultiplexing of a 4 Gbit/s TDM data stream by means of gain saturation in a semiconductor optical amplifier", Optical Amplifiers and their Applications '92, Santa Fe, paper WD4.

[41] T. Ogawa, Y. Ida, K. Hayashi, "Experimental determination of gain compression factors of a DFB laser in the presence of TM injection light", IEEE Photon. Technol. Lett., 4, pp. 345-348, 1992.

[42] R. J. S. Pedersen, B. Mikkelsen, T. Durhuus, M. J. Steinman, K. E. Stubkjaer, "Tunable DBR laser for wavelength conversion of 2.5 Gbit/s signals", 13th IEEE International Semiconductor Laser Conference, Takamatsu, paper O-7.

[43] H. Kawaguchi, K. Oe, H. Yasaka, K. Magary, M. Fukuda, Y. Itaya, "Tunable optical-wavelength conversion using a multielectrode distributed-feedback laster diode with a saturable absorber" Electron. Lett., 23, pp. 1088-1090, 1987.

[44] H. Yasaka, K. Takahata, K. Kasaya, M. Okamoto, Y. Kondo, M. Naganuma, M. Ikeda, "One-directional output optical frequency conversion device with asymmetric κ-DBR structure", ECOC '92, Berlin, paper We A10.5.

[45] S. Kobayashi, T. Kimura, "Injection locking in AlGaAs semiconductor lasers", IEEE J. Quantum Electron., 19, pp 266-270, 1983.

[46] M. Schilling, W. Idler, D. Baums, G. Laube, K. Wünstel, O. Hildebrand, "Multifunctional photonic switching operation of 1550 nm Y-coupled cavity laser (YCCL) with 28 nm tuning capability", IEEE Photon. Technol. Lett., 3, pp. 1054-1056, 1992.

[47] P. Spano, M. Tamburrini, S. Piazzolla, "Optical FSK modulation using injection-locked laser diodes" IEEE J. Lightwave Technol., 7, pp. 726-728, 1989.

[48] M. Schilling, W. Idler, D. Baums, K. Dütting, G. Laube, K. Wünstel, O. Hildebrand, "6 THz range frequency conversion of 2.5 Gbit/s signals by a 1.55 μm MQW based widely tunable Y-laser" 13th IEEE International Semicondutor Laser Conference, Takamatsu, paper O-8.

[49] T. Mukai, T. Saitoh, "Detuning characteristics and conversion efficiency of nearly degenerate four-wave mixing in a 1.5-μm travelling-wave semiconductor laser amplifier", IEEE J. Quantum Electron., 26, pp. 865-875, 1990.

[50] G. Großkopf, R. Ludwig, H. G. Weber, :140 Mbit/s DPSK transmission using an all-optical frequency converter with 4000 GHz conversion range", Electron. Lett., 24, pp. 1106-1107, 1988.

[51] N. Schunk, "All -optical frequency conversion in a travelling-wave laser amplifier", IEEE J. Quantum Electron., 27, pp. 1271-1279, 1991.

[52] A. Mecozzi, A. D'Ottavi, R. Hui, "Nearly degenerate four-wave mixing in semiconductor lasers operating above threshold", to be published in IEEE J. Quantum Electron.

[53] S. Murata, A. Tomita, J. Shimizu, M. Kitamura, A. Suzuki, "Observation of highly nondegenerate four-wave mixing (>1 THz) in an InGaAsP multiple quantum well laser", Appl. Phys. Lett., 58, pp. 1458-1460, 1991.

[54] S. Murata, A. Tomita, J. Shimizu, A. Suzuki, "THz optical-frequency conversion of 1 Gb/s-signals using highly nondegenerate four-wave mixing in an InGaAsP semiconductor laser", IEEE Photon. Technol. Lett., 3, pp. 1021-1023, 1991.

[55] R. Hui and A. Mecozzi, "Phase noise of four wave mixing of semiconductor lasers," Appl. Phys. Lett., 60, pp. 2454-2456, 1992.

ELECTRON STATES IN BIASED HETEROSTRUCTURES

R. FERREIRA and G. BASTARD
Laboratoire de Physique de la Matiere Condense
Ecole Normale Superieure
24 rue Lhomond
F - 75005 Paris
France

INTRODUCTION

In this contribution we are mainly interested in the effects of an externally applied electric field on the electronic states of semiconductors quantum wells and superlattices. Such structures are simply viewed as a sequence of different semiconductor layers grown sucessively along a well defined crystalline growth direction. The carriers' motion along this growth direction (hereafter called the z direction) is then strongly modified by the presence of the various thin (≈100Å) layers, whereas the translation symmetry (at the level of the bulk unitary cells) remains for the in-plane (x,y) directions. Single quantum wells present states which are localized in space around the well region for the z motion. Double quantum wells (DQW) consist of two well layers separated by a finite barrier layer. The localized states of the isolated quantum wells interact through the barrier region (tunnel interaction) to form the DQW eigenstates. The resulting new states are delocalized over the DQW region. For example, two identical wells (same thickness and material composition) give rise to

A. Miller et al. (eds.), Nonlinear Optical Materials and Devices for Applications in Information Technology, 207–254.

symmetrical and antisymmetrical z-dependent DQW wavefunctions (with respect to the center of the central barrier). For wells of different thicknesses (asymmetrical DQW) the tunnel coupling is less efficient and each resulting DQW state presents a preferential localization around one of the two wells, reminiscent of the localized states of the isolated wells. In other words, the tunnel coupling is less effective to mix states of distinct quantum wells wich are misaligned in energy. The application of an external electric field **F**//**z** acts on the DQW states as follows. The electrostatic potential eFz (we take its origin at the center of the intermediate barrier) raises (lowers) in energy the eigenstates of the right (left) well. Thus, by increasing the bias one is capable to externally align or bring off resonance the different states pertaining to the separated wells, and thus to force a localization or delocalization of the DQW eigenstates. Semiconductor superlattices (SL) consist of a periodic repetition of a fundamental cell composed by a sequence of wells and barriers. The simplest one consists of a well and a barrier (with eventually different thicknesses). The states of the isolated thin wells (with only one electron bound state) are all at the same energy at zero bias. The tunnel coupling lifts this (infinite) degeneracy and gives rise to an electron miniband with a finite energy width. The Bloch states are completely delocalized. In the presence of an electric field the energy levels of neighbouring wells are not at the same energy. Then, like for DQW, this energy misalignement inhibits the inter well coupling and the eigenstates of the biased SL are no longer completely delocalized. We shall show that they become localized within a Δz region which decreases continuously with increasing electric field until the wavefunction concentrates around one SL period ($\Delta z \approx$ well width if the electrostatic potential drop within a SL period becomes much greater than the tunnel coupling). This increasing spatial localization of the states with increasing applied electric field is counter-intuitive and corresponds to one of the most striking results for superlattices.

These three structures (isolated quantum wells, double quantum wells and superlattices) are considered in the following. In the first part we present briefly the theoretical formalism we use to describe the electronic states (conduction and valence bands) of such reduced dimensionality heterostructures. A brief introduction to the flat-band superlattice states is also presented. In the second part we consider the eigenstates of a single quantum well under electric field. The modifications of the nature of the carriers z-motion due to the presence of the electric field are examined both in the semi-classical and quantum regimes. The broadening of the quantum well states is evaluated via the semi-classical "ping-pong" model and the quantum description of the field induced polarizability of the quantum wells eigenstates is analysed in both the weak and high field regimes. In the third part we discuss briefly the electric field effects in double quantum wells. The latest part deals with the electric field effects in superlattices. It begins by a one band analysis of the Wannier-Stark ladders. The existence of such a ladder of states is derived by the consideration of the translation operator along the SL axis. The semi-classical (leading to the Bloch oscillators) and the quantum (leading to the Wannier-Stark ladder) analysis of the biased SL are presented and compared. Finally, both the intra-band and the inter-band optical response of the biased superlattice are discussed. In particular, the existence of an effective blue shift for the band-to-band absorption edge and its association with the increasing localization of the superlattice eigenstates with increasing field are considered in detail. Multi band effects on the Wannier-Stark quantization are then introduced. After some general remarks, we present a simple modelling of the anticrossing between two discrete Wannier-Stark levels belonging to two different bands. Finally, we discuss the consequences of such multi band interactions on the inter-subband optical transitions and on the time evolution of optically photocreated wavepackets. An appendix presents a comparison between the Bloch oscillator and Wannier-Stark wavepackets.

1 - BAND STRUCTURE OF SEMICONDUCTOR SUPERLATTICES

In the following we shall use the envelope function approximation to describe the electronic states of semiconductor heterolayers. This method is versatile and conceptually simple. It is in fact an extension of the effective mass approximation of bulk semiconductors to the multilayer case. The interested reader may find an extensive review of this method (and others) and of their use in references [1-4].

The basic idea is to write the carrier wavefuction in the energy range of interest in each host layer as a linear expansion on the hosts' periodic part of the Bloch functions at the extrema of interest. For instance in direct gap III - V materials the relevant edge is the Γ_6 (i. e. S - like symmetry with J = S = 1/2) for the conduction states and the Γ_8 (P - like symmetry and J =3/2) for the valence states. The coefficients of this expansion are functions of r. These functions, the envelope functions, are slowly varying on the scale of the hosts' unit cell. This is because they are built out of a small portion of the Brillouin zone around the extremum of interest. The in - plane translational invariance of the heterostructure allows to write these functions as product of the plane waves in x and y times functions of z. These z functions are solutions of :

$$\Sigma_{j'} \left[D_{jj'}(k_x, k_y, -i\hbar\partial/\partial z) + V(z)\delta_{jj'} \right] f_{j'}(z) = \varepsilon f_j(z) \tag{1}$$

where the summation runs over the edges one takes into consideration. $D_{jj'}$ is the bulk kinetic energy matrix in this basis. It comprises terms which are linear in $k_x, k_y, -i\hbar\partial/\partial z$ and which express the direct **k·p** coupling between the different edges retained in the analysis as well as terms which are quadratic in $k_x, k_y, -i\hbar\partial/\partial z$ and which express the indirect **k·p** coupling due to one virtual excursion from the edges under consideration to the other edges of the hosts' crystal. The latter terms are usually small when the direct **k·p** coupling is symmetry allowed (e.g. between the conduction S - like states and the valence P - like states). They are necessary when no such direct coupling exists (e.g. between the valence states). The potential energy term V(z) for a perfect heterostructure is piecewise constant and diagonal in j and j'. In each host layer it represents the energy position of the jth edge. Thus, it exhibits jumps at the interfaces whose magnitude are the energy offsets of the various edges. To be specific let us consider conduction states and assume that we restrict ourselves to a parabolic description of these states. We have then:

$$D_{jj'} = \delta_{jj'} \left[-\hbar^2\partial/\partial z(1/2m^*(z))\partial/\partial z + \hbar^2(k_x^2 + k_y^2)/2m^*(z) \right] \tag{2}$$

where j refers to the Γ_6 edge, m*(z) is the piecewise constant effective mass equal to the conduction effective mass in each host layer. V(z) for this example would be the Γ_6 conduction band offset. For the valence levels treated again in the parabolic approximation D is the Luttinger[5] matrix

$$D = \begin{bmatrix} H_{hh} & c & b & 0 \\ c^* & H_{lh} & 0 & -b \\ b^* & 0 & H_{lh} & c \\ 0 & -b^* & c^* & H_{hh} \end{bmatrix} \begin{matrix} 3/2 \\ -1/2 \\ 1/2 \\ -3/2 \end{matrix} \tag{3}$$

D has been written on the $|J = 3/2, -3/2 \leq m_J \leq 3/2>$ basis. H_{hh} and H_{lh} are the diagonal heavy hole related ($m_J=\pm3/2$) and light hole ($m_J=\pm1/2$) related kinetic energies:

$$H_{hh} = \hbar^2/2m_0 [- (\gamma_1 + \gamma_2)(k_x^2 + k_y^2) + \partial/\partial z(\gamma_1 - 2\gamma_2)\partial/\partial z]$$

$$H_{lh} = \hbar^2/2m_0 [- (\gamma_1 - \gamma_2)(k_x^2 + k_y^2) + \partial/\partial z(\gamma_1 + 2\gamma_2)\partial/\partial z]$$

$$b = \sqrt{3}\, \hbar/2m_0 (k_x - ik_y) (\gamma_3 p_z + p_z\gamma_3)$$

$$c = \sqrt{3}\, \hbar^2/2m_0 \; [\gamma_2(k_x^2 - k_y^2) - 2i\, \gamma_3 k_x k_y] \tag{4}$$

Note the so-called mass reversal effect in the diagonal terms H_{hh} and H_{lh}. Namely the motion along the growth axis is characterized by a heavy (resp. light) mass for the heavy, i.e. ±3/2 components, (resp. light, i.e. ±1/2 components) while the in - plane motion is characterized by a light mass (a heavy mass) for the heavy hole related (the light hole related) components. In principle, the γ (Luttinger) parameters are position dependent (step - like variations when going from one layer to the other).

If the b and c terms were zero, which happens at zero in - plane wavevector, the heavy and light hole states would be entirely decoupled (exactly like the two spin components ±1/2 of a conduction electron with a parabolic dispersion relation). Hence, the valence levels would be very much analogous to the conduction levels. In particular, at $k_x = k_y = 0$ there is a decoupling between the heavy hole related and light hole related edges (labelled HH_n and LH_n in the following, where n is the discrete subband index). However, as soon as one is interested in the valence subband dispersion relations upon the in - plane wavevector, the off diagonal terms can no longer be neglected. They induce two kinds of effects. Firstly, they couple the heavy and light components in the envelope wavefunctions. Secondly, the dispersion relations strongly deviate from the usual parabolic shape. In particular, for wide enough quantum wells the ground light hole subband exhibits a cammel back, i.e. a positive curvature near the zone center, followed by a negative one at large in plane wavevector. These two aspects are illustrated in figs.(1,2). Figure (1) shows the in - plane dispersion relations for the valence states of two GaAs - Ga(Al)As quantum wells of different thicknesses and clearly displays the cammel - back shape dispersion of the ground light hole subband. Figure (2) illustrates the influence of band mixing on a key property of a superlattice: the bandwidth. We show in fig.(3) the small bandwidth for heavy hole states compared to the electron one. These results were obtained by assuming a zero in - plane wavevector and arise from the heavy mass of the heavy hole along the growth axis). When k_x, k_y are non zero the heavy hole states become admixed with the light hole ones. But the light hole tunnelling is much more efficient than the heavy hole one (the light hole mass along the growth axis is indeed light). Therefore, the width of the ground miniband should change, actually increase, with the in -plane wavevector (labelled $k_\perp$ in fig.(2)) due to the heavy - light admixture in the wavefunction. Notice that this implies that the hole motions along the growth axis and in the layer plane can no longer be considered as decoupled, while this decoupling is a fairly good approximation for the electrons

A superlattice is an infinite periodic stack of two materials. The spatial period d usually ranges from a few nanometers (short period superlattices) to tens of nanometers. Because the sequence is periodic the Bloch theorem states that the eigenstates, in our formalism the envelope functions, fulfill:

$$\chi_{nq}(z + d) = e^{iqd}\,\chi_{nq}(z) \tag{5}$$

where q is the superlattice wavevector that one may restrict to the first Brillouin zone [$-\pi/d, \pi/d$ [. The dispersion relations $\varepsilon_n(q)$ associated to $\chi_{nq}(z)$ describe the motion along the superlattice axis. Note that an extra, discrete index, the subband index, is necessary to label the orbital states. There is a fairly general property regarding the dispersions $\varepsilon_n(q)$ in one dimensional periodic problems: They are necessarily of the form:

$$\cos(qd) = f(\varepsilon) \tag{6}$$

where $f(\varepsilon)$ is a function which explicitly depends on the problem under consideration. To show why this is true let us apply the Bloch theorem on both ends of a superlattice period for the problem:

$$[-\hbar^2\partial/\partial z(1/2m^*(z))\partial/\partial z + V(z)\,]\,\chi_{nq}(z) = \varepsilon\,\chi_{nq}(z) \tag{7}$$

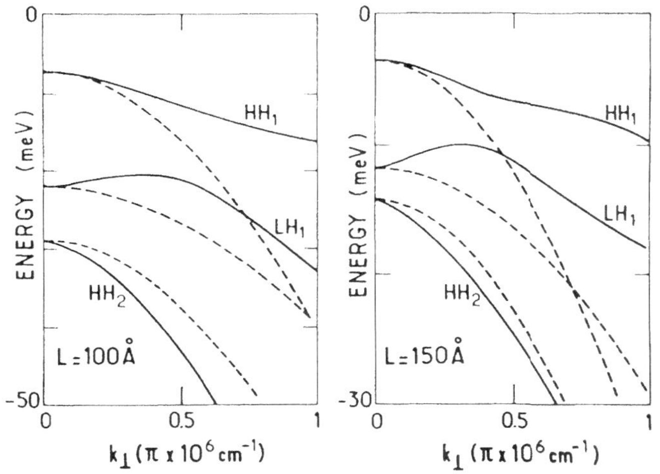

Fig.(1) Valence in - plane dispersions relations of two GaAs - Ga(Al)As quantum wells
The dashed lines correspond to the diagonal approximation of the Luttinger hamiltonian

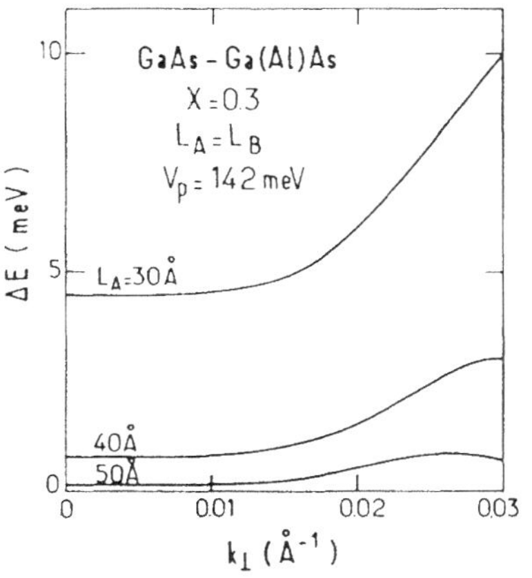

Fig.(2) $k_\perp$ dependence of the ground valence subband width in GaAs - Ga(Al)As superlattices with equal barrier and well thicknesses and d=60Å, 80Å, 100Å.

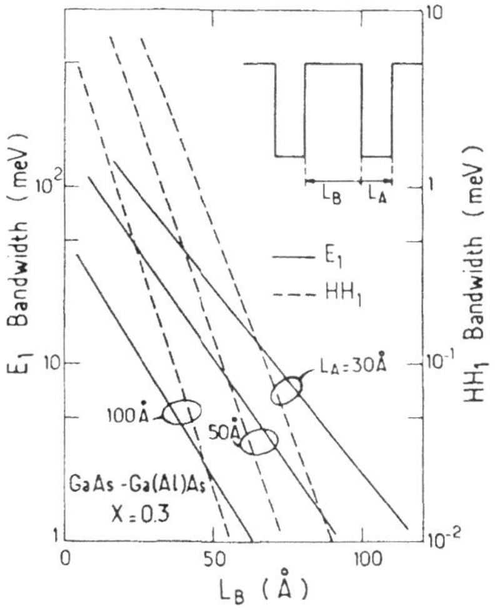

Fig.(3) Bandwidth of GaAs- Ga(Al)As superlattices for electrons and holes versus the barrier width for different well widths.

$$V(z + d) = V(z) \quad ; \quad m^*(z + d) = m^*(z)$$

Inside that period the wavefunction is a linear superposition of the two independent solutions $\alpha(z)$ and $\beta(z)$ at the energy ε of the Schrödinger equation (considered as a second order differential equation) : $\chi_{nq}(z) = a\,\alpha(z) + b\,\beta(z)$. Applying the Bloch theorem and demanding that a and b are non vanishing leads immediatly to the desired result with:

$$f(\varepsilon) = (1/2w)\,[\,-\alpha'(d)\beta(0) - \beta(d)\alpha'(0) + \alpha(d)\beta'(0) + \alpha(0)\beta'(d)\,] \tag{8}$$

where the prime means z derivative and w is the wronskian at the energy ε of the two solutions ($w(z) = \alpha\beta' - \beta\alpha'$), which is a constant. When ε varies the energy segments where $|f(\varepsilon)| \leq 1$ correspond to allowed energy states (q real): the minibands, while, under the reverse condition, there is no propagating states since q has to be pure imaginary to render the cosine larger than one. This, in turn, corresponds to evanescent states with unnormalizable wavefunctions in an infinite superlattice.

The exact form of the $f(\varepsilon)$ function is seldom known. Fortunately, the usual cas of semiconductor superlattices (which for conduction electrons corresponds to $V(z) = 0$ if $z \in$ well (thickness L_w, effective mass m_w) and $V(z) = V_b$ if $z \in$ barrier (thickness L_b, effective mass m_b), where V_b is the conduction band offset) is tractable (Kronig - Penney model) . One finds:

$$\cos(qd) = \cos(k_w L_w)\cosh(\kappa_b L_b) - 1/2(\xi - 1/\xi)\sin(k_w L_w)\sinh(\kappa_b L_b)$$

$$\xi = k_w m_b / \kappa_b m_w$$

$$k_w = (2m_w\varepsilon/\hbar^2)^{1/2} \quad ; \quad \kappa_b = [2m_b(V_b - \varepsilon)/\hbar^2]^{1/2} \qquad 0 \leq \varepsilon \leq V_b \tag{9}$$

where we have taken into account the position dependence of the effective mass in the hosts' dispersion relations as well as in the continuity conditions across the interfaces fulfilled by the envelope functions (χ and χ'/m^* continuous). In the limit of infinitely thick barriers the subband dispersions become flat and centered at the isolated quantum well bound states. The latter are the solutions of :

$$\cos(k_w L_w) - 1/2(\xi - 1/\xi)\sin(k_w L_w) = 0 \qquad (10)$$

For thick but finite barriers the variations of $f(\varepsilon)$ with ε are very large. This is because of the exponentially varying terms $\exp(\kappa_b L_b)$. These exponential variations are suppressed at the energies E_n corresponding to the isolated quantum well bound states where $f(\varepsilon)$ is nearly zero (to the extent that exp($\kappa_b L_b$) is very large and $\exp(-\kappa_b L_b)$ is negligible). Hence , in the vicinity of E_n , one may expand $f(\varepsilon)$ and find:

$$\cos(qd) = (\varepsilon - E_n) f'(E_n) \qquad (11a)$$

or $$\varepsilon = E_n + \Delta_n/2\cos(qd) \qquad (11b)$$

which is the so - called tight-binding form. Here $|\Delta_n|$ is the width of the nth subband. The Δ_n 's are alternatively negative and positive in such a way that the minimum of the miniband in the superlattice Brillouin zone alternates from $q = 0$ (ground, third,..) to $q = \pi/d$. The tight-binding form can also be obtained by direct diagonalisation of the periodic hamiltonian in the basis spanned by the isolated quantum well bound state eigenfunctions centered at the lth period $\phi_n(z - ld)$. To comply with the Bloch form one would write:

$$\chi_{nq}(z) = (2N+1)^{-1/2} \sum_l e^{iqld} \phi_n(z - ld) \qquad (12)$$

where $2N + 1$ is the (macroscopic) number of unit cells in the superlattice and the summation over l runs from $-N$ to $+N$. The tight-binding form comes from the assumption of weak coupling between the levels centered at different sites, in such a way that one may neglect all matrix elements of the periodic hamiltonian except those which are diagonal in the l index or those between nearest neighbour sites. This analysis leads to:

$$\Delta_n = -4 V_b \int_{0^{th}\ \mathrm{well}} \phi_n(z)\phi_n(z - d)dz \qquad (13)$$

which can be either negative (subbands derived from even parity quantum well states) or positive (subbands derived from odd parity quantum well states). Figure (3) shows the calculated bandwidth of the ground conduction subband in GaAs - Ga(Al)As superlattices for three GaAs slab thicknesses and versus the Ga(Al)As barrier thicknesses. One clearly see that for a given well thickness the bandwidth decreases exponentially with the barrier thicknessas as evident from eq.(13) where only the exponentially decaying tail of $\phi_n(z - d)$ in the 0th well appears in Δ_n. $|\Delta_n|$ increases with n: the tunnel coupling is more efficient when κ_b decreases, i.e. when ε approaches V_b.

Note that the $\phi_n(z - ld)$ are orthogonal for different subband index but only when centered at the same site index l. When centered at different sites they are not orthogonal. To remedy to this deficiency one may define functions $w_n(z - ld)$, called Wannier functions[10], which form an orthonormal set:

$$< w_n(z - ld) \mid w_m(z - l'd) > = \delta_{nm} \, \delta_{ll'} \tag{14}$$

The Wannier functions are localized in space (since they appear as the discrete Fourier transform of delocalized wavefunctions: the $\chi_{nq}(z)$'s). Thus, they are particularly useful to derive formal properties. They are difficult to use practically however (since they are so hard so compute from their definition). In practice, we shall often neglect the difference between the $w_n(z - ld)$'s and the $\phi_n(z - ld)$'s at the end of the calculations. This approximation is tolerable for the ground subband (n = 1) in the tight binding limit.

It is worth stressing the apparently sigular behaviour of the eigenstates versus the interwell coupling. Notice that the band dispersion, i.e. the delocalized nature of the eigenstates persists whatever faint the inter-well coupling can be. This is not the behaviour one is used to in quantum mechanics, where often if a perturbation is weak enough the perturbation calculus to the lowest order provides a correct description, with perturbed eigenstates which are of the same nature as the unperturbed ones. Here, the unperturbed states, the isolated quantum well bound states, are spatially localized around one particular cell of the superlattice. The band picture provides us, again irrespective of the inter - well coupling (as quantified by the subband width), with delocalized Bloch states: the integrated probability to find the carrier in a particular period l is independent of l. Thus, the inter - well coupling cannot be treated by conventional perturbation schemes. The answer to this difficulty comes from the fact that in a perfect superlattice the unperturbed (by tunnel coupling) energy levels of the isolated wells are degenerate: whatever the unit cell the E_n and $\phi_n(z - ld)$ are all the same. Thus the basic criterion for non degenerate perturbation calculus , which is that the matrix elements of the perturbation between the state one perturbs and the other states are small compared to their energy separation, fails in a situation where all the unperturbed levels are at the same energy. Degenerate perturbation, i. e. diagonalization of the perturbation in the subspace of the degenerate levels, are perfectly suited to our goal. This is in fact one does explicitly in the tight-binding scheme where a suitable linear superposition of the unperturbed states is found which diagonalizes the full hamiltonian. Note in particular that the coefficients of this linear expansion do not involve the strength of the coupling, another signature of the non perturbative aspect of the inter - well coupling.

More than a technical remark, we believe this point may be useful to appreciate a variety of phenomena such as localization by disorder or by an electric field in superlattices. If the levels of the various quantum wells become misaligned by any reason, then for small enough inter - well coupling it will be liable to treat its effect on the level of the nth well by nondegenerate perturbation and starting from a localized wavefunction one will end up with a perturbed wavefunction which will also be localized.

2 - STARK EFFECTS IN SEMICONDUCTOR QUANTUM WELLS

The previous paragraph has been devoted to a brief presentation of the electronic structure of heterolayers under flat band condition. Here, we consider the case where an external and constant electric field is superimposed to the band edge profile of a single quantum well structure. We confine our discussion to a theoretical description of Stark shifts. In these structures the field leads to a red shift of the ground electron and hole states and thus of the fundamental band-to-band transition. In quantum wells there is a field-induced polarization of the bound eigenstates and thus a nearly quadratic energy shift due to the interaction of the induced dipole with the field[1,4,7].

The Stark effect in isolated wells have stimulated a significant amount of academic researches. They are also at the heart of novel opto-electronic devices (e.g. fast electro-optical modulators[7]). The usefulness of these structures lays in their capability of standing large electric fields ($F \leq 10^5$ V/cm), which produce large Stark shifts, while still displaying quasi discrete bound states.

Let us assume that the field is uniform over the whole structure, as approximately realized when the quantum well is inserted in the intrinsic part of a reverse-biased p-i-n junction or in the depletion length of a reverse-biased Schottky diode. In the case of multiple quantum wells, the barrier separating two consecutive wells will be assumed thick enough to prevent any sizeable coupling between their eigenstates. Let us first recall the characteristics of the classical motion. For energies ε smaller than the lowest lying barrier (i.e. the the left hand side one in the case of a positive electric field) the classical motion is bound, the electron can only be found between $-L/2$ and $+L/2$ where L is the well thickness. Since the potential energy has a step discontinuity at the interfaces, the force experienced by the particle (besides the electrical force $-eF$, where e is the magnitude of the electronic charge) is a percussion one at $z = \pm L/2$ (: the force is the gradient of the potential energy, hence it is the derivative of a step function, i.e. it is proportional to a Dirac function $\delta(z \pm L/2)$ at the interfaces). Once this is noticed the conservation of the mechanical energy $1/2m^*v_z^2 + eFz = \varepsilon$ allows to derive the period T_c of the classical motion:

$$T_c = 2(m^*/2)^{1/2} \int (\varepsilon - eFz)^{(-1/2)}\, dz \qquad (15)$$

where the integration is between the two classical turning points where the particle reverts its tracks. We choose the electrostatic potential zero at the center of the well and measure the energies with respect to this zero. Hence the possible energies for bound classical motions are such that $-eFL/2 \leq \varepsilon \leq V_b - eFL/2$. If $-eFL/2 \leq \varepsilon \leq eFL/2$ the two turning points are $-L/2$ and $z_t = \varepsilon/eF$ while if $\varepsilon \geq eFL/2$ the two turning points are $\pm L/2$. In the former case there is a velocity jump at $-L/2$ while the velocity vanishes at the regular turning point z_t. In the latter case the velocity has a jump at both ends $\pm L/2$. Working out the integral for the classical period, one finds easily that:

$$T_c = (4/eF)\,[m^*(\varepsilon + eFL/2)/2]^{1/2} \qquad \text{if } -eFL/2 \leq \varepsilon \leq eFL/2$$

$$(16)$$

$$T_c = (4/eF)\,(m^*/2)^{1/2}\,\{(\varepsilon + eFL/2)^{1/2} - (\varepsilon - eFL/2)^{1/2}\} \qquad \text{if } \varepsilon \geq eFL/2$$

It is worth noticing that the classical period is not a monotonic function of the energy but instead exhibits a maximum value when $\varepsilon = eFL/2$. To explain this feature we note that at very large energies compared to eFL (but still compatible with our requirement of a bound motion) the period goes to zero because the electrostatic potential energy becomes negligible with respect to the kinetic energy and therefore, the motion is analogous to that of a free particle bouncing between two walls separated by a distance L . The period T_c of the latter motion decreases owing to the speed increase of the particle with the energy (like $\sqrt{\varepsilon}$). On the other hand, on the low energy side the motion becomes more and more sensitive to the electrostatic potential energy. The period now decreases when ε approaches $- eFL/2$ because the spatial extent $L/2 + z_t$ of the classical motion shrinks to zero with the energy.

To study the quantum motion we have to investigate the solutions of:

$$[- \hbar^2/2m^* d^2/dz^2 + V_b(z) + eFz] \chi(z) = \varepsilon \chi(z) \tag{17}$$

In eq. (17) the in-plane motion has been dropped and band mixing effects have been neglected, $V_b(z)$ is the potential energy profile of a single rectangular quantum well. A constant electric field leads to pathological behaviour of the eigenstates of eq.(17) versus F. At zero field the Schrödinger equation admits at least one bound state ($\varepsilon < V_b$) while an arbitrarily small F is sufficient to transform the allowed energy spectrum into a continuum. This is because the potential energy is arbitrarily large and negative at large and negative z. This situation is not specific of quantum well physics but is found in many other instances (e.g. the hydrogen atom in an electric field)

Despite the lack of true bound states we expect, if F is not too large, that there will exist in the continuous spectrum some particular energies where the carrier wavefunction piles up in the quantum well. Moreover, these particular energies will smoothly extrapolate to the quantum well true bound states when $F \rightarrow 0$. In fact, we do know from experiments that, over a significant field range (typically $F \leq 10^5$V/cm), the quantum well structures support states which behave as if they were truly bound. Thus, it is worth trying to convince ourselves that some stationary eigenstates of eq.(17) are indeed peculiar in that they display an accumulation of their wavefunctions in the well. An alternative description, which is often more revealing, is to consider them as metastable (i.e. non stationary) solutions of the time dependent tilted quantum well problem. This kind of description is relevant when the decay time of the quasi bound states is long. Hereafter, we shall also denote the peculiar stationary solutions of eq.(17) as the metastable state, since it can be shown that the real part of the complex eigenenergies of the time-dependent problem coincide with the (real) energies of the peculiar solutions of eq.(17).

Apart from the exact, Airy-like solutions, there are various ways to find the metastable states of eq.(17). Firstly, one may cut the electric field at some large distance from the investigated quantum well and impose the existence of an infinite barrier. The problem becomes that of finding true bound states in a complicated band edge profile. The outcome of such calculations is that there exists a large number of states which are essentially localized in the large triangular well and show a very small probability of being in the well. A small number of states are found to display an enhanced probability of being in the well. These are the metastable states whose energy positions extrapolate smoothly to those of the zero field bound states of the well.

One may also cut the electric field at some large distances on both sides of the well and investigate the transmission coefficient $T(\varepsilon) = |t(\varepsilon)|^2$ of a plane wave with unity amplitude impinging at $z = -\mathbb{L}$ on the barrier and being finally transmitted at $z = +\mathbb{L}$. $T(\varepsilon)$ is very low (because $\mathbb{L}$ is large) except in the vicinity of some energies which belong to the segment $[h_1, h_2]$ where it displays sharp peaks ($T \leq 1$). As usual, these transmission resonances correspond to the trapping of the particle inside the the quantum well[6]. If the resonances are narrow this means that the corresponding trapping times τ_{trap} are long : $\tau_{trap}\,\Delta E > \hbar/2$ where ΔE is the width of the resonance.

Another way to depict these resonances is to consider them as virtual (or metastable) bound states, i.e. as bound states of the quantum well with a complex energy $E_n - i\hbar/2\tau_{trap}$. The reason why the energies of these states have to be complex and not purely real is that the solutions of the Schrödinger equation eq.(17) with the boundary conditions corresponding to a piling up of the wavefunction at $t = 0$ in the quantum well and to an outgoing wave at $z = -\infty$ do not fulfill the requirement of the probability current conservation, i.e. are not stationnary. On the other hand, it can be shown that such solutions are approximate eigenstates of eq.(17) provided their energies are complex.

Clearly, the metastable nature of the bound states in a biased quantum well comes from the non zero tunnelling probability across the left hand side (right hand side) barrier for conduction (valence) electrons. In order to set up a criterion allowing us to neglect, in a first approximation, the escape of the particle outside the quantum well, the condition

$$eF\kappa_b^{-1} << V_b - E_1 \tag{18}$$

was proposed some time ago[8], where E_1 is the confinement energy in the quantum well and κ_b^{-1} the characteristic decay length of the bound state wavefunction at zero field. The meaning of eq.(18) is that the lowering of the barrier by the field over the characteristic decay length of the zero field wavefunction must remain negligible with respect to the effective barrier height $V_b - E_1$ at $F = 0$.

One can derive a similar criterion by looking at the problem from a different angle. Suppose we know that we have built up at $t = 0$ a quasi bound state E_1, which, if the left hand side (l.h.s.) barrier were flat, would be truly bound. Correspondingly, the particle would (classically) oscillate back and forth in the well for ever with a period $T_c(E_1)$ where $T_c(\varepsilon)$ has been evaluated previously. Since, however, the l.h.s. barrier is tilted we know that the electron which oscillates in the well progressively escapes to infinity. Now, on a semi - classical basis, the probability that a carrier impinging the l.h.s. barrier with energy E_1 escapes towards infinity is simply $D(E_1)$, where $D(E_1)$ is the transmission of the tilted barrier. The escape frequency $1/\tau_{esc}$ will therefore be the number of collisions on the l.h.s. barrier per unit time $1/T_c(E_1)$ times the probability that the particle be transmitted through that barrier $D(E_1)$:

$$1/\tau_{esc} = 1/T_c(E_1) \times D(E_1) \tag{19}$$

In the semiclassical approximation $D(E_1)$ is given by:

$$D(E_1) = D_0 \exp\{ -(4/3eF)\, (2m^*/\hbar^2)^{1/2}\, (V_b - E_1)^{3/2} \} \tag{20}$$

with $D_0 \approx 1$. The lifetime $\tau_{esc}(E_1)$ will be long if the argument of the exponential in eq.(20) is large and negative, which means :

$$V_b - E_1 >> 3/4\, eF\kappa_b^{-1} \tag{21}$$

which (apart from the factor 3/4) is just the criterion we derived previously on the smallness of the barrier lowering by the field over the distance κ_b^{-1} where the wavefunction is important in the barrier. Note that the criterion physically means that $\tau_{esc}(E_1) >> T_c(E_1)$ or, in other words, that the particle oscillates many times in the well before escaping. This is realized if the triangular barrier is opaque ($D << 1$).

Assuming that the lower bound for >> is 10 and taking $V_b - E_1 = 0.125$ eV, $m^* = 0.07m_0$ we find that eq.(21) is satisfied if $F < 7.93 \times 10^4$ V/cm. If, in addition, $L = 100$ Å, $D_0 \approx 1$ and $E_1 = 70$ meV, we get $\hbar/2\tau_{esc}(E_1) \approx 2.5\ 10^{-4}$ meV and $\tau_{esc}(E_1) \approx 1.3$ ns. For these typical sample parameters the level broadening is still small with respect to the confinement energy, or in semi - classical terms the classical period is short compared to the escape time . Notice that having well defined quantum levels does not necessarily imply that an optical experiment will reveal them. In particular, photoluminescence experiments (spontaneous emission of photons through a radiative transition) may be deceiptive in biased quantum wells. This is because the magnitude of the luminescence signal is not directly related to the density of states but rather depends on the ratio τ_r / τ_{nr} of the radiative to the non radiative lifetimes. The latter, in good structures, can be assimilited to τ_{esc} but the former is considerably longer than the classical period T_c. For the previous example the escape lifetime is relatively long but not much longer than a recombination lifetime ($\tau_r \approx 1$ ns). Notice finally the strong dependence ($F \exp(-F_c/F)$) of τ_{esc}^{-1} with the field. It arises from the transparency coefficient eq.(20). If instead of applying 8×10^4 V/cm we only apply 50 kV/cm, $\tau(E_1)$ increases by several decades and the field-induced tunnelling becomes a completely negligible effect with respect to recombination or capture phenomena. Figure (4) shows the calculated τ_{esc} versus F for E_1, HH_1 and LH_1 in GaAs-Ga(Al)As quantum wells with thicknesses 30 Å, 60 Å and 90Å. τ_{esc} is found to vary considerably (from 1ps to 1s) in the investigated field range. Moreover, fig.(4) clearly shows that the field-induced escape of the carrier outside the quantum well is practically negligible when $L \geq 100$Å.

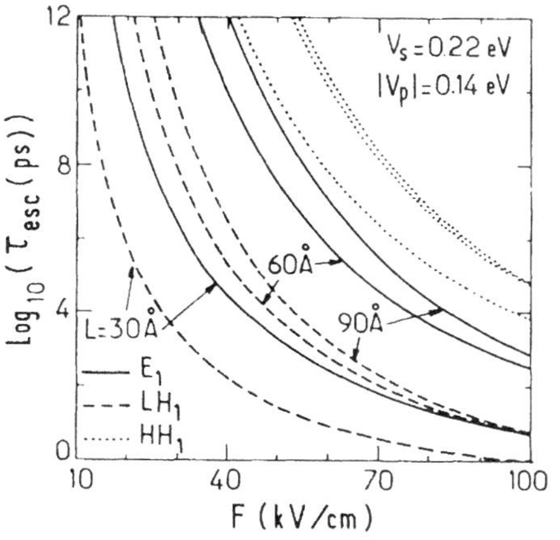

Fig.(4): Semi-classical estimates of the escape time of an electron (E_1), a heavy (HH_1) and light (LH_1) hole in biased GaAs-Ga(Al)As quantum wells with L = 30Å, 60Å and 90Å. $k_\perp = 0$ for holes.

Most of the quantum well structures which have been so far investigated (e.g. in electro - absorption) have shown excellent performances (i.e. pronounced exciton peaks) in field up to $\cong 10^5$ V/cm.The previous considerations have aimed to point out why it could be so in spite of the field-induced tunnelling. From now on we shall neglect this effect and investigate the second useful feature of the electric field effects on quantum well states, which is the existence of large Stark shifts.

The zero-field eigenstates of eq.(17) may be classified according to the parity operator. Since the eFz term is odd in z it only couples the zero-field eigenstates of opposite parities. Since we forget about field-induced tunnelling we may have recourse to perturbation (second order) or variational approaches to calculate the field dependence of the eigenenergies. The basic physics is that the quantum well bound states become polarized by the electric field which in turn lead to a shift of their eigenenergies by $-(1/2)\, D_n \cdot F$ where D_n is the average value of the dipole operator in the nth state. In the low field limit $D_n = \alpha_n F$ where α_n is a c-number. This results in a quadratic (Stark) energy shift upon F. For larger fields α_n becomes F-dependent displaying some saturation : $|D_n| \approx eL/2$. This is the carrier accumulation regime where the wavefunction piles up near the l.h.s. of the well for electrons and the right hand site (r.h.s.) for holes. This carrier accumulation is the useful feature for electro-optical devices : by inhibiting the carrier escape towards $\pm\infty$ (which is unavoidable in bulk materials) the quantum well walls give access to large ($\approx$ 30 meV) energy shifts while preserving the carrier localization and thus the beneficial action of enhanced excitonic binding . This effect has led to a number of electro-optical devices, e.g. fast modulators[7].

In the small field limit we can use a second order perturbation approach to obtain:

$$E_n(F) = E_n(0) + e^2F^2\Sigma_{m\neq n} |<\phi_m |z| \phi_n>|^2/(E_n-E_m) \quad (22)$$

where $|\phi_n>$ is the nth zero field eigenstate with energy $E_n = E_n(0)$ and where the summation over m runs over the zero field bound and unbound states. The latters give a small contribution and are usually neglected. From eq.(22) we see that the ground state (n=1) experiences a red shift, as always. Since $E_n - E_m \approx L^{-2}$ and $< n | z | 1 > \approx L$, $E_1(F) - E_1(0)$ scales like $L^4F^2m^*$. But the domain of validity of eq.(22), which is that the field-induced shift remains small with respect to the unperturbed energy splittings, narrows in a way that m^*FL^3 = constant. At the same approximation the perturbed wavefunctions are:

$$|\chi_n> = |\phi_n> + eF\Sigma_{m\neq n} <\phi_m |z| \phi_n>|\phi_m>/(E_n - E_m) \quad (23)$$

The average induced dipole (as measured from the well center) in the nth perturbed state is:

$$D_n = -e <\chi_n |z| \chi_n> = -2e^2F\Sigma_{m\neq n} |<\phi_m |z| \phi_n>|^2/(E_n-E_m) \quad (24)$$

as expected from the previous reasoning.

Once the field is too large to use eq.(24), one may use variational approaches. A linear variational treatment consists of expanding $\chi(z)$ on the uncomplete basis spanned by the zero-field bound eigenstates of eq.(17). In this way, one obtains the field dependences of all the bound states. If one is interested in the ground state only (as often in device applications), a non linear variational wavefunction like :

$$\chi(z) = \phi_1(z) \exp(-\beta z) \quad (25)$$

is quite accurate, as it describes the tendency towards accumulation ($\beta > 0$ for electron; $\beta < 0$ for holes), and of simple use. In eq.(25) $\phi_1(z)$ is the ground bound solution of eq.(17) at zero field. The wavefunction given in eq.(25) also contains the implicit signature of significant field-induced tunnelling : as $|\beta|$ increases

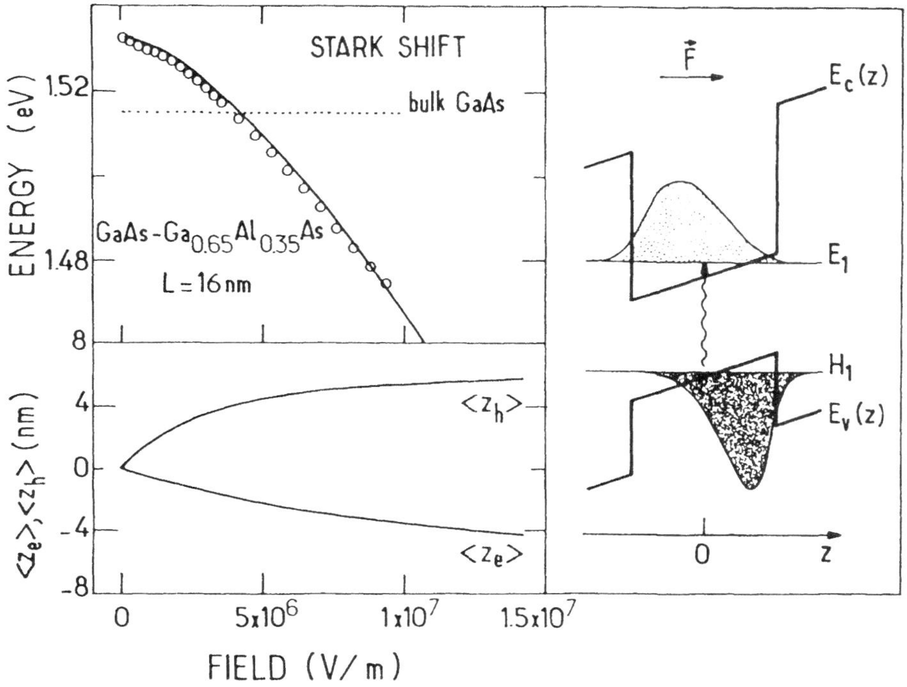

Fig.(5) A summary of the Stark effect in quantum wells. Right pannel: schematic representation of the conduction and valence band edge profiles of an isolated quantum well in presence of an external electric field and the envelope functions associated with the ground electron and hole states. Left upper pannel : variation with the electric field strength of the calculated (solid line) and measured[9] (open cercles) inter-band transitions energies in a 16 nm GaAs-Ga(Al)As quantum well. Left lower pannel : calculated mean values of the electron and heavy-hole positions in the biased well versus the electric field strength.

with F it happens that it becomes larger than $\kappa_b(0)$ the zero-field wavevector characterizing the evanescent wing of the ground bound state. The minimization procedure becomes impossible and one may rightfully consider that the very notion of quasi discrete bound state fades away. A summary of the Stark effect in quantum wells is depicted in fig.(5) where we have presented the calculated and measured transitions energies in a quantum well together with the calculated field dependence of the mean positions of the electron and the hole in the quantum well. Generally speaking there is a good quantitative understanding of the Stark effect in GaAs quantum wells.

The electric field-induced polarization of the carrier wavefunctions suppresses their parity properties. Thus, optical transitions which were parity-forbidden at zero field become allowed at non vanishing F. Their growth occurs at the expense of the F=0 parity-allowed transitions. Miller et al[7] have very nicely discussed the sum rules associated with optical transitions in biased quantum wells as well as the progressive evolution of the optical absorption lineshape from that of biased quantum wells to the Franz-Keldysh effect in thick, bulk-like, structures ($L \geq 500$Å).

3 - ELECTRIC FIELD EFFECTS IN DOUBLE QUANTUM WELLS

Double quantum wells (DQW) are formed by two wells separated by an intermediate barrier layer, these three layers beeing confined by two thick (symmetrically disposed) terminal barriers. These systems are of interest since thei correspond to the simplest semiconductor heterostructures presenting (i) an inter-well coupling of the eigenstates and (ii) field induced localization and/or delocalization of the wavefunctions within the two wells in presence of an external electric field. Let us consider firstly the zero field eigenstates of a DQW. In a tight-binding approch, each quantum well contributes with N_w bound states which interact due to the finite intermediate barrier thickness. When only the two levels at the same energy are taken, the tight-binding eigenvalue problem reads :

$$\begin{bmatrix} E_i - \varepsilon & \lambda_i \\ \lambda_i & E_i - \varepsilon \end{bmatrix} \qquad (26)$$

where λ_i is the nearest neighbour transfer integral between the wells corresponding to the i-th isolated well states (energy E_i). In eq.(26) we have taken for simplicity the two isolated wavefunctions as orthogonal. The wavefunctions of the DQW are then the symmetric ($\varepsilon_{i,S} = E_i - |\lambda_i|$) and antisymmetric ($\varepsilon_{i,A} = E_i + |\lambda_i|$) linear combinations of the two isolated wells wavefunctions. For large wells ($N_w>1$) the tight-binding matrix becomes more difficult to deal with when all the inter-well tunnel couplings (and not only the resonant ones) are taken into account. In any case, however, the wavefunctions have a definite parity , due to the spatial eveness of the DQW potential profile ($V_{DQW}(-z) = V_{DQW}(+z)$) with respect to the center of the intermediate barrier: z=0.

The eigenstates of biased double quantum wells can accurately be obtained by diagonalizing the eFz term between all the zero field bound states of the problem. Let E_{iS} and E_{iA} be the exact zero field DQW energies and $\varphi_{iS}(z)$ and $\varphi_{iA}(z)$ the corresponding wavefunctions. Let us consider a DQW structure with four bound states (i=1,2). In this basis the diagonalization of the electrostatic potential gives:

$$\begin{bmatrix} E_{1S}-\varepsilon & eF(z)_{1S,1A} & 0 & eF(z)_{1S,2A} \\ eF(z)_{1A,1S} & E_{1A}-\varepsilon & eF(z)_{1A,2S} & 0 \\ 0 & eF(z)_{2S,1A} & E_{2S}-\varepsilon & eF(z)_{2S,2A} \\ eF(z)_{2A,1S} & 0 & eF(z)_{2A,2S} & E_{2A}-\varepsilon \end{bmatrix} = 0 \qquad (27)$$

since the eFz potential can couple only states of different z-parity. If one neglects the off-diagonal blocks, which amounts to performing two diagonalization separately, one gets immediately:

$$\varepsilon_{1,\pm} = (E_{1A}+E_{1S})/2 \pm \sqrt{\{[(E_{1A}-E_{1S})/2]^2 + [eF(z)_{1S,1A}]^2\}}$$

$$\varepsilon_{2,\pm} = (E_{2A}+E_{2S})/2 \pm \sqrt{\{[(E_{2A}-E_{2S})/2]^2 + [eF(z)_{2S,2A}]^2\}} \qquad (28)$$

At low fields the four DQW levels present a quadratic variation with the field (in particular the ground state presents a quadratic red shift at low fields, like in a single quantum well ; see also below), whereas a linear variation appears at high fields. Note that the high field slopes are of opposite sign for the two ($i,\pm$) states with same i, but they differ in magnitude for i=1 and i=2. For thick barriers, however, they are very similar. Indeed, in this case we can approximate :

$$\varphi_{i,S}(z) \approx (1/\sqrt{2})\,[\,\phi_i(z-d/2) + \phi_i(z+d/2)]$$

$$\varphi_{i,A}(z) \approx (1/\sqrt{2})\,[\,\phi_i(z-d/2) - \phi_i(z+d/2)] \tag{29}$$

and show that $(z)_{1S,1A} \approx (z)_{2S,2A} \approx d/2$. Also, in that tight-binding limit we have $(E_{iA}+E_{iS})/2 \approx E_i$ and $(E_{iA}-E_{iS})/2 \approx |\lambda_i|$. Thus eqs.(28) becomes

$$\varepsilon_{1,\pm} \approx E_1 \pm \sqrt{\{\,|\lambda_1|^2 + [eFd/2]^2\,\}}$$

$$\varepsilon_{2,\pm} \approx E_2 \pm \sqrt{\{\,|\lambda_2|^2 + [eFd/2]^2\,\}} \tag{30}$$

If eFd exceeds the coupling terms between the wells (λ_1 for the ground state, λ_2 for the excited state), the tunnel effect between the wells becomes effectively non-resonant and thus weakens, to the extent that if F becomes very large there is no more coupling between the wells : the eigenenergies respectively converge towards $E_1 \pm eFd/2$ and $E_2 \pm eFd/2$ while their associated wavefunctions converges towards $\phi_1(z \pm d/2)$ and $\phi_2(z \pm d/2)$ respectively. Thus, the physics behind eqs.(28,30) is the field induced turning off of the zero field resonant tunnel coupling between the wells and the concomitant localization of the eigenstates.

The second important physical feature we would like to stress arises from the effect of the off-diagonal blocks on the eigenstates. When these blocks are absent there exists a crossing between ε_{1+} and ε_{2-} taking place at $F=F_c$ (where $e|F_c|d \cong (E_2 - E_1)$ if $e|F_c|d/2 >> |\lambda_1|, |\lambda_2|$). This crossing is actually replaced by an anticrossing when the off-diagonal blocks are taken into account. If the field was strong enough to have localized both the E_1 and E_2 related states (i.e. if $e|F_c|d/2 >> |\lambda_1|, |\lambda_2|$) in either wells, the eigenstates again delocalize in a field range ΔF around F_c which strongly depends on the intermediate barrier thickness: the thicker the barrier the smaller are the anticrossing and the departure of the eigenstates from a linear dependence upon F. In the field range $F_c \pm \Delta F/2$ a carrier assumed to be in the upper branch will make a very efficient relaxation towards the lower branch by emitting acoustical or optical phonons (energy conservation permitting) or by being elastically scattered in the lower branch and converting some of its longitudinal energy into transverse kinetic energy, this process being followed by a subsequent intra-lower branch desexcitation. Finally, the off diagonal terms induce a deviation from a straight line dependence at large F ($F>>F_c$) of the eigenstates which is quadratic upon F. This is nothing but the intra-well Stark effect (note that the quadratic variation at low fields is associated mostly to an inter-well effect). Figure (6) displays the calculated results of the field dependent energies for a 100Å-20Å-100Å GaAs- $Ga_{0.7}Al_{0.3}As$ symmetric double quantum well (barrier height: 213 meV, $m^{*}= 0.07m_0$). The electrostatic potential is taken to vanish at the center of the middle barrier and the energy zero is taken at the bottom of the GaAs conduction band edge at F = 0. The states moving up (down) in energy are those which are mostly localized in the right hand side (left hand side) well. When F increases, the eigenstates become progressively localized in a given well, except near F_c where an anticrossing take place between ε_{1+} and ε_{2-}. When this anticrossing is passed through there is an interchange between the main spatial localizations of the second and third states respectively.

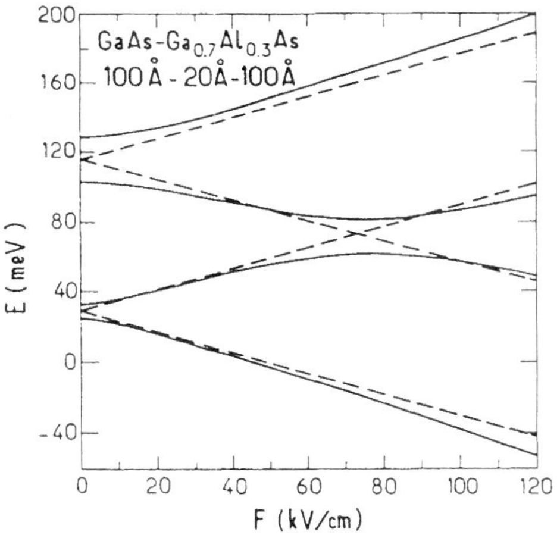

Fig.(6) Calculated conduction eigenstates (solid lines) of a biased 100Å-20Å-100Å GaAs-Ga(Al)As double quantum well versus the field strength F. The dashed lines are the strong field asymptotes of the decoupled E_1 and E_2 levels in each well.

4 - ELECTRIC FIELD EFFECTS IN SUPERLATTICES

One of the original motivation for the growth of semiconductor superlattices was the hope of achieving a negative differential resistance by realizing a Bloch oscillator. The Bloch oscillator, suggested in 1927, results from a semiclassical analysis of the electron motion in a potential which is the sum of a periodic term V(z) and a linearly varying term eFz. F is considered as so small that its effects can be treated semi-classically on the band structure defined by V(z). The fully quantum-mechanical version of the Bloch oscillator was derived by Wannier in the late 50's[10]. Wannier showed that a one band analysis of the problem leads to the replacement of the quasi continuous band spectrum of the crystalline solid (F=0) by a ladder of bound levels evenly spaced by eFd where d is the period of the lattice along the electric field (Wannier-Stark ladder)[10]. Wannier's results were long disputed on the ground that the electric field-induced interband mixing (Zener breakdown) would unavoidably wash out the ladders. Careful theoretical and numerical analysis however showed that Wannier's findings were essentially correct, the interband couplings leading to a broadening of the bound levels into virtually bound ones. For a recent review of theoretical aspects see Nanciu[11]. In actual bulk samples, however, it was quickly realized that the time needed to complete a period of the semiclassical bound motion was considerably longer than any realistic collision time on impurities, defects, phonons etc..., thus leaving only a faint hope to observe this fascinating effect. In fact, this is only recently that the Wannier - Stark ladders were experimentally evidenced; not in bulk materials but in semiconductor superlattices (see [12] for a recent introduction) where, as we shall see below, the electric field effects are considerably enhanced compared to the bulk case. In the following we shall present an introduction to the Wannier-Stark quantization in semiconductor superlattices. Efforts will be made to provide a comparison between the fully quantum treatment and the semiclassical one. We shall first discuss the simplest case which corresponds to the one band analysis made by Wannier[10] and which is applicable to a good approximation to the conduction subbands of superlattices whose wells, if they were isolated, would only support a single bound state. This section will also include a short discussion of the intraband and interband optical properties associated with the Wannier-Stark states. Multiband effects on the Wannier-Stark quantization will be discussed for electrons.

The Wannier-Stark Ladders

We consider an infinite semiconductor superlattice subjected to a constant electric field. For simplicity we assume that the conduction states are built out of a single parabolic band which has the same symmetry (say Γ_6) for both host materials. The hamiltonian is written as:

$$H = T + \Sigma_n V(z - nd) + eFz \qquad (31)$$

where V(z-nd) is the "atomic" square well potential felt in the superlattice period centered at z = nd. Consider the translation operators

$$T_{jd} = \exp(ip_z jd/\hbar) \tag{32}$$

We find:

$$[\,T_{jd}\,, H\,] = T_{jd}\,H - H\,T_{jd} = jeFd\,T_{jd} \tag{33}$$

Equation (33) states that the translation operators are no longer constants of motion if $F\neq0$, a result physically arising from the potential energy drop eFjd between two superlattice periods separated by jd. However, the fact that $[T_{jd}\,,\,H]$ is proportional to T_{jd} itself implies that this operator evolves harmonically with time. In particular, suppose that at time t = 0 the electron state was also an eigenstate of T_{jd} with the eigenvalue $\exp(ik_0jd)$. Then, eq.(33) shows that the electron wavefunction will remain an eigenstate of T_{jd} with the eigenvalue $\exp[ik(t)jd]$ where:

$$k(t) = k_0 - eFdt/\hbar \tag{34}$$

which is a rigourous justification of the "Newton" law of the semi-classical mechanics:

$$\hbar\; dk/dt = -\, dV_{ext}/dz = -\, eF \tag{35}$$

Notice that this law is exact only because the potential is linearly varying with z. Any other variation would not allow the generalization of eq.(35) to be exactly derivable.

When eq.(35) is used in conjunction with the definition of the velocity for a semi-classical electron:

$$v = (1/\hbar)\; \partial\varepsilon_n/\partial k \tag{36}$$

the semi-classical equations of motion becomes entirely prescribed. In eq.(36) $\varepsilon_n(k)$ is the dispersion relation of the nth subband (at zero field) and n, the subband index, is a constant of motion in the semi-classical analysis. Assuming a decoupling between the z and (x,y) motions, we obtain:

$$v_z(t) = \mathbf{v}.\mathbf{F}/F = (1/\hbar)\,\partial/\partial k_0[\varepsilon_n(k_0 - eFt/\hbar)] \quad (37)$$

which proves that both $v_z(t)$ and $z(t)$ are periodic function of time with a period T_B equal to:

$$T_B = 2\pi\hbar/eFd = 2\pi/\omega_B \quad (38)$$

since $\varepsilon_n(q)$ is periodic upon q with periodicity $2\pi/d$. The oscillatory motion described by eqs.(37,38) is the Bloch oscillator and ω_B is the Bloch angular frequency. The Bloch oscillator is independent upon the exact shape of the dispersion relations $\varepsilon_n(q)$. To go one step further and obtain more explicit results, we restrict our attention to the ground subband $\varepsilon_1(q)$ derived from the ground bound state E_1 of the isolated wells. We adopt the simple tight binding expression for $\varepsilon_1(q)$:

$$\varepsilon_1(q) = E_1 - 2|\lambda|\cos(qd) \quad (39)$$

The subband width Δ_1 is thus equal to $4|\lambda|$. From eqs.(37) we obtain:

$$v_z(t) = (2|\lambda|d/\hbar)\sin[(k_0 - eFt/\hbar)d] \quad (40)$$

$$z(t) = z_0 + (2|\lambda|/eF)\cos[(k_0 - eFt/\hbar)d] \quad (41)$$

where z_0 is a constant of integration. The Bloch oscillator executes an harmonic motion with period T_B and amplitude $2|\lambda|/eF$ around its equilibrium position z_0. This oscillator is however peculiar in that its amplitude is unrelated to its total energy E_{tot} since E_{tot} is equal to $E_1 + eFz_0$ whereas for a regular harmonic oscillator (e.g. a mass m* attached to a spring) one knows that the amplitude a is related to E_{tot} by $E_{tot} = 1/2m^*a^2\omega^2$, where ω is the angular frequency of the oscillator. The difference between both kinds of oscillator ultimately rests on the difference between the dependence of their kinetic energies upon the momentum. While the conventional oscillator can display an arbitrarily large kinetic energy, the Bloch oscillator one is bound by $E_1 + 2|\lambda|$. The latter property is closely associated with the counterintuitive response of the electron to an external field in the crystalline solid. While our intuition associates a constant acceleration to the electron motion in a constant field, the electron in the solid actually oscillates. The band structure (or the Bragg reflection effect) is still overwhelming in the semiclassical equations and falsifies our intuitive representation. Notice finally that the amplitude of the Bloch oscillator blows up when F vanishes. This raises the question of the applicablity of eqs.(40,41) to real superlattice structures, which are finite. The finite size effects could be incorporated in eq.(35) by adding, say, infinitely repulsive potential walls corresponding to both ends of the structure. We immediately see that, for such a finite structure comprising 2N + 1 periods, one should get $|z_0 \pm 2|\lambda|/eF| < (N + 1/2)d$ to ensure that eqs.(40,41) are still relevant. This criterion is better fulfilled for oscillators centered as far as possible from the edges ($z_0 = 0$) or for strong fields. Otherwise, one should numerically solve eqs.(35,36) to get the "edge-interacting Bloch oscillator" motions.

Let us now examine the quantum treatment of eq.(31) restricting ourselves to a one band tight binding expansion. The general form of the spectrum can be obtained without any approximation. Suppose that ψ_ν is a solution of eq.(31) corresponding to the eigenvalue ε_0. Consider the function $T_d \psi_\nu(z) = \psi_\nu(z + d)$ and apply H on it. We find:

$$HT_d \psi_\nu(z) = (\varepsilon_0 - eFd)\, T_d \psi_\nu(z) \qquad (43)$$

which shows that $T_d \psi_\nu(z)$ is an eigenstate of H corresponding to the energy ε_0 - eFd. This procedure can be iterated as well as the application of T_d^+ ($= T_{-d}$) which translates the energies upward. Finally through the successive translations T_{jd}, j a relative integer, one generates the ladder:

$$\varepsilon_\nu = \varepsilon_0 + \nu eFd \;,\quad \nu \text{ a relative integer} \qquad (44)$$

Since we have restricted our analysis to a one band tight binding description of the problem there are as many available states as periods in the superlattice. Thus, the spectrum given by eq.(44) exhausts all the possible states of the biased superlattice and we may conclude that in the one band analysis the Wannier-Stark ladders are evenly spaced. Notice that the one band assumption is here crucial insofar as the unknown ε_0 in eq.(44) is necessarily unique. Had we included several bound states per well or, a fortiori, the continuum states, it would have been impossible to draw any conclusion on the even spacing of the level and, in fact, when several bound states are included in the calculations, the spectrum is unevenly spaced. What remains however is the existence of a periodicity in eFd on the energy scale because eq.(43) always holds, but the "elementary cell" of this periodic "lattice" instead of containing a single eigenvalue contains as many of them as the number of states per period which have been retained.

The one band result is called a Wannier-Stark ladder. In its derivation, Wannier was able to prove that ε_0 is actually the center of gravity of the band at zero field[10]:

$$\varepsilon_0 = d/2\pi \int_{BZ} \varepsilon_1(q)dq \tag{45}$$

which coincides with E_1 in the simplest tight binding scheme. The spacing between two consecutive eigenvalues is $\hbar\omega_B$ as one could have anticipated from the semiclassical result.

At this point it is useful to again derive eq.(44) by means of an expansion of $\psi_v(z)$ on the complete basis spanned by the Bloch functions $\psi_{nq}(z)$ of the crystal at zero electric field:

$$\psi_v(z) = \Sigma_{n,q}\, \Phi_{nv}(q)\, \psi_{nq}(z) \tag{46}$$

In this representation the Schrödinger equation is:

$$[\varepsilon_{nq} - \varepsilon + ieF\partial/\partial q\,]\Phi_{nv}(q) + eF\,\Sigma_{n'}\, X_{nn'}(q)\Phi_{n'v}(q) = 0 \tag{47}$$

$$X_{nn'}(q) = i\int_{\Omega} u^*_{nq}\, \partial/\partial q'\, u_{n'q'} \tag{48}$$

and the u's are the periodic part of the Bloch functions ($\psi_{nq}(z) = u_{nq}(z)\, e^{iqz}$). Notice that $X_{nn'}$ is diagonal in q. Thus, in the Bloch representation the electric field leaves the wavevector unchanged but admits an intraband term $ieF\partial/\partial q$ as well as interband ones $X_{nn'}$. The one band assumption amounts to keeping only $n = n'$ in eqs(47,48). This immediatly leads to:

$$\Phi_{n\nu}(q) = \Phi_0 \exp[- (i/eF)\int^{q} \{ \varepsilon - \varepsilon_{nq'} - eFX_{nn}(q') \} dq'] \qquad (49)$$

where the lower bound of the integral is arbitrary. Note the non analytical behaviour of the $\Phi_{n\nu}$'s versus F. From eq.(49) we see that there is no hope to retrieve the F = 0 (band) spectrum by a smooth extrapolation of the $F\neq0$ expressions. The Wannier quantization arises because of the periodicity of $\Phi_{n\nu}(q)$ in the reciprocal space. To be meaningful eq.(46) should be completed with the periodic boundary conditions $\Phi_{n\nu}(q + 2\pi/d) = \Phi_{n\nu}(q)$. This is because two Bloch states which differ by a reciprocal lattice vector are physically identical (in particular they correspond to the same eigenvalue of the translation operator). The periodicity condition on $\Phi_{n\nu}(q)$ leads immediatly to:

$$\varepsilon = \nu eFd + (d/2\pi) \int_{(q)}^{(q+2\pi/d)} [\varepsilon_{nq'} + eFX_{nn}(q')] dq' \qquad (50)$$

The afore mentionned periodicity ensures that the integral in eq.(50) is independent of q. Thus the Wannier - Stark spectrum eqs.(44,45) is recovered apart from the small F dependent term $eF <X_{nn}>$.

The Wannier-Stark spectrum recalls in many respects the harmonic oscillator one. There exist several differences however. Firstly, the Wannier-Stark spectrum has no ground state, which implies that the usual rules on the nodes of the eigenfunctions do not apply. In fact each eigenstate of the Wannier-Stark problem has an infinite number of nodes and the ψ_ν 's are all but the same wavefunction whose argument is translated by an integer times d (since $T_{jd}\, \psi_\nu = \psi_{\nu-j}$) when going from one state to the other. The equivalent of the raising (c^+) and lowering operator (c) of the harmonic oscillator are the translation operators T_d^+ and T_d respectively. However, T_d and T_d^+ commute while c and c^+ do not. This has some important implications when considering the perturbation of a Wannier-Stark ladder by a harmonically varying electric field, i.e. when one addresses the question of light absorption or emission by an electron moving in a linearly biased superlattice. We shall return to this point later on.

To know more about the actual shape of the ψ_ν's we expand the solutions of H on the basis spanned by the w(z-nd) which are the Wannier functions centered at z = nd:

$$\psi_\nu = \Sigma_n \, c_{n,\nu} \; w(z\text{-}nd) \tag{51}$$

We neglect couplings between all the neighbours but the nearest and absorb the shift integral into a redefinition of E_1 to finally end up with the secular equation:

$$(E_1 - \varepsilon_\nu + eFnd)c_{n,\nu} - |\lambda|(c_{n+1,\nu} + c_{n-1,\nu}) = 0 \tag{52}$$

where the identification of the $< w(z - nd)|\ eFz\ |\ w(z - md)>$ matrix elements to $eFnd\delta_{n,m}$ has been used. One recognizes in eq.(52) the recursion relations of the Bessel functions. If the superlattice is infinite the $c_{n,\nu}$'s should not blow up, which implies that the divergent Bessel functions (I_n) should be eliminated . This leaves us with:

$$c_{n,\nu} = J_{\nu-n}(-2/f) \tag{53}$$

$$\varepsilon_\nu = E_1 + \nu eFd \tag{54}$$

where f is the dimensionless strength of the electric field:

$$f = eFd/|\lambda| = 4eFd/\Delta_1 \tag{55}$$

In finite superlattices one should take the c_n's as linear combinations of the J_n's and I_n's and determine the eigenenergies by writing appropriate boundary conditions at the superlattice edges

Equations (51-55) call for several remarks. Firstly, the one band tight binding model provides a universal description of the electric field effects on a superlattice: one only has to scale eFd and ε_ν - E_1 to the bandwidth . This scaling allows us to readily understand why the Wannier-Stark quantization is far more easy to observe in superlattices than in the bulk materials: a typical bandwith in a bulk crystal is 2eV corresponding to a lattice periodicity of 6Å while the corresponding $\Delta_1$and d are typically 70 meV and 60Å in a superlattice. Assuming an electric field strength of 10^5 V/cm, we get $f_{bulk} \cong$

0.012 and $f_{sl} \cong 3.43$. In other words, the electric field has a small effect in a bulk crystal compared to that produced by the periodic potential while in a superlattice the two effects are comparable. Secondly, the replacement of the extended minibands spectrum by a set of discrete localized levels as evidenced by eqs.(54) calls for a physical interpretation. This is the field-induced turning off of the tunnel effect between consecutive wells which is at the heart of the Wannier-Stark quantization and which again explains why the semiconductor superlattices are ideal candidates for its observation. Consider the biased superlattice (or bulk material) as a collection of localized potential wells whose bound states are misaligned by the electric field and which are nearest neighbour coupled with an interaction strength $-|\lambda|$. By extension of the familiar H_2^+ molecular ion or the double well problems, we know that the tunnel coupling between the wells 0 and n will remain operative if the misalignement between the two levels neFd remains smaller than $2|\lambda|$. This means that the number of sites visited by an eigenstate ψ_ν on each side of the νth site is $2|\lambda|/eFd = 2/f$. Thus, for a typical bulk material, a Wannier-Stark state extends over 2x170 periods while for the same field strength ψ_ν hardly extends beyond the νth site in a superlattice. This is because the superlattice periods are larger that the minibands are narrower and that the misalignement of consecutive energy levels are bigger than the corresponding quantities in bulk materials that the tunnel coupling between the wells of the superlattice can be so effectively turned off. An extra benefit is gained by the 10 fold increase between consecutive eigenstates of the ladder, making the ratio $eFd/(\hbar/\tau)$ ($= \omega_B\tau$, where τ is the scattering time) 10 times longer in a superlattice than in a bulk material if the τ's are identical. Notice finally that the argument on the number of sites visited by an electron in the νth Wannier-Stark state on each side of the νth site leads to a result (2/f) which exactly coincides with the amplitude of the Bloch oscillator.

Another striking property of the Wannier-Stark states (when compared with our intuitive expectation) is the even probability distribution of the ψ_ν around the νth period (since $J_{n-\nu}(x) = (-1)^{n-\nu} J_{\nu-n}(x)$). In the one band approximation, there is as much probability for a Wannier-Stark state to be found in regions of lower than in regions of higher electrostatic potential energy, while one might have expected a piling up of the eigenstates where this energy is the lower.

To complete our survey of the one band analysis of the Wannier-Stark ladders let us summarize the comparison between the quantum and the semi-classical approaches (see also the appendix). We have already mentionned that the semi-classical motion occurs with an angular frequency ω_B which is exactly that expected from the quantum spacing $\hbar\omega_B$. Moreover, we found that a rough evaluation of the spatial extent of a quantum eigenstate is just equal to the amplitude of the Bloch oscillator. This agreement can be further quantified by calculating the root mean square deviation Δ_z of the particle position around its equilibrium position:

$$\Delta_z = [<(z-z_0)^2> - <z-z_0>^2]^{1/2} \tag{56}$$

where the symbols <..> either denote quantum averages over ψ_ν (with $z_0 = \nu d$) or over the period $2\pi/\omega_B$ of the semi-classical motion. We find:

$$[\Delta_z]_{Bloch} = 2d\sqrt{2}/f \tag{57}$$

$$[\Delta_z]_n = < \phi_0 \mid z^2 \mid \phi_0> + 2d\sqrt{2}/f \tag{58}$$

Both expressions are nearly identical in the limit of small f (where one indeed expects the semi-classical analysis to be the more valid). For strong fields the field dependent uncertainty decrease in both cases as both the semi-classical and quantum motions become increasingly restricted. This shrinking can reach a complete collapse in the case of the Bloch oscillator, since both the position and velocity of the Bloch oscillator can be defined with an arbitrary accuracy, whereas in the quantum case there always remains a residual uncertainty $< \phi_0 \mid z^2 \mid \phi_0>$ associated with the finite extension of the isolated well eigenstates in the well and barrier layers.

We have previously mentionned that the Wannier-Stark ladder has no ground state and that T_d^+ and T_d, the equivalent of the harmonic oscillators raising and lowering operators, commute. These two properties lead to an interesting consequence with respect to the response of the Wannier-Stark state to a sinusoidally varying electric field (parallel to the static one). Let the electromagnetic (e.m.) perturbation be described by $eF_{em}z\cos(\omega t)$. If F_{em} is small one can investigate by perturbation the transition rate that an electron in the νth state makes a transition to the μth one. This rate is proportional to

$|\langle \nu |z| \mu \rangle|^2$. By using the tight binding expansion (eq.(51)) we get:

$$\langle \nu |z| \mu \rangle = \Sigma_{ll'} \; c_{l,\nu} \, c_{l',\mu} \; \langle w(z - ld) \, |z| \, w(z - l'd) \rangle \tag{59}$$

By writing that $z = (z+ld) - ld$ and by neglecting any integral of the form $\langle w(z - ld) \, |z-ld| \, w(z - l'd) \rangle$ one finally gets:

$$\langle \nu |z| \mu \rangle = (\nu d)\delta_{\nu,\mu} + (d/f)\delta_{\nu,\mu\pm 1} \tag{60}$$

The diagonal term is of no relevance here since it leads to no absorption or emission of energy by the electron. The off-diagonal term displays the expected shape. Namely, a given Wannier-Stark state is coupled by the e.m. wave to its nearest neighbour. The key point is that the coupling is ν-independent, a feature clearly associated with the fact that all the Wannier-Stark states are characterized by wavefunctions which are isomorphical. If the transitions rates (α) between ν and $\nu\pm 1$ are the same, it is not very difficult to show that no net energy is absorbed by the electron due to its interaction with the oscillating electric field. Let f_ν be the (arbitrary) distribution function of the νth state. The net power absorbed due to the $\nu \rightarrow \nu\pm 1$ transitions is $\hbar\omega\alpha\, [f_\nu(1- f_{\nu+1}) + f_{\nu-1}(1- f_\nu) - f_\nu(1- f_{\nu-1}) - f_{\nu+1}(1- f_\nu)]$. This is equal to $\hbar\omega\alpha\, [f_{\nu-1} - f_{\nu+1}]$, i.e. does not depend on f_ν. Instead of considering only the $\nu \rightarrow \nu\pm 1$ transitions, one may calculate the net power absorbed due to all the transitions between the states $-N$, $-(N-1)$,..., $N-1$, N and find that it is equal to $\hbar\omega\alpha\, [f_{-N} - f_N]$. Thus, there is no bulk absorption. The only possible absorption inside a Wannier-Stark ladder is therefore due to edge effects, i.e. to the fact that any real superlattice is finite[9]. Notice that we have neglected the spontaneous emission in the previous analysis. Had we considered the net power absorbed by a regular harmonic oscillator (say a ladder of Landau levels) that the results would have been entirely different. The transition rates $n \rightarrow n+1$ ($n-1$) would have been proportional to $n+1$ (n), where n is the Landau level index, and the net power absorbed by the electron due to the $n \rightarrow n\pm 1$ transitions would have been proportional to $[(n+1)(f_n - f_{n+1}) + n(f_{n-1} - f_n)]$. The total power absorbed due to all the possible transitions would be proportional to $\Sigma_n (n+1-n) f_n$, i.e. to the concentration of electrons as it should for a bulk effect.

The Wannier-Stark states are thus of little interest for the intraband transitions. They display however rich electro-optical properties which are associated with interband transitions, i.e. to transitions between the valence and conduction ladders Let us only sketch here the salient features.

Consider the possible optical interband transitions in a finite but thick superlattice. The electromagnetic wave is assumed to propagate along the growth axis. The in-plane wavevector has to be conserved for allowed optical transitions in the dipole approximation. If there is no electric field the superlattice wavevector q is conserved as well. If the superlattice is biased, the electrostatic potential breaks the q conservation rule, but we have seen that the superlattice minibands are destroyed and replaced by the Wannier-Stark ladders. For single, parabolic and non degenerate conduction and valence bands, the optical matrix elements involving the coupling between the light and the electron breaks into:

$$M_{cv} = \langle u^c \mid \varepsilon \cdot p \mid u^v \rangle \langle \psi^{(c)} \mid \psi^{(v)} \rangle \tag{61}$$

where u^c (u^v) is the periodic part of the conduction (valence) zone center Bloch function, ε is the light polarization vector and $\psi^{(c)}$, $\psi^{(v)}$ are the envelope wavefunctions of the conduction and valence subbands, in our case the Wannier-Stark states $\psi_\nu{}^{(c)}$, $\psi_\mu{}^{(v)}$. Notice that although both the valence and conduction ladders are evenly spaced by the same eFd and parallel, the eigenstates $\psi_\nu{}^{(c)}$, $\psi_\nu{}^{(v)}$ are not the same, for their projection on the Wannier functions $w^{(c)}(z\text{-}nd)$, $w^{(v)}(z\text{-}nd)$, namely $J_{\nu-n}(-2/f_c)$, $J_{\nu-n}(2/f_v)$, are different due to the different conduction and valence subband widths. The overlap integrals $\langle \psi^{(c)} \mid \psi^{(v)} \rangle$ are thus, in general, non vanishing and all the possible transitions between ν and μ become allowed. This result, again, contrasts with the case of the magneto-absorption (i.e. the interband transitions between Landau levels) where, in spite of different ladder spacings ($\hbar\omega_c$ and $\hbar\omega_v$) the conduction and valence eigenstates are the same for a given n, which leads to the selection rule on the envelope functions $\Delta n = 0$.

It is not very difficult to calculate the transition rates for $\mid v, \mu\rangle \rightarrow \mid c, \nu\rangle$ transitions, to integrate over $k_\perp$, taking into account its conservation in the optical process and to finally end up with an absorption coefficient which, in the limit of a thick sample and for a type I superlattice, where electrons and holes are mostly localized in the same layer, is given by:

$$\alpha(\omega) = (2N+1)\alpha_0 \, \Sigma_p J_p^{\,2}(2/f_{cv})\, Y[\hbar\omega-(\varepsilon_g+E_1+HH_1+peFd)] \qquad (62)$$

where 2N+1 is the (large) number of periods in the superlattice, Y(x) the step function, ε_g the bandgap of the well-acting material, E_1 and HH_1 the confinement energies of the ground electron and hole states in the quantum wells when they are isolated, p a relative integer ($-N \leq p \leq +N$) and f_{cv} the dimensionless electric field strength:

$$f_{cv} = eFd/(|\lambda_c| + \lambda_v) \qquad (63)$$

Finally, α_0 is the absorption coefficient related to the $HH_1 \rightarrow E_1$ optical transition in the isolated quantum well ($\alpha_0 \cong 0.6\%$ in III-V quantum wells). We see that the absorption coefficient is just the sum of staircase absorption edges taking place at the energies $\varepsilon_g + E_1 + HH_1 + peFd$. These staircases have a simple physical explanation: they are associated with the oblique transitions in real space which promote an electron in the 0th valence Wannier-Stark state to the pth conduction one. These oblique transitions are symmetrically placed on the energy scale with respect to the p=0 vertical transition. There is therefore no absorption edge for the band-to-band transitions (if NeFd > $\varepsilon_g + E_1 + HH_1$). However, the oscillator strength of the oblique transitions sharply drop with p, to such an extent that they quickly become unmeasurably small. We illustrate these considerations in fig.(7) where the absorption coefficient of a 41 periods superlattice is plotted versus the reduced photon energy $\mathcal{E}$ ($\mathcal{E} = [\hbar\omega-(\varepsilon_g+E_1+HH_1)]/(|\lambda_c| + \lambda_v)$) for several field strengths f_{cv}. At zero field the unperturbed superlattice absorption coefficient is drawn and displays the well-known Arccosine shape. When f_{cv} increases, steps corresponding to the oblique ($p \neq 0$) and vertical (p=0) transitions develop. It is seen that their amplitudes are not monotonic functions of f_{cv} (because $J_p^{\,2}$ has an infinite

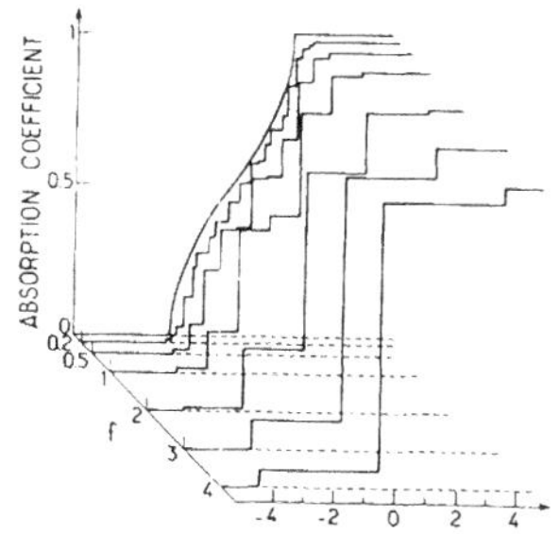

Fig.(7) Calculated band to band absorption in a biased 41 periods thick type I superlattice.

number of nodes). However, when $f_{cv} > f_{p}$, where $2/f_p$ is the smaller zero of J_p^2, the pth oblique transition fades away. For $f_{cv} > 4$ (which corresponds to a potential energy drop over a period equal to the sum of the conduction and valence subband widths), one is left with a dominant p = 0 vertical step (nearly 90% of the total absorption coefficient) and two small p = ±1 steps, evenly sharing the remaining 10% of the absorption. There is therefore an effective blue shift of the band-to-band absorption edge[13]. This blue shift can be quite large since $(\Delta_c+\Delta_v)/2$ can easily reach 30 meV. Notice that when the blue shift is significant, the well and barrier thicknesses are usually small (≅ 30Å), which implies that the intra-well Stark shift discussed previously is small (≅ 1-2 meV). On the other hand, the excitonic effects become larger with increasing f_{cv} because the structure continuously evolves from a quasi three dimensional material ($f_{cv} \approx 0$) to a quasi bi dimensional material ($f_{cv} \geq 4$), which enhances the exciton binding energy. Thus, the measurable blue shift Δ_{BS} is limited to:

$$\Delta_{BS} = (\Delta_c+\Delta_v)/2 - (R^*_{QW} - R^*_{SL}) - E_{SQW} \qquad (64)$$

where $R^*_{QW} - R^*_{SL}$ is the increase of the exciton binding energy when going from the coupled to the electric field-isolated quantum well situations. We show in fig.(8) an estimate of the periods where a useful blue shift can be obtained in a superlattice assuming equal well and barrier thicknesses. The criteria used to define the rectangle in fig.(8) are that $(1/2)\Delta_c$ should be larger than 10 meV and that the electric field required to achieve an almost complete Wannier-Stark localization (f = 4) should be smaller than 10^5V/cm. It is seen that a period d = 70 Å , which is not exceedingly demanding from the growth point of view, fulfills both criteria. The blue shift in a (Ga,In)As-(Ga,Al,In)As superlattice is also shown in fig.(8). Finally, fig.(8) presents a "fan" diagram (transition energies versus electric field strength) showing the field dependences of the vertical and oblique transitions in a GaAs-(Ga,Al)As superlattice. Oblique transitions from p=-5 up to p = +3 have been observed, which demonstrates both the relevance of the Wannier-Stark description of a biased superlattice and the fact that the coherence of the conduction states extends at least up to 7 periods (the hole states are quickly field-localized in a given period and, actually, act as markers of the conduction envelope functions).

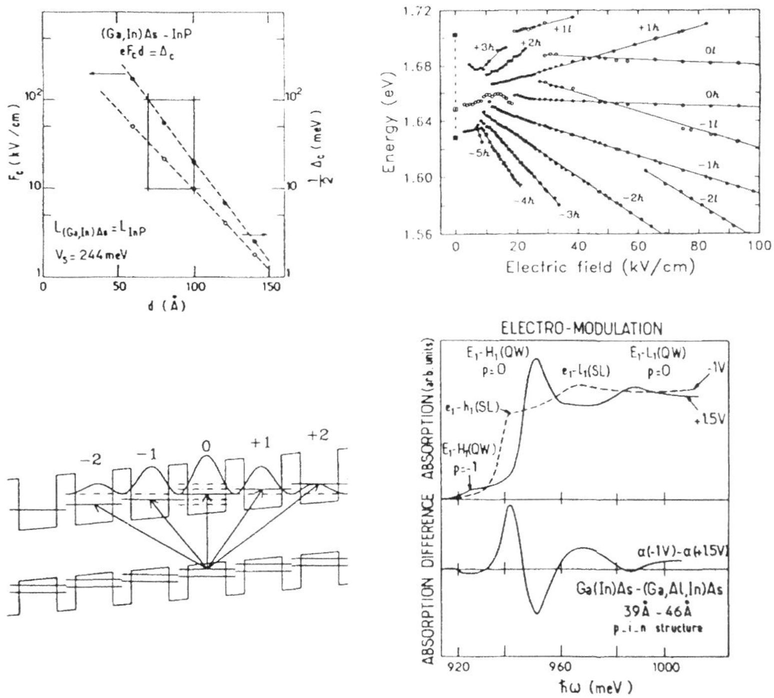

Fig.(8) A summary of the Wannier-Stark effect in superlattices. Left lower pannel : schematic representation of different inter-band optical transitions between the valence and conduction ladders[14]. Left upper pannel: the critical field needed to achieve f=4 (F_c, left scale) and the maximum blue shift of the E_1 conduction states ($(1/2)\Delta_{c1}$, right scale) are plotted versus the superlattice period d in GaAs-$Ga_{0.7}Al_{0.3}As$ superlattices with equal layer thicknesses. The rectangle defines the area where $F_c<10^5$V/cm and $(1/2)\Delta_{c1} > 10$ meV. Right upper pannel : electric field dependence of interband transitions between Wannier-Stark states in a 60Å period GaAs-$Ga_{0.7}Al_{0.3}As$ superlattice[14]. Right lower pannel : measured absorption line shape versus photon energy at two voltages in a 39Å-46Å $Ga_{0.47}In_{0.53}As$-(Ga,Al,In)As superlattice (upper scale). Absorption difference versus photon energy (lower scale).

Multiband Effects on the Wannier - Stark Quantization

The previous paragraph has been devoted to a one band tight binding analysis of the Wannier-Stark quantization. In reality, any bulk or superlattice band structure displays an infinite number of bands. Let n be the subband index of the superlattice minibands. The zeroth order of approximation consists of constructing a Wannier-Stark ladder attached to each subband:

$$\varepsilon_{n\nu} = \langle\varepsilon_{n0}\rangle + \nu eFd \ , \ \nu \text{ a relative integer} \tag{65}$$

For discrete values of the field there will exist crossings between $\varepsilon_{n\nu}$ and $\varepsilon_{m\mu}$ ($edF_{\mu\nu} = (\langle\varepsilon_{n0}\rangle - \langle\varepsilon_{m0}\rangle)/(\mu-\nu)$). In reality, these crossings are replaced by anticrossings since there is no reason why the matrix elements $\langle w_n(z-id) \mid eFz \mid w_m(z-jd) \rangle$ should all be zero if $m \neq n$. In particular, there exists a broadening of the $|n\nu\rangle$ Wannier-Stark state due to its interaction with the infinity of ladders arising from the excited minibands situated in the energy continuum of the unbiased superlattice. These interactions have long cast doubts on the very existence of the Wannier-Stark ladders, for it may have happened that the escape could have been faster than the Bloch period T_B, invalidating Wannier's approach. It took some time to establish that these effects were often small, a feature which recalls the findings in the problem of the intra-well Stark effect where one also deals with virtual bound states but where many physical quantities are accurately calculated by models which neglect the interaction with the continuum.

If one thus forget about finite lifetime effects, there are still possibilities of interactions between the ladders attached to different subbands generated by the hybridization of different quantum well bound states. In the vicinity of the $F_{\mu\nu}$'s defined above, the eigenstates which were field-localized in the one band approximation delocalize again. This delocalization, similar to the one described above for double wells, is very important for the carrier relaxation and transport along the growth axis. Notice however that its spatial extent is limited: if only nearest neighbour couplings are significant the larger achievable delocalization, out of completely localized states at the zeroth order of approximation, takes place over two superlattice periods at the maximum of the anticrossing between the two interacting ladders.

The reasoning we made previously about the linear variation of the Wannier-Stark states with F is no longer operative when several bound states per period are admixed by the field. Instead of dealing with a single $\langle\varepsilon_0\rangle$ appearing as an additive constant to all the energy levels of a ladder, and thus liable to be absorbed in a redefinition of the zero of energy or to be evaluated by some direct calculations, there are as many $\langle\varepsilon_{n0}\rangle$ as bound states per period. They cannot be all eliminated. Moreover, the previous reasoning on the T_d

(eq.(43)) is unable to tell us a priori if the ε_{n0} are field dependent or not, which precludes us to assert that the eigenstates vary linearly with F in the multiband situation. In fact they do not, as shown by the numerical diagonalizations of the superlattice hamiltonian. What remains true however, is the fact that the sequence $\varepsilon_V^{(n)}$, where n =1,2,..,M labels the various bound states per period, repeats itself periodically on the energy scale and that the wavefunctions $\psi_V^{(n)}(z)$ generate all the eigenfunctions of the problem by successive applications of T_d or $T^+{}_d$.

Interaction between two Wannier-Stark Ladders

In the following, we present a simple, yet accurate enough, modelling of the anticrossings between two discrete Wannier - Stark levels belonging to two different bands. The stationnary electron eigenstates of a semiconductor superlattice in the presence of a longitudinal electric field are the solutions of eq.(31). As discussed above, without electric field the superlattice eigenstates are for a wide class of materials well described by nearest neighbour tight binding models, as witnessed by the excellent fit of the exact dispersion relations to a single cosine law. We restrict our considerations to the two lower minibands. We call Δ_1 and Δ_2 their widths and $w_1(z - nd)$, $w_2(z - md)$ the Wannier functions of these two bands, where n and m are relative integers. The electric field term admits diagonal and off diagonal matrix elements in the Wannier basis. For the intra - band terms there is:

$$< w_{1,2}(z - nd) | \, z \, | w_{1,2}(z - n'd) > = nd\delta_{n,n'} \qquad (66)$$

As for the interband terms we have found that the approximation:

$$< w_1(z - nd) | \, z \, | w_2(z - n'd) > = Z_0\delta_{n,n'} + Z_1\delta_{n,n'\pm 1} \qquad (67)$$

works well for a variety of superlattices, with $Z_N = <\phi_1(z)|z|\phi_2(z\text{-}Nd)>$. Note that the Z_0 contribution is essentially an intra- well term: it represents for a single quantum well the polarisation of the isolated wells' wavefunctions by the field (the one which gives rise to the Stark effect). The intra - well term is dominant : for a GaAs-Ga(Al)As structure and period d ≈ 12 nm (80Å well width), $|Z_1|$ is ≈ 7 times smaller than $|Z_0|$. It remains however important to keep it in the analysis in order to get an accurate description of the splitting at the anticrossings for the following reason: Without interband coupling each miniband gives rise to an evenly spaced (by eFd) ladder with energies

$$E_{1\nu} = \langle\varepsilon_1(q)\rangle + \nu eFd \; ; \; E_{2\mu} = \langle\varepsilon_2(q)\rangle + \mu eFd \tag{68}$$

where μ and ν are relative integers and where $\langle\varepsilon_1(q)\rangle$ and $\langle\varepsilon_2(q)\rangle$ are the average energies of the E_1 and E_2 subbands respectively. $\langle\varepsilon_1(q)\rangle$ and $\langle\varepsilon_2(q)\rangle$ are close in energy from the isolated quantum well bound states. The corresponding wavefunctions are:

$$\psi_{1\nu} = \sum_n J_{\nu-n}(-\Delta_1/2eFd)\, w_1(z - nd); \tag{69}$$

$$\psi_{2\mu} = \sum_m J_{\mu-m}(\Delta_2/2eFd)\, w_2(z - md) \tag{70}$$

where J_n is the Bessel function of integer order n. The uncoupled ladders intersect at fields $F_{\mu\nu}$ equal to $[\langle\varepsilon_1(q)\rangle - \langle\varepsilon_2(q)\rangle]/(\mu-\nu)ed$. The interband couplings transform these crossings into anticrossings: The anticrossing between $|1,\nu\rangle$ and $|2, \mu = \nu - p\rangle$ gives rise to the energies $\varepsilon_\pm(\nu, p)$. For $F \approx F_{\mu\nu}$ the $\varepsilon_\pm(\nu, p)$ are approximatly the solutions of the secular equation:

$$\begin{bmatrix} E_{1\nu} - \varepsilon_\pm(\nu, p) & \delta_p \\ \delta_p & E_{2\mu} - \varepsilon_\pm(\nu, p) \end{bmatrix} \begin{bmatrix} c_1^\pm(p) \\ c_2^\pm(p) \end{bmatrix} = 0 \tag{71}$$

where we have only retained the two resonant contributions in the analysis and:

$$\delta_p \approx eF\,\{ Z_0 J_{-p}\,[(\Delta_1 + \Delta_2)/2eFd] + Z_1 \sum_\pm J_{-p\pm1}\,[(\Delta_1 + \Delta_2)/2eFd]\,\} \tag{72}$$

Thus, the difference between Z_0 and Z_1 can be offset by the magnitude of the J' s appearing in δ_p. In strong fields, when the arguments of the Bessel functions become small and if p = 1 the magnitude of Z_1J_0 becomes comparable or even larger than that of Z_0J_{-1} (and both terms have the same sign). At resonance the anticrossing gap is $2|\delta_p(F_{\mu\nu})|$. Let $\hbar\omega_p(F) = \varepsilon_+(\nu, p) - \varepsilon_-(\nu, p)$ be the field dependent energy difference between the two levels anticrossing at $F \approx F_{\mu\nu}$. In particular we have $2|\delta_p(F_{\mu\nu})| = \hbar\omega_p(F_{\mu\nu})$. Note also that $\varepsilon_\pm(\nu+1, p) = \varepsilon_\pm(\nu, p) + eFd$ and that the coefficients c_1 and c_2 are ν-independent. This complies with the general properties.

The non resonant contributions can be inserted into eq.(71) up to the second order. This amounts to adding diagonal contributions of the form $e^2F^2\Sigma_{\mu'\neq\mu} |<\psi_{2\mu'}|z|\psi_{1\nu}>|^2 \{\varepsilon_\pm - E_{2\mu'}\}^{-1}$ to $E_{1\nu}$ and a similar term to $E_{2\mu}$ where 2, μ and μ' are interchanged with 1, ν and ν' respectively. We have checked that the inclusion of those terms are not essential : they hardly affect the magnitude of the anticrossing gaps and merely amounts to shifting the $F_{\mu\nu}$' s towards high fields. We summarize in fig.(9) our modelling of the two band anticrossing for a 8nm - 4nm GaAs - Ga(Al)As superlattice. Since we do not know the w_1 and w_2 functions explicitly, we have replaced them by the ϕ_n' s defined previously in the evaluation of Z_0 and Z_1. We note a fair

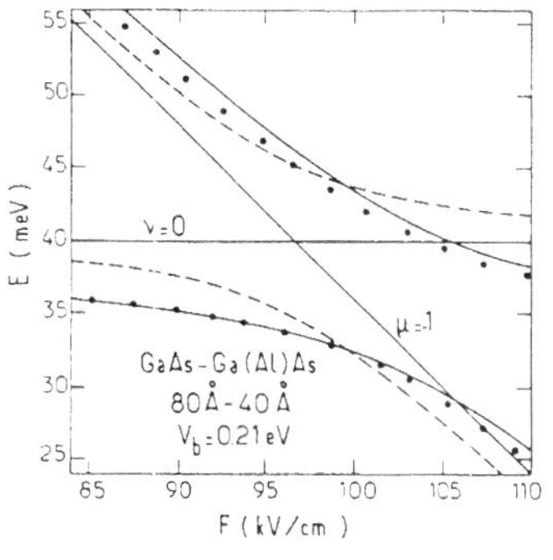

Fig.(9) Calculated field dependence of the Wannier Stark levels $\nu = 0$ and $\mu = -1$ belonging to two subbands of a 80Å - 40Å GaAs-Ga(Al)As superlattices. Straight lines: uncoupled ladders. Dashed (solid) lines: interacting ladders without (with) diagonal corrections. Large dots: numerical solution (central levels of a 15 periods SL).

description of the $\varepsilon_\pm(\nu, p)$ versus F curves by eq.(71) when compared with an "exact" (i.e. numerical) calculation of the central eigenstates (to avoid edge effects) of a 15 periods superlattice.

Inter-Subband Optical Transitions

In the following, we consider briefly the consequences that such anticrossings may have on the frequency of the absorbed or emitted light between the levels of such interlaced ladders. The oscillator strength of the optical transitions between the levels $|-,\nu, p\rangle$ and $|\pm,\nu',p\rangle$ of such an hybrid ladder is proportionnal to $|\langle -,\nu,p| z | \pm,\nu',p\rangle|^2$. The energies of such transitions are $\varepsilon_{\pm}(\nu', p) - \varepsilon_{-}(\nu, p)$. Clearly, all the transitions $|-,\nu, p\rangle \rightarrow |-,\nu', p\rangle$ occur at the energy $(\nu'-\nu)eFd$, i. e. at the fundamental and harmonics of the Bloch energy. However, like in the one band case, the oscillator strengths of the absorption is exactly equal to that of the induced emission. Thus, there is no net absorption of the electromagnetic wave in an infinite superlattice due to $|-,\nu, p\rangle \rightarrow |-,\nu', p\rangle$ transitions. Clearly, the same conclusion is reached if $|+,\nu, p\rangle \rightarrow |+,\nu', p\rangle$ transitions are involved. In actual, i. e. finite materials, absorption or emission may take place due to edge effects. Their signature would be their independence upon the superlattice thickness. Possible stimulated emission or absorption are associated with $|-,\nu, p\rangle \rightarrow |+,\nu', p\rangle$ transitions. There is an absorption corresponding to $\nu = \nu'$, at the energy $\hbar\omega_p(F)$, which is not compensated by an emission at the same energy and with the same initial state. This solitary line is accompanied by an absorption sideband at the energy $eFd + \hbar\omega_p(F)$ and by an emission line at the energy $eFd - \hbar\omega_p(F)$. Note that all these uncompensated (because the energies are different) absorption / emission processes are of the inter - subband type (like $E_1 \rightarrow E_2$ in a single quantum well) rather than of the Wannier - Stark type. They should produce net absorption or emission of an electromagnetic wave depending of the steady state (or transient) occupancy of the $|\nu, \pm\rangle$ hybrid levels.

Time Dependent Aspects of Interactions between Ladders

Recent time resolved optical experiments have been used to study the dynamical behaviour of electrons in superlattices under an electric field. In particular, the time varying polarization that accompanies the tunnelling back and forth of an electron between the wells of a superlattice has been detected in the form of a tunable coherent emission between 0.5 and 2 THz[15]. In the following, we discuss the consequences that the energy anticrossings may have on the frequency of the motion of wavepackets of the two interacting Wannier - Stark levels.

The time dependent counterpart of the anticrossings studied in eq.(71) is a

Rabi precession between the two interacting states if the system is prepared in an initial state which is not an eigenstate of eq.(71). This is realized if a short light pulse (characteristic width $\Delta T \leq 1/\omega_p \leq \hbar/2|\delta_p|$) is shined onto the crystal. Assume for simplicity that $|\psi(t=0)> = |1, \nu>$. Then, the probability P(t) to find the system in the state $|2, \mu = \nu - p>$ is equal to

$$P(t) = |<2, \mu = \nu - p| \exp(-iHt/\hbar) | 1, \nu >|^2 \tag{72}$$

Working with the same approximation as in eq.(71) and since:

$$<\pm, \nu', p | 2, \mu > = c_2^{\pm}(p)\delta_{\nu\nu'} \tag{73}$$

$$<\pm, \nu', p | 1, \nu > = c_1^{\pm}(p)\delta_{\nu\nu'} \tag{74}$$

we find easily that:

$$P(t) = [c_1^+ c_2^+]^2 + [c_1^- c_2^-]^2 + \\ 2c_1^+ c_1^- c_2^+ c_2^- \cos[\omega_p(F)\, t] \tag{75}$$

This Rabi precession is accompanied by an oscillatory displacement of the electron between the νth and the μth period of the superlattice. We calculate the average position of the electron at time t ($z(t) = < \psi(t) | z | \psi(t) >$) with the same assumptions as previously ($| \psi(0) > = |1, \nu >$) and find:

$$z(t) = (c_1^+)^2 < +, \nu, p | z | +, \nu, p > + (c_1^-)^2 < -, \nu, p | z | -, \nu, p > \\ + 2\, c_1^+ c_1^- < +, \nu, p | z | -, \nu, p > \cos[\omega_p(F)\, t] \tag{76}$$

In addition, there is:

$$< \pm, \nu, p | z | \pm, \nu, p > = (c_1^{\pm})^2 < 1, \nu | z | 1, \nu > + \\ (c_2^{\pm})^2 < 2, \mu | z | 2, \mu > + 2c_1^{\pm} c_2^{\pm} < 1, \nu | z | 2, \mu > \\ = (c_1^{\pm})^2 \nu d + (c_2^{\pm})^2 (\nu - p) d + 2c_1^{\pm} c_2^{\pm} \delta_p / eF \tag{77}$$

$$< +, \nu, p \mid z \mid -, \nu, p > = c_1^+ c_1^- \nu d + c_2^+ c_2^- (\nu - p) d +$$
$$(c_1^+ c_2^- + c_1^- c_2^+) \delta_p / eF \tag{78}$$

We show in fig.(10) the calculated time dependence of z(t) for a 8nm / 4nm GaAs/Ga(Al)As superlattice. The anticrossing between the $|1, 0>$ and $|2, -1>$ levels takes place at $F_r = 96.8$ kV/cm. The different curves correspond to different field detuning $\Delta F = F - F_r$, from $\Delta F = -20$kV/cm to +20kV/cm from the resonance field. The (F dependent) time scale is the period of the Rabi precession, i. e. $2\pi/\omega_p(F)$. It is only at exact resonance that the particle excursion extends over one period. Off resonance there is an increasing spatial localization with increasing detuning. There exists a slight asymmetry between the $+\Delta F$ and $-\Delta F$ curves. It is due to the fact that δ_p in eq.(71) is field dependent. Note that the z(t) curves which corrrespond to the $|1, 0>$ and $|2, -p>$ anticrossings with p>1 would qualitatively be the same as those shown in fig.(10) with, however, much sharper resonance like profiles and, of course, at the resonance field an excursion down to the - p period. As has been shown by Schneider et al[16] these resonant delocalizations of the eigenstates provide an efficient way for the vertical transport of the carrier.

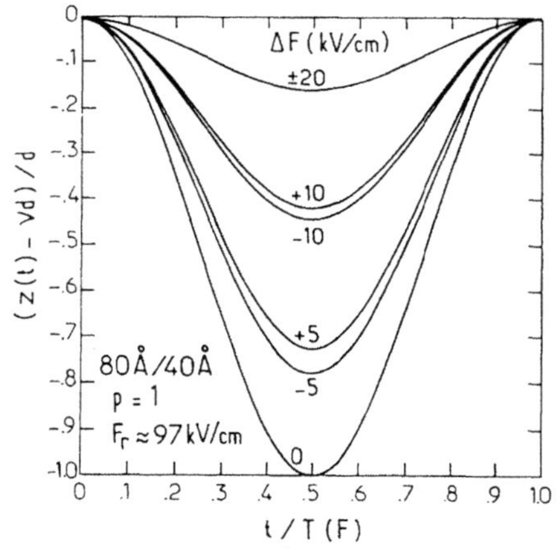

Fig.(10) Calculated time dependence of the position of the center of the wavepacket in a 80Å - 40Å GaAs-Ga(Al)As superlattice for several field detunings $\Delta F = F - F_r$. Here, $F_r \approx$ 97 kV/cm corresponds to the crossing of the unperturbed $|1, \nu = 0>$ and $|2, \mu = -1>$ levels.

Appendix. Comparison between the Bloch Oscillator and Wannier - Stark Wavepackets

Both descriptions are one band models. The width of the superlattice subband is Δ_1. Consider a wavepacket $\psi(t)$ of Wannier - Stark states which at t = 0 is:

$$\psi(t=0) = \Sigma_\nu \, g_\nu \, \psi_\nu \tag{79}$$

where we take g_ν as real and normalized:

$$\Sigma_\nu \, g_\nu{}^2 = 1 \tag{80}$$

At time t there is:

$$|\psi(t)> = \exp(-iHt/\hbar) \, |\psi(0)> = \Sigma_\nu \, g_\nu \, \psi_\nu \exp(-i\varepsilon_\nu t/\hbar) \tag{81}$$

The matrix elements of the position operator z and of the electron velocity $v_z = [z, H]/i\hbar$ are:

$$< \psi_\nu \, | \, z \, | \, \psi_{\nu'} > = \nu d + (\Delta_1/4eF)(\delta_{\nu',\nu+1} + \delta_{\nu',\nu-1}) \tag{82}$$

$$\begin{aligned} < \psi_\nu \, | \, v_z \, | \, \psi_{\nu'} > &= (\nu' - \nu)eFd < \psi_\nu \, | \, z \, | \, \psi_{\nu'} > /(i\hbar) \\ &= (\Delta_1 d/4i\hbar)(\delta_{\nu',\nu+1} - \delta_{\nu',\nu-1}) \end{aligned} \tag{83}$$

where we have assumed the nearest neighbour tight binding expansion to be valid. Thus, we find from eqs.(81-83) the average position $<z>(t)$ ($= < \psi(t) | \, z \, | \, \psi(t) >$) and velocity $<v_z>(t)$ ($= < \psi(t) | \, v_z \, | \, \psi(t) >$) of the wavepacket at time t:

$$< v_z>(t) = -(\Delta_1 d/2\hbar) \sin(\omega_B t) \, [\, 1 + \Sigma_\nu \, g_\nu \, (g_{\nu+1} - g_\nu)] \tag{84}$$

$$<z>(t) = \Sigma_\nu \, \nu d g_\nu{}^2 + (\Delta_1/2eF) \cos(\omega_B t) \, [\, 1 + \Sigma_\nu \, g_\nu \, (g_{\nu+1} - g_\nu)] \tag{85}$$

The initial velocity is zero (which could have been anticipated because both the wavefunctions and the wavepacket coefficients are real and the ψ_ν are localized). The first term in $<z>(t)$ can be seen by direct inspection to be equal to $(1/eF)<\psi(t=0) | \, H \, | \, \psi(t=0) >$. For the same initial condition (zero initial velocity) the semi - classical equations of motion provide us with:

$$v_z(t) = -(\Delta_1 d/2\hbar)\sin(\omega_B t) \tag{86}$$

$$z(t) = z_0 + (\Delta_1/2eF)\cos(\omega_B t) \tag{87}$$

where z_0 is indeed the total mechanical energy ("kinetic" + potential) of the oscillator (i. e. $eFz(t) + \varepsilon_1[\, q(t)\,]$). There is therefore an amazing closeness between an exact quantum treatment and the semi - classical equations of motions. A point however should be stressed. The relative difference between the quantum and semi - classical equations, viz $\Sigma_\nu\, g_\nu\, (\, g_{\nu+1} - g_\nu\,)$, can be made small by chosing an arbitrarily extended wavepacket. The trouble is that this choice is F independent. In contrast with the usual semi - classical treatments of quantum problems (e.g. the WKB approximation) the quality of the semi - classical solutions of the Wannier - Stark problem does not require any condition regarding the smallness or the spatial smoothness of the perturbing potential. In some sense the exact Wannier - Stark states coincide with their semi - classical limit.

5 - ACKNOWLEDGEMENTS

We thank Pr. E. Burstein and Dr. C. Weisbuch for giving us the possibility to participate in this summer school. We are pleased to thank Drs E. E. Mendez and P. Voisin for fruitful collaboration. The Laboratoire de Physique de la Matière Condensée, Ecole Normale Supérieure is "URA 1437 au CNRS".

6 - READING GUIDE

The following are review articles or textbooks which present broader and more detailed analysis than the ones presented in this article:

[1] An excellent introduction to quantum mechanics can be found in "Quantum theory" by D. Bohm, Prentice Hall, Englewood Cliffs, N. J. (1951).

[2] An introductory textbook to the electronic structure of semiconductor heterolayers is "Wave mechanics applied to semiconsuctor heterostructures" by G. Bastard, Les Editions de Physique, Les Ulis (1988).

[3] Two detailed reviews of the electronic states in semiconductor heterolayers are "Theory of semiconductor superlattices" by D. L. Smith and C. Mailhiot, Rev. Mod. Phys.**62**, 173 (1990) and "Electronic properties of two-dimensional systems" by T. Ando, A. B. Fowler and F. Stern, Rev. Mod. Phys.**54**,437(1982).

7 - REFERENCES

The following list of references is by no mean exhaustive. Most of the quoted papers are reviews. The reader is urged to consult also the references cited therein.

[1] G. Bastard, J. A. Brum and R. Ferreira, Electronic states in semiconductor heterostructures, Solid State Physics **44**, 229(1991).

[2] M. Altarelli, Band structure, impurities and excitons in superlattices, *in* : "Heterojunctions and Semiconductor Superlattices", G.Allan, G. Bastard, N. Boccara, M. Lanoo and M. Voos, Eds. Springer Verlag, Berlin (1986)

[3] D. L. Smith and C. Mailhiot, Theory of semiconductor superlattice electronic structure, Rev. Mod. Phys. **62**, 173 (1990)

[4] L. J. Sham and Yan - Ten Lu,Theory of electronic structure in superlattices, Journ. of Lumin. **44**, 207 (1989).

[5] J. M. Luttinger,Quantum theory of cyclotron resonance in semiconductors .: general theory, Phys. Rev. 102, 1030 (1956). See also G. L. Bir and G. E. Pikus, "Symmetry and Strain Induced Effects in Semiconductors", Wiley, New York (1974)

[6] D. Bohm, "Quantum theory", Prentice-Hall, Englewood Cliffs, N.J. (1951)

[7] D. A. B. Miller, J. S. Weiner and D. S. Chemla, Electric-field dependence of linear optical properties in quantum well structures : waveguide electroabsorption and sum rules, IEEE J. Quantum Electron. **QE22**, 1816 (1986)

[8] G. Bastard, E. E. Mendez, L. L. Chang and L. Esaki, Variational calculations on a quantum well in an electric field, Phys. Rev. **B28** 3241 (1983).

[9] L. Vina, E. E. Mendez, W. I. Wang, L. L. Chang and L. Esaki, Stark shifts in GaAs/GaAlAs quantum wells studied by photoluminescence spectroscopy, J. Phys. C. **20**, 2803 (1987).

[10] G. H. Wannier, "Elements of Solid State Theory", Cambridge U. P., Cambridge, England (1959) see also Dynamics of band electrons in electric and magnetic fields, Rev. Mod. Phys. **34**, 645 (1962)

[11] G. Nenciu, Dynamics of band electrons in electric and magnetic fields: rigorous justification of the effective Hamiltonians, Rev. Mod. Phys. **63**, 91 (1991).

[12] E. E. Mendez and G. Bastard, Wannier-Stark ladders and Bloch oscillations in superlattices, Physics Today **46**,34 (1993).

[13] J. Bleuse, G. Bastard and P. Voisin, Electric -field - induced localization and oscillatory electro - optical properties of semiconductor superlattices,

Phys. Rev. Lett. **60**, 220 (1988).

[14] E. E. Mendez and F. Agullo - Rueda, Optical properties of quantum wells and superlattices under electric fields, Journ. of Lumin. **44**, 223 (1989).

[15] C. Waschke, H. Roskos, R. Schwedler, K. Leo, H. Kurz and K. Köhler, Coherent submillimeter -wave emission from Bloch oscillations in a semiconductor superlattice, Phys. Rev. Lett. **70**, 3319 (1993)

[16] H. Schneider, H. T. Grahn, K. von Klitzing and K. Ploog, Resonance - induced delocalization of electrons in GaAs - AlAs superlattices, Phys. Rev.. Lett. **65**, 2720 (1990).

QUANTUM WELL OPTICAL SWITCHING DEVICES

DAVID A. B. MILLER
Rm. 4B-401, AT&T Bell Laboratories
Holmdel, NJ07733-3030
USA

ABSTRACT These lecture notes summarize the basic physics of quantum wells for optical switching devices, the principles of quantum well optical modulators and self-electrooptic-effect devices, and the current state of the art in such devices for systems experiments.

1. Introduction to Quantum Wells

First we will introduce quantum wells by discussing their basic physics, their structure, fabrication technologies, and their elementary linear optical properties.

1.1. SEMICONDUCTOR BAND STRUCTURE AND HETEROSTRUCTURES

All of the physics and devices that will be discussed here are based on properties of direct gap semiconductors near the center of the Brillouin zone. For all of the semiconductors of interest here, we are concerned with a single, S-like conduction band, and two P-like valence bands. The valence bands are known as the heavy and light hole bands. Importantly for quantum wells, the electrons in the conduction band, and the (positively charged) "holes" in the valence band behave as particles with effective masses different from the free electron mass. The simplest "k.p" band theory says that the electron effective mass, m_e, and the light hole effective mass, m_{lh}, are approximately equal and proportional to the band gap energy. For GaAs, which has a band gap energy of ~ 1.5 eV, the actual values are $m_e \sim 0.069\ m_o$ and $m_{lh} \sim 0.09\ m_o$, where m_o is the free electron mass. The heavy hole effective mass, m_{hh}, is typically more comparable to the free electron mass ($m_{hh} \sim 0.35\ m_o$ for the most common situation in quantum wells), and does not vary systematically with the band gap energy.

Quantum wells are one example of heterostructures - structures made by joining different materials, usually in layers, and with the materials joined directly at the atomic level. When two semiconductors are joined, it is not clear in advance how the different bands in the two materials will line up in energy with one another, and their is no accurate predictive theory in practice. Hence, an important experimental quantity is the "band offset ratio"; this is the ratio of the difference in conduction band energies to the difference in valence band energies. For GaAs/AlGaAs heterostructures, for example, approximately 67% of the difference in the band gap energies is in the conduction band offset, and 33% is in the valence band offset, giving a ratio 67:33. In this particular material system, both electrons and holes see higher energies in the AlGaAs than in the GaAs, giving a so-called "Type I" system. Heterostructures in which

A. Miller et al. (eds.), Nonlinear Optical Materials and Devices for Applications in Information Technology, 255–284.

electrons and holes have their lowest energies in different materials are called "Type II", but such structures are not currently of device interest and will not be considered further here.

Heterostructures in general have many uses. They can be used for advanced electronic devices (e.g., modulation-doped field-effect transistors, heterojunction bipolar transistors, resonant tunneling devices), optical components (e.g., waveguides, mirrors, microresonators), and optoelectronic devices and structures (e.g., laser diodes, photodetectors, quantum well and superlattice optical and optoelectronic devices). Although heterostructures may be useful in electronics, they are crucial in many optoelectronic devices (e.g., lasers). Perhaps their most important technological aspect may be that they can be used for all of these electronic, optical, and optoelectronic purposes, and hence may allow the integration of all of these.

1.2. QUANTUM WELL STRUCTURES AND GROWTH

A quantum well is a particular kind of heterostructure in which one thin "well" layer is surrounded by two "barrier" layers. Both electrons and holes see lower energy in the "well" layer, hence the name (by analogy with a "potential well"). This layer, in which both electrons and holes are confined, is so thin (typically about 100 Å, or about 40 atomic layers) that we cannot neglect the fact that the electron and hole are both waves. In fact, the allowed states in this structure correspond to standing waves in the direction perpendicular to the layers. Because only particular waves are standing waves, the system is quantized, hence the name "quantum well".

There are at least two techniques by which quantum well structures can be grown, molecular beam epitaxy (MBE), and metal-organic chemical vapor deposition (MOCVD). Both can achieve a layer thickness control close to about one atomic layer. MBE is essentially a very high vacuum technique in which beams of the constituent atoms or molecules (e.g., Ga, Al, or As) emerge from ovens, land on the surface of a heated substrate, and there grow layers of material. Which material is grown can be controlled by opening and closing shutters in front of the ovens. For example, with a shutter closed in front of the Al oven, but open shutters in from of the Ga and As ovens, GaAs layers will be grown. Opening the Al shutter will then grow the alloy AlGaAs, with the relative proportion of Ga and Al controlled by the temperatures of the ovens. With additional ovens and shutters for the dopant materials, structures of any sequence of GaAs, AlAs, and AlGaAs can be grown with essentially arbitrary dopings. MOCVD is a gas phase technique at low pressure (e.g., 25 torr). In this case the constituents are passed as gasses (e.g., trimethylgallium and arsine) over a heated substrate, with the resulting composition being controlled by the relative amounts of the appropriate gasses. Hybrid techniques, using the gas sources of MOCVD in a high vacuum molecular beam system, also exist, and are known variously as gas-source MBE or chemical beam epitaxy (CBE). Which technique is best depends on the material system and the desired device. Typical structures grown by these techniques might have total thickness of microns, and could have as many as hundreds of layers in them.

There are many different materials that can be grown by these techniques, and many of these have been used to make quantum well structures. One significant restriction is that it is important to make sure that the lattice constants (essentially, the spacing between the atoms) of the materials to be grown in the heterostructure are very similar. If this is not the case, it will be difficult to retain a well-defined crystal structure throughout the layers - the growth will not be "epitaxial". The growth is simplest when the lattice constants are identical. Fortunately, AlAs and GaAs have almost identical lattice constants, which means that arbitrary structures can be grown with high quality in this materials system. Another commonly used system is InGaAs with InP; in this case, the proportions of In and Ga are adjusted to give a lattice constant for the ternary (three-component) InGaAs alloy that is equal to InP. Use of four component (quaternary) alloys (e.g., InGaAsP) allows sufficient degrees of freedom to adjust both the lattice constant and the bandgap energy. Up to a certain critical thickness, which depends on the degree of lattice constant mismatch, it is possible to grow structures with materials that naturally have different lattice constants. In this case, the materials grow in a highly strained state but can adopt the local lattice constant and retain good epitaxial crystal structure. Such strained materials are of increasing technological importance, although we will not discuss them further here.

A partial list of materials used for quantum well structures includes: III-V's - GaAs/GaAlAs on GaAs (Type I), GaSb/GaAlSb on GaSb (Type I), InGaAs/InAlAs on InP (Type I), InAs/GaSb (Type II), InGaAs/GaAs (Type I, strained); II-VI's - HgCdTe/CdTe, ZnSe/ZnMnSe (semimagnetic), CdZnTe/ZnTe (Type 1, strained); IV-VI's - PbTe/PbSnTe; IV - Si/SiGe (strained).

1.3. PARTICLE-IN-A-BOX QUANTUM WELL PHYSICS

We can understand the basic properties of a quantum well through the simple "particle-in-a-box" model. Here we consider Schrödinger's equation in one dimension for the particle of interest (e.g., electron or hole)

$$\frac{-\hbar^2}{2m}\frac{d^2\phi_n}{dz^2}+V(z)\phi_n=E_n\phi_n \tag{1}$$

where V(z) is the structural potential (i.e., the "quantum well" potential) seen by the particle along the direction of interest (z), m is the particle's (effective) mass, and E_n and ϕ_n are the eigenenergy and eigenfunction associated with the n'th solution to the equation..

The simplest case is shown in Fig. 1. In this "infinite well" case, we presume for simplicity that the barriers on either side of the quantum well are infinitely high. Then the wavefunction must be zero at the walls of the quantum well.

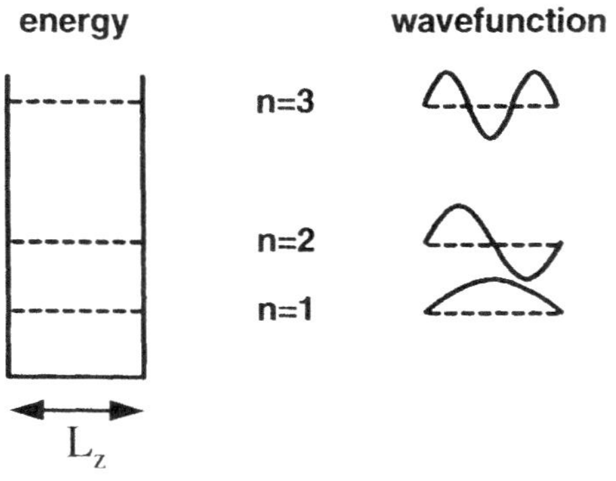

Fig. 1 "Infinite" quantum well and associated wavefunctions.

The solution is then particularly simple:

$$E_n = \frac{-\hbar^2}{2m}\left[\frac{n\pi}{L_z}\right]^2 \quad n = 1, 2, ... \quad \phi_n = A\sin\left(\frac{n\pi z}{L_z}\right) \tag{2}$$

The energy levels (or "confinement energies") are quadratically spaced, and the wavefunctions are sine waves. In this formula, the energy is referred to the energy of the bottom of the well. Note that the first allowed energy (corresponding to n=1) is above the bottom of the well.

We see that the energy level spacing becomes large for narrow wells (small L_z) and small effective mass m. The actual energy of the first allowed electron energy level in a typical 100 Å GaAs quantum well is about 40 meV, which is close to the value that would be calculated by this simple formula. This scale of energy is easily seen, even at room temperature.

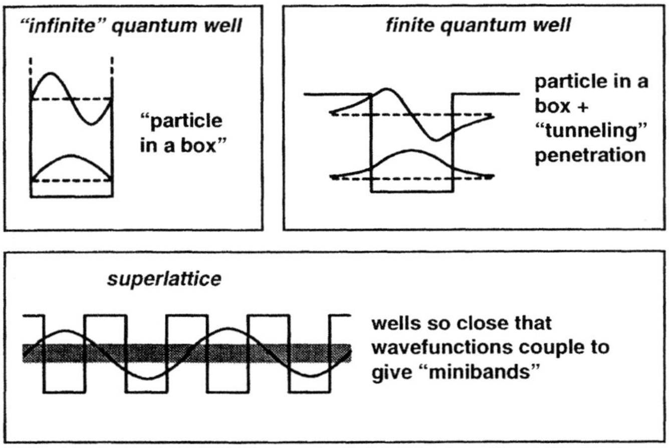

Fig. 2 Comparison of "infinite" quantum well, "finite" quantum well, and superlattice behavior. For the superlattice, a wavefunction for one of the possible superlattice miniband states is shown (actually the state at the top of the miniband).

The solution of the problem of an actual quantum well with finite height of barriers is a straightforward mathematical exercise. It does, however, require that we choose boundary conditions to match the solutions in the well and the barriers. One boundary condition is obvious, which is that the wavefunction must be continuous. Since the Schrödinger equation is a second order equation, we need a second boundary condition, and it is not actually obvious what it should be. We might think that we would choose continuity of the wavefunction derivative across the boundary; we cannot do so because the masses are different on the two sides of the boundary in general, and it can be shown that such a simple condition does not conserve particle flux across the boundary when the masses are different. One that does conserve particle flux is to choose (1/m) $d\phi/dz$ continuous; this is the most commonly used one, and gives answers that agree relatively well with experiment, but there is no fundamental justification for it. This lack of fundamental justification should not worry us too much, because we are dealing anyway with an approximation (the "envelope function" approximation). If we were to use a proper first principles calculation, we would have no problem with boundary conditions on the actual wavefunction.[4] The solution of the finite well problem does not exist in closed form [5], requiring numerical solution of a simple equation to get the eigenenergies. The wavefunctions of the bound states are again sine waves inside the quantum well, and are exponentially decaying in the barriers. The energies are always somewhat lower than those we would calculate using the infinite well. It can be shown from the solution [5] that there is always at least one bound state of a finite quantum well. Fig. 2 illustrates the differences between the idealized "infinite" quantum well, the actual "finite" well.

Also in Fig. 2, we have illustrated a superlattice. It will be useful here to define the difference between quantum wells and superlattices. The simplest, crystallographic, definition of a superlattice is a "lattice of lattices". With that definition, any regular sequence of well and barrier layers would be a superlattice. A more useful definition here is the "electronic" definition; in this definition, we call such a regular structure a superlattice only if there is significant wavefunction penetration between the adjacent wells. Otherwise, the physics of the multiple layer structure is essentially the same as a set of independent wells, and it is more useful to call the structure a multiple quantum well (MQW). If there is significant wavefunction penetration between the wells, we will see phenomena such as "minibands", and the structure is then usefully described as a superlattice. The "minibands" arise when quantum wells are put very close together in a regular way, just as "bands" arise in crystalline materials as atoms are put together. Just as with quantum wells, simple models for superlattices can be constructed using envelope functions and effective masses, and such models are also good first approximations. As a rule of thumb, for well and barrier layers thicker than about 50 Å in the GaAs/AlGaAs system, with a typical Al concentration of about 30 % in the barriers, the structure will probably be best described as a multiple quantum well.

2. Linear Optical Properties of Quantum Wells

To understand the linear interband optical absorption in quantum wells, we will first neglect the so-called "excitonic" effects. This is a useful first model conceptually, and explains some of the key features. For a full understanding, however, it is important to understand the excitonic effects. In contrast to bulk semiconductors, excitonic effects are very clear in quantum wells at room temperature, and have a significant influence on device performance.

2.1. OPTICAL ABSORPTION NEGLECTING EXCITONS

The simplest model for absorption between the valence and conduction bands in a bulk semiconductor is to say that we can raise an electron from the valence band to a state of essentially the same momentum in the conduction band (a "vertical" transition) by absorbing a photon. The state in the conduction band has to have essentially the same momentum because the photon has essentially no momentum on the scale usually of interest in semiconductors. In this simple model, we also presume that all such transitions have identical strength, although they will have different energies corresponding to the different energies for such vertical transitions. The optical absorption spectrum therefore has a form that follows directly from the density of states in energy, and in bulk (3D) semiconductors the result is an absorption edge that rises as the square root of energy, as shown in Fig. 3.

In quantum wells, for the direction perpendicular to the layers, instead of momentum conservation we have a selection rule. The rule states that (to lowest order) only transitions between states of the same quantum number in the valence and conduction bands are allowed.

This rule follows from the fact that the optical absorption strength is proportional to the overlap integral of the conduction and valence (envelope) wave functions. For sinusoidal standing waves, as shown for the infinite quantum well in Fig. 2, there is only a finite overlap between identical standing waves. (This rule is somewhat weaker for finite wells, although it is still a very good starting point.) We can if we wish still view this as conservation of momentum, since we can regard the standing waves as states of specific momenta. (The converse is also true, in that even in bulk semiconductors we can equally well view the momentum conservation rule as following from allowed overlap integrals, in that case between plane wave wavefunctions.)

In a quantum well, the electrons and holes are still free to move in the directions parallel to the layers; hence, we do not really have discrete energy states for electrons and holes in quantum wells; we have instead "subbands" that start at the energies calculated for the confined states. The electron in a given confined state can in addition have any amount of kinetic energy for its in-plane motion in the quantum well, and so can have any energy greater than or equal to the simple confined-state energy for that subband. The density of states for motion in the plane of the quantum well layers turns out to be constant with energy, so the density of states for a given subband really is a "step" that starts at the appropriate confinement energy. Optical transitions must still conserve momentum in this direction, and just as for bulk semiconductors, the optical absorption must still therefore follow the density of states. Hence, in this simple model, the optical absorption in a quantum well is a series of steps, with one step for each quantum number, n. It is easily shown, from the known densities of states, that the corners of the steps "touch" the square root bulk absorption curve (when that curve is scaled to the thickness of this infinite quantum well). Thus, as we imagine increasing the quantum well thickness, we will make a smooth transition to the bulk behavior, with the steps becoming increasingly close until they merge into the continuous absorption edge of the bulk material.

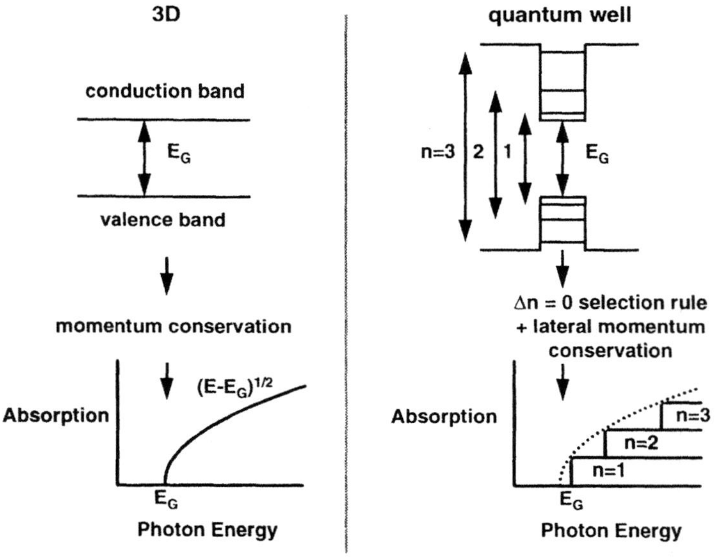

Fig. 3. Optical absorption in bulk (i.e., 3D) semiconductors and in quantum wells, in the simplest model where excitonic effects are neglected.

2.2. CONSEQUENCES OF HEAVY AND LIGHT HOLES

As we mentioned above, there are two kinds of holes that are relevant here, corresponding to the heavy and light hole bands. Since these holes have different masses, there are two sets of hole subbands, with different energy spacings. The light hole subbands, because they have lighter mass, are spaced further apart. The consequence for optical absorption is that there are actually two sets of "steps". The heavy-hole-to-conduction set starts at a slightly lower energy and is more closely spaced than the light-hole-to-conduction set. The heavy hole set is usually dominant when we look at optical absorption for light propagating perpendicular to the quantum well layers.

If, however, we look in a waveguide, with light propagating along the quantum well layers, there are two distinct optical polarization directions: one with the optical electric vector parallel to the quantum well layers (so-called "transverse electric" or TE polarization); and the other with the optical electric vector perpendicular to the quantum well layers (so-called "transverse magnetic" or TM polarization). The TE case is essentially identical to the situation for light propagating perpendicular to the layers, where the optical electric vector is always in the plane of the quantum wells; the optical electric vector is always perpendicular to the direction of propagation for a plane wave. The TM case is substantially different from the TE case, however.

Because of microscopic selection rules associated with the unit cell wavefunctions, for the TM polarization, the heavy-hole-to-conduction transitions are forbidden, and all of the absorption strength goes over to the light-hole-to-conduction transitions. Hence, at least with this simple model, there is now only one set of steps in the absorption. The reason for loss of the heavy hole transitions is not a special property of quantum wells; this kind of selection rule is a consequence of defining a definite symmetry axis in the material, in this case the growth direction of the quantum well layers. Exactly the same selection rule phenomenon will result if we apply a uniaxial stress to a bulk semiconductor. One practical consequence of this selection rule effect is that quantum well waveguide lasers essentially always run in TE polarization; there are more heavy holes than light holes in thermal equilibrium, and hence the gain associated with heavy holes is larger, and hence the gain is larger for the TE polarization. In general, this simple classification of holes into "heavy" and "light" is only valid near the center of the Brillouin zone. The detailed structure of the valence bands is particularly complicated since the various subbands would actually appear to cross one another. In fact, such crossing are avoided, and the resulting subband structure is quite involved. For many devices, such effects are not very important (at least for a basic understanding of the devices), and we will not discuss such valence band effects further here.

2.3. OPTICAL ABSORPTION INCLUDING EXCITONS

Fig. 4 shows an actual absorption spectrum of a quantum well sample.

We see from Fig. 4 that the quantum well absorption is indeed a series of steps, and simple calculations based on the particle-in-a-box models will correctly give the positions of the steps. But it is also clear that there are sets of peaks in the spectra not predicted or explained by the simple "non-excitonic" model described above. These peaks are quite a strong effect, and will be particularly important near the band-gap energy, here at about 1.46 eV photon energy. Most of the devices we discuss here operate in this region.

To understand these peaks, we need to introduce the concept of excitons. The key point missing in the previous discussion is that we have neglected the fact the electrons and holes are charged particles (negative and positive respectively) that attract each other. Hence, when we have an electron and a hole in a semiconductor, their wavefunctions are not plane waves; plane waves correspond to the case of uniform independent motion of the electron and the hole. Instead, we should expect that the electron and hole wavefunction should correspond to the case where the electron and hole are close to one another because of their Coulomb attraction. The formal error we made in the analysis above is therefore that we did not use the correct eigenfunctions for the electrons and holes, and hence we formally got an incorrect answer when we used those eigenfunctions to calculate the optical absorption.

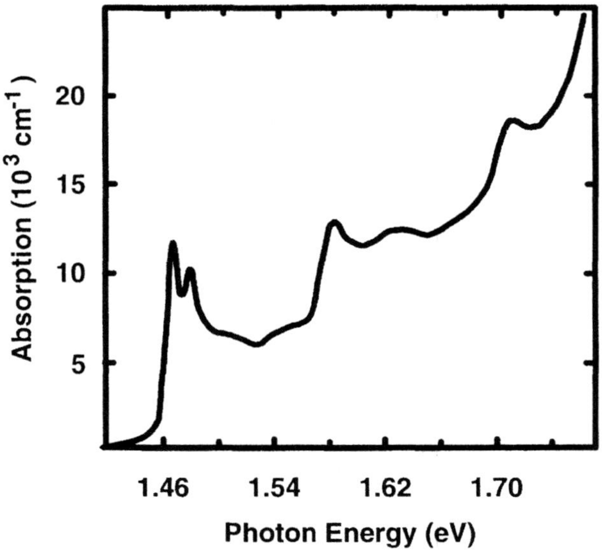

Fig. 4. Absorption spectrum of a typical GaAs/AlGaAs quantum well structure at room temperature.

Unfortunately, to include excitonic effects properly we need to use a different picture, since the whole band structure picture is a single particle picture; it essentially describes the energies seen by either a single electron or a single hole, but cannot handle both at once. Fortunately, there is a simple picture that allows us to understand the resulting excitonic effects.

The correct approach is not to consider raising an electron from the valence band to the conduction band, but rather to consider the creation of an electron-hole pair. In this picture, we find the eigenfunctions of the electron-hole pair in the crystal, and base our calculation of optical absorption on those pair states. First, then, we must understand what are the states of an electron-hole pair in a crystal. Fortunately, at least for the case where the attraction is not too strong, this problem is already solved; it is essentially the same problem as the states of the hydrogen atom, corrected for the different effective masses and dielectric constants in the semiconductor. Using this model, we find, for example, the binding energy of lowest, 1S, exciton is

$$E_B = \frac{\mu e^4}{8h^2\varepsilon_R^2\varepsilon_o^2} \quad (3)$$

where h is Planck's constant, ε_R is the relative permittivity, ε_o is the permittivity of free space, and μ is the reduced mass

$$\mu = \frac{m_e m_h}{m_e + m_h} \quad (4)$$

For bulk GaAs (where excitons can be clearly seen at low temperature), $E_B \sim 4$ meV.

Hence, we find the first remarkable property of excitons compared to the non-excitonic model. It is possible to create an exciton with an energy E_B less than that required to create a "free" electron-hole pair. A free electron-hole pair is analogous to an ionized hydrogen atom, and the energy required to create such a free pair in the semiconductor is actually the simple band-gap energy of the single particle picture. Hence we expect some optical absorption at photon energies just below the band-gap energy. This point is illustrated in Fig. 5, on the left, showing a possible transition at an energy $E_{exciton} = E_{bandgap} - E_B$.

It is not only the absorption that creates the 1S exciton that is important, although under most conditions it is the only one that we see clearly as a distinct peak. In fact the entire absorption spectrum of these kinds of semiconductors is properly described in terms of the complete set of hydrogenic states. For example, the absorption above the band-gap energy results from the creation of excitons in the unbound hydrogenic states. In the classical sense, such states correspond to hyperbolic orbits. There is also additional absorption just below the bandgap energy from the other, bound excitonic states.

Note now that we are explaining the optical absorption in terms of the creation of a particle, the exciton (or, exactly equivalently, an electron-hole pair). It is important to understand that the absorption we see is not that associated with raising an existing hydrogenic particle to an excited state, as would be the case with conventional atomic absorption; we are instead creating the particle. An analogy that may help understand this distinction is the absorption in the vacuum that creates electron-positron pairs. This is illustrated in Fig. 5, on the right.

By absorbing photons in the vacuum, we can in principle create positronium atoms, which are simply hydrogenic systems composed of an electron and a positron (instead of a proton). In this case, we are simultaneously creating a positron in the Fermi sea (the analog of the valence band), as well as an electron. This happens to be a two-photon transition, and is therefore a rather weak effect in practice, but it illustrates the difference between the creation of an atom and the absorption between levels of the atom. A very important difference between these is that in the creation case, we have "excitonic" absorption even when we have no excitons in the material. In the positronium case, it is very clear that there need be no positronium atoms there since we are starting with a perfect vacuum!

In the simple model where we neglected excitonic effects, the strength of a particular transition was determined by the square of the overlap integral between the "initial" (valence subband) state and the "final" (conduction subband) state. In principle, we have a similar result for the present excitonic model, although in this case the initial state is the "empty" crystal, and the final state is the crystal with an exciton added. We will not formally derive the consequences of this change in model, but the net result is that, in the exciton creation case, the strength of the absorption to create an electron-hole pair in a given state is proportional to the probability that the resulting electron and hole will be in the place (strictly, the same unit cell). Now we can understand why the excitonic absorption gives such a strong peak for creation of 1S excitons. In the 1S exciton, the electron and hole are bound closely to one another (within a diameter of about 300 Å in bulk GaAs), and the 1S wavefunction actually peaks at zero relative displacement. Hence the probability of finding the electron and the hole in the same place is actually very large, and so the resulting absorption is strong.

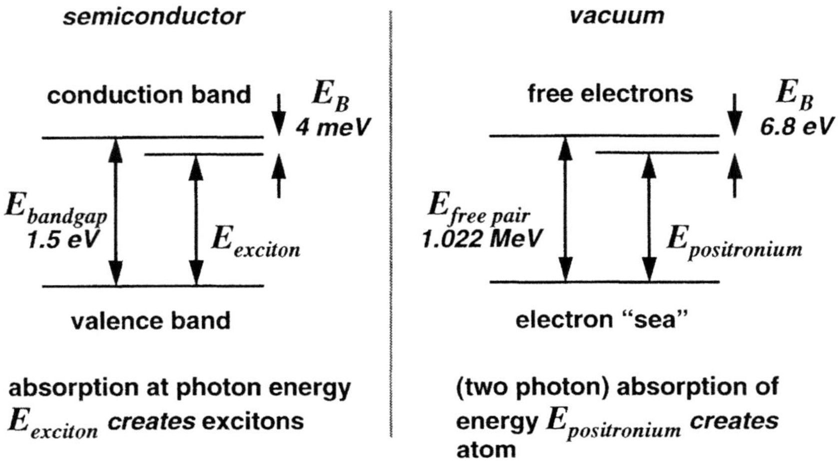

Fig. 5. Analogy between excitonic absorption and the optical creation of positronium atoms.

Thus far, we have discussed excitons in general, without explaining why they are particularly important in quantum wells. Excitonic effects are clear in many direct gap semiconductors at low temperature. At room temperature, however, although excitonic effects are still very important in determining the shape and strength of the interband absorption, the actual peaks corresponding to the bound states are difficult to resolve. The main reason for this lack of resolution is that the bound states are rapidly ionized by collisions with optical phonons at room temperature. In fact, they are typically ionized in times short compared with a classical orbit time. By a Heisenberg uncertainty principle argument, the associated optical linewidth is broadened, to such a point that the line is no longer clearly resolvable.

The quantum well offers two important differences compared to the bulk material, both of which stem from the confinement in the quantum well. Fig. 6 shows, in a semiclassical picture, an exciton in bulk GaAs and an exciton in a quantum well.

Hydrogenic theory for the GaAs exciton gives a diameter of about 300 Å, as mentioned above. When we create an exciton in a quantum well that is only 100 Å thick, the exciton must become smaller, at least in the direction perpendicular to the quantum wells. Remarkably, however, it also becomes smaller in the other two directions (in the plane of the quantum well). This surprising conclusion can be checked, for example, by variational calculations. We can rationalize it by saying that nature prefers to keep the exciton more nearly spherical to minimize energy overall. If, for example, we allowed the exciton to become a flat "pancake" shape, it would acquire high kinetic energies from the large second derivatives at the edges of the pancake. It is also the case that the exact solution of the two-dimensional hydrogen atom, corresponding to a very thin quantum well with very high walls, has a diameter one quarter that of the three-dimensional hydrogen atom. The two consequences are: (1) the electron and hole are even closer together than in the three-dimensional case, so the absorption strength to create such an exciton is even larger; (2) the exciton has a larger binding energy because the electron and hole are closer together, and hence it orbits "faster". As a result of the faster orbiting, the exciton is able to complete a classical orbit before being destroyed by the optical phonon, and hence it remains a well-defined resonance. Equivalently, although the linewidth of the quantum well exciton is comparable to that of the bulk exciton, the binding energy is larger, and the peak is still well resolved from the onset of the "interband" absorption at the band-gap energy. These two reasons explain both why the quantum well excitons are relatively stronger and also better resolved than the bulk excitons. The practical consequence is that in quantum wells we may be able to make some use of the remarkable excitonic peaks since we can see them at room temperature.

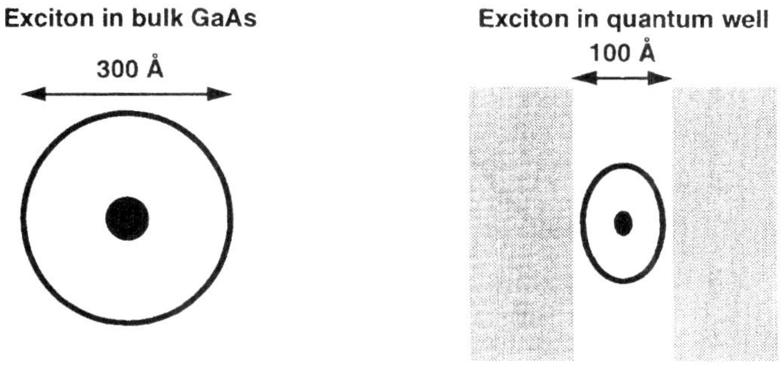

Fig. 6. Comparison of bulk and quantum well exciton sizes and shapes.

Note that in practice we frequently refer to the peaks in the spectrum as the exciton peaks or the exciton absorption peaks. It is important to remember always that these peaks represent the absorption that *creates* the (1S) exciton, and also that all of the rest of the interband absorption is also excitonic, although not to create the 1S state. Loosely, we often talk of "the exciton" as meaning only the 1S exciton.

Incidentally, the two strong peaks that we typically see near the band-gap energy are the 1S exciton absorption peaks associated with the first heavy-hole-to-conduction transition (the stronger peak at lower photon energy) and with the first light-hole-to-conduction transition (the weaker peak at slightly higher photon energy).

3. Nonlinear Optics in Quantum Wells

There are many possible nonlinear optical effects in quantum wells. Here we will discuss only one class of effects, namely those related to optical absorption saturation near to the band-gap energy. This particular set of effects has been strongly considered for various different kinds of devices, and is a serious candidate for applications in laser modelocking.

In the simplest case, we shine a laser beam on the material so that the resulting optical absorption generates a significant population of electrons and holes, either creating "excitons" (i.e., electron-hole pairs initially in bound hydrogenic states) or "free carriers" (i.e., electron-hole pairs initially in unbound hydrogenic states). Absorbing directly into the exciton (absorption) peaks will generate excitons, whereas absorbing at higher photon energies will generate free carriers. For steady state effects at room temperature with continuous laser beams, it makes little difference which we generate, since the excitons will ionize rapidly (e.g., in ~ 300 fs) and in thermal equilibrium we will essentially have only free carriers anyway. Fig. 7 shows the effect on the absorption spectrum of generating a significant density (e.g., ~ 10^{17} cm^{-3} free carrier density) in the quantum wells.

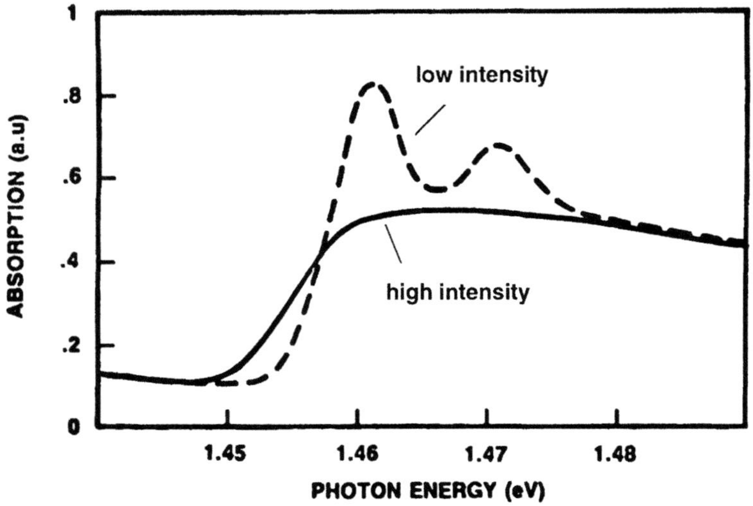

Fig. 7. Absorption spectrum without (dashed line) and with (solid line) optically created free carriers.

As can be seen from Fig. 7, with the presence of such a free carrier density, the exciton peaks have become saturated. The detailed physics of such phenomena is very complicated, and can only be analyzed using many-body theory. We can, however, understand qualitatively various of the mechanisms [6] [7]. It is important in understanding these mechanisms to remember that the excitonic absorption peaks are associated with the creation of excitons. The simple absorption saturation mechanism that we would see in, for example, an atomic vapor, in which the absorption saturates when half of the atoms have been excited to their upper state, does not exist here; there is no density of atoms to start with since there are no excitons present and hence we cannot deduce any density of excited atoms at which saturation occurs by such an argument.

There is, however, one mechanism that is particularly easy to understand. Our physical intuition tells us that we cannot create two similar excitons in the same place; this is exactly like trying to create two atoms in the same place. Hence, as we begin to fill up space with excitons, we will start to run out of space to create more. Consequently, the probability of being able to create more excitons must reduce, and so the optical absorption associated with creating them must decrease. Therefore the exciton absorption line will saturate. This mechanism is illustrated on the left of Fig. 8. For such a mechanism, we will get saturation with about one created exciton per exciton area, which works out to an exciton density of about 10^{17} cm^{-3}.

The real reason for this saturation mechanism is Pauli exclusion; we cannot have two electrons in the same state in the same place. It is also true that free carriers, if they are "cold" enough to be in the states near the band center from which the exciton is comprised, can also prevent creation of more excitons, again by the Pauli exclusion principle. This is illustrated on the right of Fig. 8.

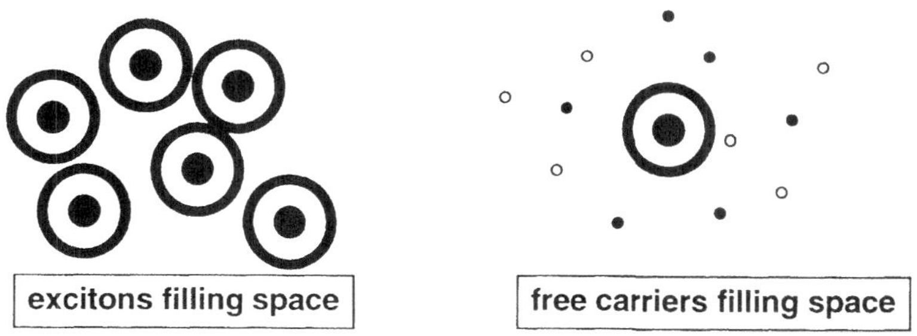

Fig. 8. Illustration of saturation by filling space with excitons (left) and filling space with free carriers. At a density of one exciton per exciton volume, or one "cold" free carrier per exciton volume, it becomes essentially impossible to create any more excitons. This can be visualized geometrically in terms of the probability of trying to "throw" one more exciton onto the plane without it landing on any of the other excitons or free carriers.

Although we have discussed the saturation mechanisms so far in a qualitative and simplistic way, it turns out that these mechanisms are rigorously correct, with the only detail being the effective area of the exciton to use in the argument [7]. There are also other mechanisms that are somewhat less obvious, and some of these are of comparable size [7]. A second class of mechanisms that will change the absorption is screening effects. If we create a density of free carriers, they will tend to change the dielectric constant, and hence change the size of the exciton, typically increasing it. If the size of the exciton increases, then the probability of finding the electron and hole in the same place decreases; hence the optical absorption strength decreases, giving an effect that also behaves like saturation. This direct, classical Coulomb screening is thought to be weak in quantum wells because the walls of the wells prevent the movement of charge necessary for effective screening. There is, however, an even more subtle effect that is actually of comparable size to the other "saturation" mechanisms, which is exchange screening. Essentially, when we include the effect of Pauli exclusion in our calculations, we find that the results we would calculate using our simple classical Coulomb effects are not correct at high density, because the Pauli exclusion forces the electrons to be further apart than we had thought (and similarly for the holes). This is described as if it were a screening effect, although it is actually a Pauli exclusion effect that causes us to correct our previous Coulomb calculation. This exchange screening also tends to increase the size of the exciton, reducing absorption.

Although in the steady state at room temperature, we see primarily the effect of free carriers on excitons, at lower temperatures or at short time scales we can see the effect of excitons on excitons. These latter effects are actually typically somewhat stronger (e.g., a factor of 1.5 - 2). Hence, if we initially create excitons and monitor the saturation of the exciton peaks, we can see a fast transient absorption saturation associated with the exciton-exciton effects, followed by a somewhat weaker saturation as the excitons ionize. This actually allows us to measure the exciton ionization time, which is about 300 fs at room temperature.

These saturation effects associated with the exciton peak in quantum wells are relatively sensitive. They have been explored for a variety of nonlinear optical switching devices. Even with the additional benefits of the strong quantum well excitons, these effects are not large enough to make low enough energy devices for current practical systems interest for information processing. The effects are important, however, for two reasons. One reason is that they set limits on the operating power of other devices, such as the modulators and electroabsorptive switches discussed below. Secondly, the effects are large enough for serious use as mode locking saturable absorber elements for lasers, and this is becoming practical now [8].

4. Quantum Well Electroabsorption Physics

When electric fields are applied to quantum wells, their optical absorption spectrum near to the band-gap energy can be changed substantially, an effect we can call electroabsorption. Such effects have been extensively investigated for optical modulators and switches [9]. There are two very distinct directions in which we can apply electric fields to quantum wells, either with the electric fields parallel to the layers or with the electric field perpendicular to the layers. The case of fields perpendicular to the layers is the one most peculiar to quantum wells, and it is called the Quantum-Confined Stark Effect (QCSE). Here, we will first discuss the case of fields parallel to the layers.

4.1. ELECTRIC FIELDS PARALLEL TO THE LAYERS

For the case of electric fields parallel to the layers, we get effects that are qualitatively similar to those seen in bulk semiconductors. The benefit of the quantum well here is that we can exploit the excitonic electroabsorptive effects at room temperature. The main effect we see is that the exciton absorption peak broadens with field.

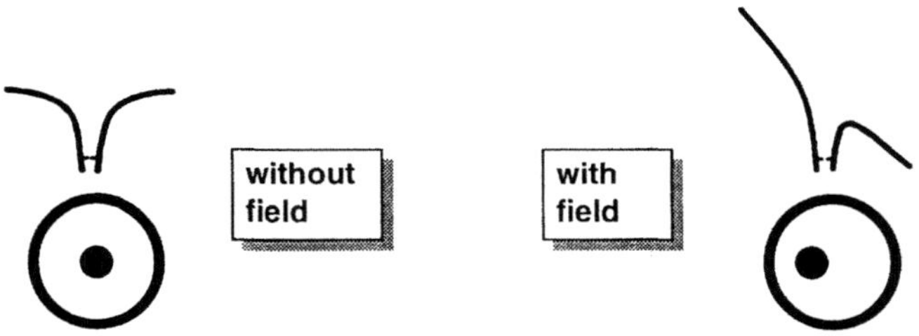

Fig. 9. Effect of field on quantum well excitons for fields parallel to the layers.

Fig. 9 illustrates qualitatively the Coulomb potential of the electron in the presence of the hole, and sketches how the exciton is deformed as the field is applied. In principle, when we apply an electric field to a hydrogenic system, we can shift the energy levels and the resulting transition energies, an effect known in atoms as the Stark effect. For a symmetrical state, such as the ground state of a hydrogenic system, the Stark effect can be viewed as the change in electrostatic energy caused by polarizing the atom with the field, **E**. The change in energy is therefore - (1/2) **P.E**, where **P** is the induced polarization. This corresponds to a reduction in the energy of the hydrogenic system. Hence, we might expect that as we apply an electric field to a semiconductor, we should see the 1S exciton absorption peak move to lower energies, because the energy of the resulting exciton we create is lower by this Stark shift.

It is true that there is such a Stark shift, but it is not the dominant effect that we see. The reason is that the Stark shift of a hydrogenic system is limited to about 10% of the binding energy. Since the binding energy of the exciton is only about 10 meV, the shift is therefore limited to about 1 meV, and hence it is not a very large effect. When we try to shift the energy by more than this, the hydrogenic system becomes field ionized (i.e., the electron and hole are "ripped" apart by the field) so rapidly that the particle cannot complete even a substantial fraction of a classical orbit before being destroyed, and the whole concept of a bound state loses any useful meaning. In fact, what we see primarily as we apply the field is the broadening of the exciton absorption resonance caused by the shortening of the exciton's lifetime - again, a "Heisenberg uncertainty principle" broadening. (It is just possible to see the shift at low fields in a carefully controlled experiment.)

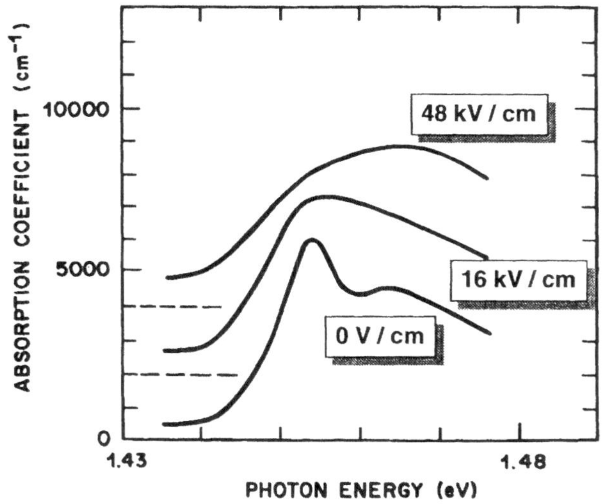

Fig. 10. Absorption spectra for a quantum well sample at room temperature with various electric fields applied in the plane of the quantum well layers. The spectra are shifted vertically for clarity. The exciton peaks broaden with field.

It is worth noting that the fields we are capable of applying to the exciton are gigantic in a relative sense. Applying 1V / μm (= 10^4 / cm) corresponds to a field of one binding energy (~ 10 meV) over one exciton diameter (~ 100 Å). Such a field corresponds to a massive perturbation, certainly much larger than can readily be achieved with a hydrogen atom itself. It is also not surprising that such a field should cause the exciton to be field-ionized in less than a classical orbit time.

Fig. 10 shows a typical set of spectra for parallel field electroabsorption in quantum wells. The broadening and disappearance of the peaks is clearly seen. One consequence is the appearance of a weak absorption tail at lower photon energies. The appearance of this tail is often referred to as the Franz-Keldysh effect. This is somewhat misleading, however, since the Franz-Keldysh effect is really a non-excitonic effect, whereas excitonic effects dominate near to the band-gap energy; the Franz-Keldysh effect, for example, cannot model the exciton broadening at all.

4.2. ELECTRIC FIELDS PERPENDICULAR TO THE LAYERS

The behavior of the electroabsorption for electric fields perpendicular to the quantum well layers is quite distinct from that in bulk semiconductors. Fig. 11 shows a typical set of spectra.

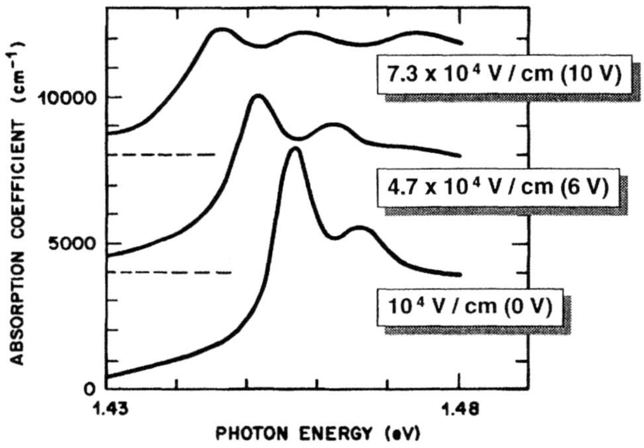

Fig. 11. Absorption spectra for electric fields applied perpendicular to the quantum well layers. The spectra are shifted vertically for clarity. The voltages correspond to those applied to a diode structure containing quantum wells.

Here we can see that, instead of being broadened by the electric field, the exciton absorption peaks are strongly shifted by the field. The shifts can be 10's of meV, and the applied fields here can be much larger than those shown above for the parallel field case, while still preserving the exciton peaks.

The reason for the difference in the electroabsorption in the perpendicular field case is a straightforward consequence of the quantum well. As we apply an electric field perpendicular to the layers, we pull the electron in an exciton in one direction (towards the positive electrode) and the hole in the other direction, just as we would expect. The difference is that the walls of the wells prevent the exciton from field ionizing. Instead, the electron becomes squashed against one wall of the quantum well and the hole against the other, as illustrated in Fig. 12 for the lowest electron and hole states (n=1). Because the electron and hole are still relatively close to one another, they are still strongly attracted by their Coulomb attraction, and they still orbit round one another in the plane of the quantum wells, albeit in a somewhat displaced orbit. Hence the exciton can still exist as a particle for times longer than a classical orbit time, and the exciton absorption peak is not greatly broadened. Because the particle still exists even with very strong fields, we can obtain very large Stark shifts. In fact, the Stark shifts can be many times the binding energy (this is not unphysical since we are decreasing the energy, not increasing it). To see this effect, it is of course important that the quantum well is significantly smaller than the bulk exciton diameter; otherwise, the exciton can effectively be field-ionized simply by separating them by a sufficient distance within the well. For obvious reasons, this shift of the exciton absorption peaks is called the Quantum-Confined Stark Effect (QCSE). In principle, we could see similar effects with a hydrogen atom itself, but to do so we would need to confine the hydrogen atom within a distance less than 1 Å, and apply fields of ~ 10^{11} V/cm, both of which are currently impractical.

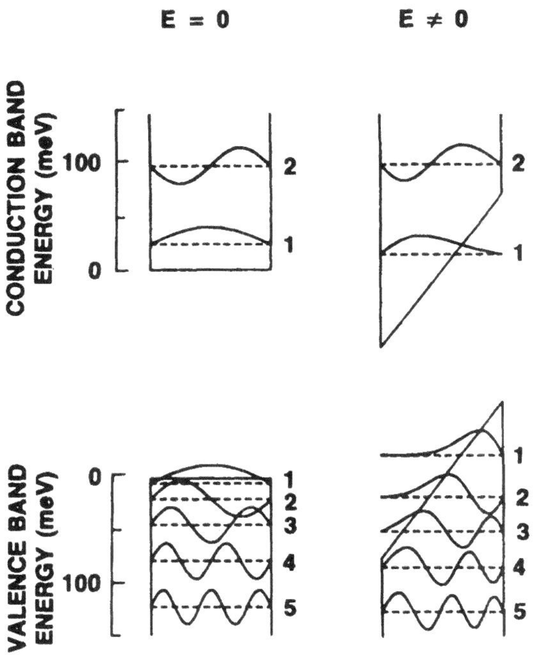

Fig. 12. Electron and hole wavefunctions for the first few states in an "infinite" quantum well. Without field, the wavefunctions are sinusoidal; with field they are Airy functions.

We can also see from Fig. 11 that, as the exciton absorption peaks are shifted to lower energies, they do become weaker. The reason for this is that the electron and hole are being separated from one another by being pulled in opposite directions, and hence there is less probability of finding them in the same place. Consequently, the absorption strength decreases. It is worth mentioning here that, as absorption strength is lost in this way by the so-called "allowed" transitions (e.g., the first hole state to the first electron state), it is picked up by formerly "forbidden" transitions (e.g., the second hole state to the first electron state) [10]. With the applied field, the electron and hole wavefunctions in all of the levels are distorted; instead of being sinusoidal, they are Airy functions, as illustrated in Fig. 12. In this case there tend to be finite overlap integrals between all possible states. The strengths of these various "forbidden" transitions are bounded by sum rules - essentially, the electric field cannot change the total amount of absorption in the system.

We will not give details of the quantum mechanical calculation of the shifts of the peaks. It is, however, relatively straightforward. Although the above explanations for the reasons for the continued existence of the excitons are important for the effect, and the excitonic effects are also important for the strength of the absorption, the dominant part of the shift comes from the underlying shift of the single particle states. We can see this shift in Fig. 12, where we see the n=1 electron level moving down and the n=1 hole level moving up; the net result is that the energy separation between the n=1 hole and electron levels is reduced with field. The shift in the exciton binding energy itself is relatively unimportant by comparison (typically a few meV). The calculation of the shift of individual electron or hole levels with field can be done by various means. For a tilted well, the wave equation becomes Airy's differential equation, and hence the exact solutions (e.g., as in Fig. 12) are Airy functions.

It is also, incidentally, quite correct to view the quantum-confined Stark effect shifts as resulting from the polarization of the exciton by the electric field, hence giving a (1/2) **P.E** shift in the exciton energy. To lowest order, the polarization is proportional to the field (i.e., the induced separation of the electron and hole is initially linear with field), and so the QCSE is a quadratic effect to lowest order.

Although we have discussed effects here only for the case of simple "rectangular" quantum wells, many other forms of quantum wells are also possible, such as coupled quantum wells, graded quantum wells, stepped quantum wells. These other structures also show related effects induced by applied electric fields, although we will not discuss these here. Also, superlattices show a class of effects induced by fields, known as Wannier-Stark localization. This effect and effects in coupled quantum wells are closely related.

5. Quantum Well Modulators

The QCSE is particularly attractive for optical modulators. A simple device structure is shown in Fig. 13. A quantum well region, typically containing 50 to 100 quantum wells, is grown as the undoped "intrinsic" ("i") region in a p-i-n diode. This quantum well region will therefore have a thickness of about 1 - 2 microns. In operation, the diode is reverse biased. 1 V across 1 micron corresponds to a field of 10^4 V/cm, so substantial QCSE shifts and changes in the absorption spectrum can be made with applied voltages of the order of 5 - 10 V. The reverse biased diode is convenient because there is no conduction current that needs to flow in the diode in order to apply these relatively substantial fields.

The "p" and "n" regions of the diode are made out of a material that is substantially transparent at the wavelength of interest. In the case of GaAs quantum wells, they would most likely be made out of AlGaAs, the same material as used for the quantum well barriers. For the particular case of GaAs/AlGaAs quantum wells, the usual substrate material is GaAs. Unfortunately, GaAs is opaque at the wavelengths used with such modulators, hence the substrate has to be removed for a transmissive modulator, such as the one in Fig. 13.

There are two very important features of the quantum well modulator. The first is that the electroabsorptive mechanisms (the QCSE) in the quantum well are strong enough to make a modulator that can work for light propagating *perpendicular* to the surface of the semiconductor. Despite the fact that this modulator will typically only be a few microns in thickness altogether, it can make changes in transmission of a factor of 2 - 3 in a single pass of a light beam. This modulation is large enough to make usable systems. This feature has the very important consequence that one can therefore make two-dimensional arrays of devices. This opens up many possibilities for novel highly-parallel optoelectronic systems.

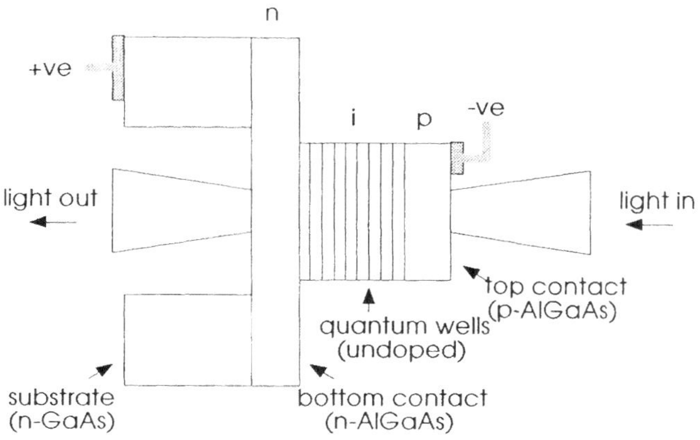

Fig. 13. Quantum well modulator diode structure.

A second key feature is that the operating energy of this modulator is small. The amount of energy required to change the optical properties in this modulator is essentially the stored electrostatic energy at the operating field. In other words, one has to charge up the capacitance of the device. The capacitance of a 1 micron square area of a semiconductor device 1 micron thick is approximately 10^{-16} F. Hence the total energy required to charge the device is (1/2) CV^2 ~ 1 - 2 fJ/μm^2. This energy density is comparable to the energy density in switching electronic devices, and is much smaller (e.g., by a factor of ~ 100) than the energy per unit optical area required to turn on a laser diode or saturate an optical absorption. Hence this device is very attractive as a potentially highly efficient optical output device for electronic circuits.

In addition, the speed limit on the use of the QCSE is probably in the range of ~ 100 fs or less. Practical modulator devices will not be limited by the electroabsorption mechanism itself, seeing instead only the usual resistive/capacitive limits in applying voltage to the diodes. Because the diodes can be small, the capacitance can also be small, and hence these devices are attractive for high speed modulators.

Although we will discuss here primarily only devices where the light propagates perpendicular to the surface, the QCSE can also be used effectively for waveguide modulators (where the light propagates along the surface). In this case, a useful modulator can be made with only a few quantum wells because the propagation distance can be long (e.g., 100 microns). As a result, waveguide quantum well modulators can operate with low voltages (e.g., < 1 V).

The removal of the substrate, as shown in Fig. 13, is usually undesirable since it leaves a very thin structure. An alternative is to grow the quantum well modulator on top of a mirror [11]. Such a mirror can be made directly in the semiconductor growth machine by making a multilayer dielectric stack mirror from the semiconductor materials themselves. Such a technique leaves a crystalline surface suitable for the epitaxial growth of the quantum well device. The semiconductor layers do not have a large index contrast between them, however, so a relatively large number of pairs of quarter-wave layers is usually required (e.g., 12 - 20 pairs of layers, each layer ~ 800 Å thick). A second consequence of the small index contrast is that the mirror has a relatively narrow wavelength band for reflection (e.g., < 100 nm at 850 nm wavelength). As a result, the control of the thickness of the mirror layers often imposes tighter tolerance on the growth rate than does the thickness of the quantum well layers themselves. (The exciton peak shifts relatively little for a change of a few percent in the layer thickness, whereas such a percentage change in the mirror thicknesses could move the reflection band away from the operating wavelength region.)

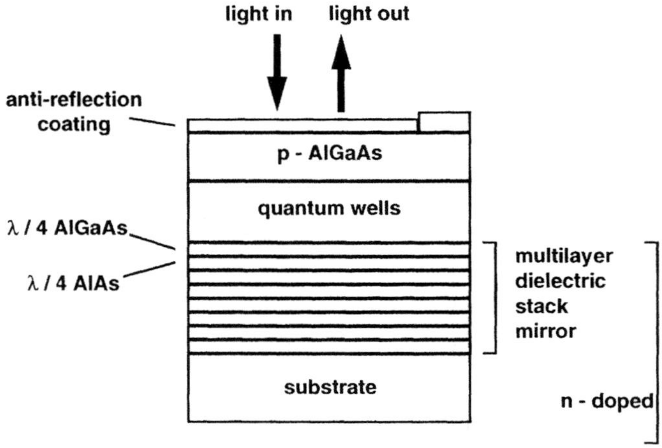

Fig. 14. Quantum well reflection modulator.

The reflection modulator also brings other benefits. It improves the contrast ratio (actually squaring it compared to a simple modulator), because the light beam makes two passes through quantum well material. The structure also permits conventional electronic chip mounting and substrate heat sinking because we do not have to pass light through the bottom of the chip.

Some Quantum Well Modulator Material Systems	
Materials	*Typical Wavelength*
GaAs/GaAlAs	850 nm
InGaAs/InP	1.5 μm
GaSb/GaAlSb	1.5 μm
InGaAs/GaAs	980 nm, 1.06 μm
InAsP/InP	1.06 μm
AlGaAs/AlAs	680 nm
InGaAsP/InP	1.3 μm
CdZnTe/ZnTe	610 nm

Table 1

An extension of the reflective modulator concept is to put the modulator inside a resonant cavity by adding a reflecting surface on top. One interesting design is the asymmetric Fabry-Perot resonator [12]. In this case, the bottom reflector is made more reflecting than the top reflector. When the right amount of absorption is created inside the resonator, the bottom reflector "looks" as if it has the same reflection as the top reflector. In any Fabry-Perot resonator with equal front and back mirror reflectivities, in the on-resonance condition there is exactly no reflection from the resonator. Hence by increasing or reducing the absorption in the quantum well layers to exactly the right amount, the modulator structure has exactly no reflection. This can give a modulator with very high contrast ratio and also lower drive voltage since typically less absorbing material is required than in the non-resonant case. The major practical problem with this concept is that, since the idea only works on resonance, the thicknesses of the layers have to be very precisely controlled during fabrication, a formidable task for current growth techniques.

Quantum well modulators have been demonstrated in many material systems and at many wavelengths. Table 1 shows a list of some of the materials and the operating wavelengths.

The quantum well modulators can be run in two different modes, illustrated in Fig. 15. In the so-called "λ_0" mode, the absorption decreases with increasing reverse bias voltage. The operating wavelength is then typically chosen to coincide with the position of the exciton absorption peak near zero bias. In the "λ_1" mode, the absorption increases with increasing reverse bias. Note, incidentally, that by applying enough voltage, e.g., 10 V in Fig. 15, the operation at the indicated "λ_1" wavelength ends up by also showing a decrease of absorption with reverse voltage once the exciton peak has moved past the operating wavelength.

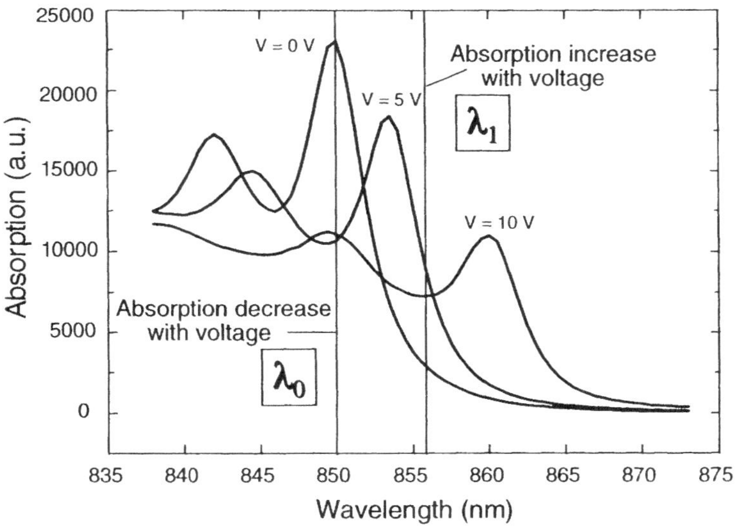

Fig. 15. Illustration of the two characteristic operating wavelengths of quantum well modulators. At λ_0, the modulator absorption decreases with increasing reverse bias voltage. At λ_1, the absorption increases with voltage.

Most obviously the choice of mode affects whether the modulator increases or decreases transmission (or reflection) as voltage is applied. There are, however, other consequences of the choice of mode. One relates to the maximum power that can be applied to the modulators without saturation. When running at λ_1, whenever there is absorption in the modulator there is also substantial electric field over the quantum wells. The electric field tends to cause the carriers to escape from the wells rapidly [13][14], and so they tend not to build up in the wells. Carrier build-up in the wells can cause saturation of the absorption. Hence this λ_1 operation has higher saturation intensities [15]. This can be important in practice; for example, the present design of modulators for the FET-SEED devices (discussed below) is intended to handle 500 μW incident power at 850 nm wavelength in a 3 μm diameter spot, corresponding to intensities of the order of 5 - 10 kW/cm^2.

The second important point about the choice of operating mode relates to the feedback induced in the modulator circuit by the photocurrent generated in the quantum well diode. Depending on the details of the electrical circuit, the photocurrent generated in the modulator tends to change the voltage across the modulator, hence changing the absorption, and hence in turn changing the photocurrent. Operating at λ_0 tends to give positive feedback that can lead to bistability. Operating at λ_1 tends to give negative feedback that can lead to linearized modulation. These feedback effects will be discussed below in the section on self-electrooptic effect devices (SEEDs).

The quantum well modulator can also be used to make changes in refractive index. As is well known, changes in absorption spectrum generally lead to changes in the refractive index spectrum, the two being related by the Kramers-Krönig relations. As a result, there is a refractive index change associated with the QCSE [16]. This effect can also be used to make devices, such as electro-optic modulators [17]. One problem with using refractive effects is that we cannot normally use the refractive effects in those spectral regions where they are largest, such as close to the exciton absorption peaks. The reason is that refractive devices usually require that we can make more than half a wavelength of path length change in less than one absorption length, and the absorption is too strong near to the exciton peaks. Consequently, it is difficult to make useful refractive devices that can be operated for light propagating perpendicular to the surface (e.g., with ~ 1 μm of optical thickness as in the absorptive p-i-n modulator structure). As we move to longer wavelengths, however, the refractive effects fall off more slowly than the absorption effects. Consequently, waveguide devices of about 100 μm or more of length can be made, and these are operated at wavelengths somewhat longer than the band-edge wavelength. Even these lengths are significantly shorter than those of other waveguide electro-optic modulators, and such devices offer the additional advantage that they can be integrated with other semiconductor devices, such as lasers, photodetectors and optical amplifiers.

6. Self-Electrooptic-Effect Devices

The Self Electrooptic Effect Device (SEED) principle is to combine a quantum well modulator (or a set of modulators) with a photodetector (or a set of photodetectors) to make an optically controlled device with an optical output (or outputs) [18] [19]. A major reason for thinking about such a class of devices is an opportunity for efficient integration. Quantum well modulators driven directly by external electrical connections are limited in speed and operating energy by the external electrical parasitics, such as capacitance and the need to drive with impedance matched lines at high speeds. The modulators themselves can be driven with very low total energies if the drive is electrically local and not brought in by some external connection. Furthermore, since they are semiconductor devices, the prospects for integration with various optoelectronic and electronic devices are good.

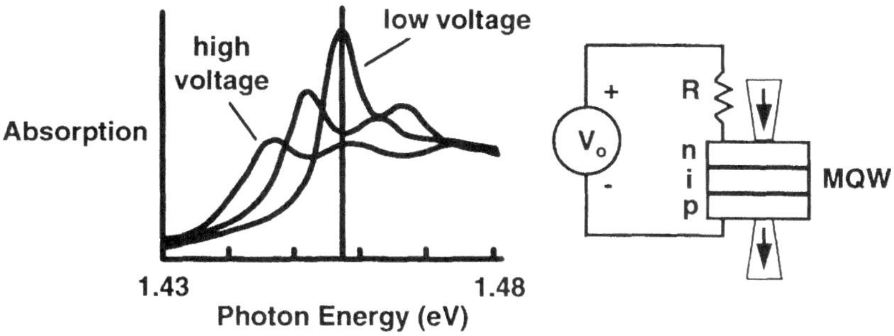

Fig. 16. Illustration of operating photon energy (or wavelength) point and circuit for simple resistive optical bistability.

Although it is often the case that the conversion from optics to electronics and back to optics is inefficient and costly, this need not be the case if the devices are well integrated. In practice, the quantum well devices can be integrated effectively, and allow two-dimensional arrays of smart optoelectronic units or "smart pixels". Such devices offer new possibilities in information processing and switching architectures.

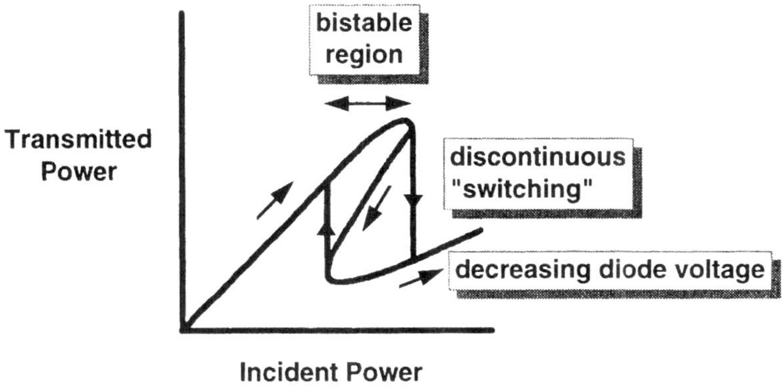

Fig. 17. Characteristic bistable curve for the simple SEED circuit of Fig. 16.

There are two general classes of SEEDs, those based on diodes alone and those also integrating transistors. The simplest SEED configuration is shown in Fig. 16. This is a so-called "resistor-biased" SEED (R-SEED), and it is optically bistable. It is operated at λ_0 where the absorption increases as the reverse bias voltage is reduced.

This device is optically bistable as shown in Fig. 17. The bistability results from a positive feedback mechanism. Suppose initially there is little light shining on the diode. Then there is little photocurrent, and consequently little voltage drop across the resistor. Hence the full supply voltage appears across the diode, and the absorption of the diode is relatively low. As we start to shine more light on the diode, we get more photocurrent. Because of the voltage drop now across the resistor, we get now less voltage across the diode, and hence more absorption, and hence more photocurrent, and so on. As a result of this mechanism, we can get switching into a highly absorbing state. Once in this state, the device requires less input power to hold it in this state because the diode is now so highly absorbing. Decreasing to some lower power eventually leads to a similar but opposite mechanism that causes the diode to switch back to its high-voltage, low transmission state, as shown in Fig. 17.

As an alternative to using a resistor as the "load" in the bistable circuit of Fig. 16, we could use another reverse-biased photodiode. The photodiode could then perform the function of the load resistor, but with an effective "resistance" value controlled by the light shining on this second, "load" photodiode. Such a circuit is called a "diode-biased" SEED (D-SEED), and it has various interesting analog modes that we will discuss below. A further extension of this concept is to use a second quantum well diode as the "load", giving the so-called "symmetric" SEED (S-SEED), as shown in Fig. 18 [20].

The S-SEED is unusual in that it is bistable in the ratio of the two input light beam powers rather than switching at any particular absolute power. We can analyze this device in a similar fashion to the R-SEED bistability; it has the same kind of intrinsic positive feedback. In the case of the S-SEED, it switches simply when the photocurrent in one diode exceeds that in the other, and hence we can run the S-SEED over a very wide range of powers. We can run it at very low absolute powers, in which case it will switch slowly, or at high powers, in which case it will switch rapidly. (The switching speed is determined by the charging and discharging of the diode capacitances by the photocurrent.)

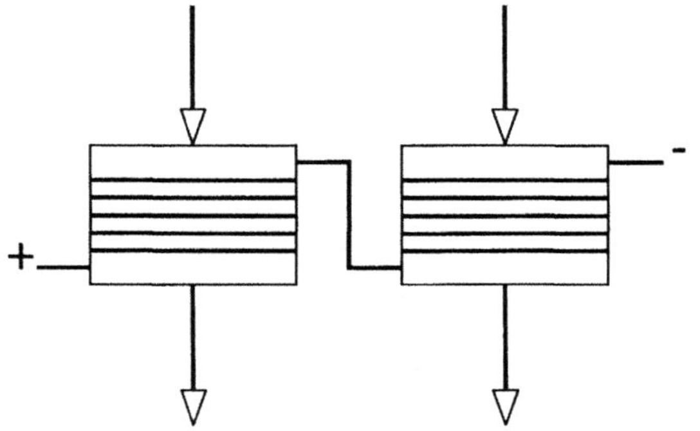

Fig. 18. Schematic of symmetric SEED (S-SEED)

The S-SEED is a particularly useful device for systems experiments. One very useful feature is "time-sequential gain". Put simply, we can switch the device when running it low powers and read it out at high powers, hence achieving effective gain, which is always required in logic systems. Importantly, this "gain" does not require that we set the operating power particularly carefully. Attempts to make logic systems with simple bistable devices have been very difficult because of the necessity of biasing very close to the switching point in order to get gain; the S-

SEED avoids this problem. Another important feature of the S-SEED is that it provides complementary and differential inputs and outputs. The logic value is represented not by any particular beam power, but by the ratio of two beam powers, a number that is independent of attenuation. Attenuation of various kinds is unavoidable and sometimes unpredictable in optical systems, but with such a two-beam representation the logic value is not changed by such attenuation, provided only that both beams are attenuated equally. The S-SEED is also insensitive to even gross power supply fluctuations. A further benefit of the use of two beams is that high contrast modulation not needed. The S-SEED has been used successfully in various systems experiments [21].

The S-SEED concept can be extended to other logical functions. It is possible to perform elementary logic functions with the S-SEED itself. One extension is to use more than two diodes in series, the so-called "multistate" SEED (M-SEED) [22]. Depending on the biasing mode, this can show a variety of different stable states. For example, under voltage biasing, N diodes in series can have N stable states (i.e., only one diode "on"). Under current biasing, all of the diodes can be independently bistable. A related class of S-SEED extensions is the "logic" SEED (L-SEED) [23]. In this case families of circuits can be constructed that are exactly analogous to NMOS or CMOS silicon logic families.

The SEEDs discussed so far are all based on operation at λ_0, and all rely on positive feedback effects. Operation at λ_1 gives another class of devices with unusual linear analog modes. The simplest of these is the self-linearized modulator [19], illustrated in Fig. 19.

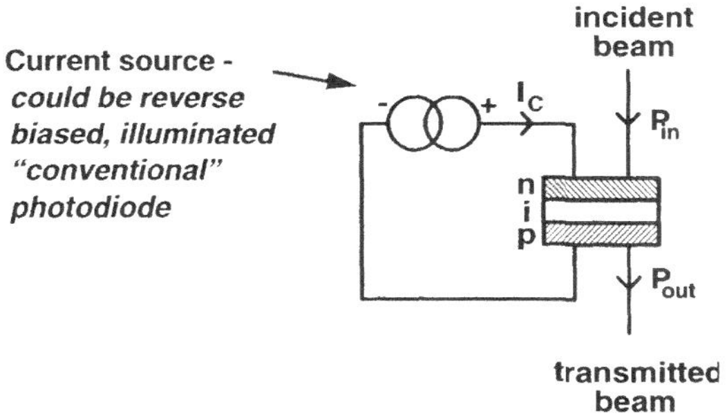

Fig. 19. Circuit for self-linearized modulator operation.

Over most of the reverse bias voltage range, the quantum well diode produces one electron of photocurrent for every photon absorbed. Now suppose we have the circuit as in Fig. 19. The current source is presumed to give the same current independent of voltage. One simple way of making such a current source is to connect a large resistor in series with a large voltage supply. Another more useful current source here is an illuminated conventional photodiode in series with a reverse-biasing voltage supply. A conventional diode can readily be designed to give one electron of photocurrent for every photon incident on it. Regardless of the specific form of the current source, we can analyze the basic mechanism. Suppose that the photocurrent is less than the current source current. Then the diode capacitance will charge up, increasing the voltage across the diode. Because we are operating at λ_1, the absorption and hence the photocurrent will increase. Conversely if the photocurrent is too large, the diode will discharge, reducing absorption and photocurrent. Hence the diode voltage adjusts itself to set the photocurrent equal to the source current. Because the diode current is only photocurrent, we now have an exactly linear modulator that absorbs one photon for every electron of current drive.

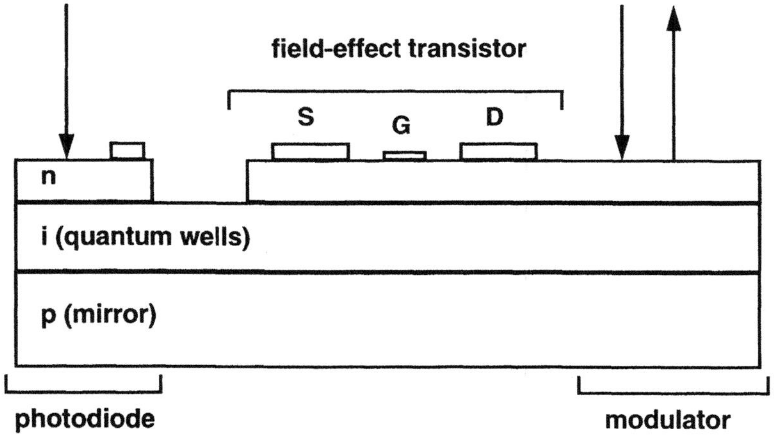

Fig. 20. Concept of the field-effect transistor SEED (FET-SEED). S - source; G - gate; D - drain.

When using the conventional photodiode as the current source, such a circuit can therefore give a linear light-by-light modulator. An interesting recent extension is to differential circuits [24] that can perform image differentiation, subtraction, multiplication, and other analog functions.

Quantum well diodes, operating as both photodetectors and modulators, can also be integrated with transistors. There are at least two reasons to do such integration. One problem found in systems experiments with S-SEED devices is that, although they are relatively low energy switching devices by optical standards, they need too much optical power to allow large, fast systems to be built using present laser technology. Using transistors for electronic gain allows the optical energy requirements to be reduced. A second reason is that electronics is very good at performing complex logic functions, at least locally. Combining electronics with the abilities of optics for interconnection may allow the best of both worlds. Indeed there are many advantages of optics for interconnection that become apparent once a good integration technology is available (see, e.g., ref. [25]).

Fig. 20 shows the concept of one integration, the field-effect transistor SEED (FET-SEED) [26]. Here a quantum well diode is grown on top of a mirror as usual, in this case with the n-layer at the top. Then field effect transistors can be fabricated in the top layer. Hence this concept allows photodetectors for optical inputs, transistors for gain and logic, and quantum well modulators for optical outputs.

Fig. 21 shows the actual layer structure for some current circuits [27]. The present state of the art in this technology is that, in the laboratory, small circuits have been operated with 22 fJ input optical energy, and at speeds up to 650 Mb/s. In actual multistage systems, larger "smart pixels" are operating with 100 fJ optical input energy at 155 Mb/s. A smart pixel array with 96 optical beams and 400 transistors was used in this system [28]. Such operating energies and speeds are of serious interest. For example, with 100 fJ input energy and a factor of 10 loss overall in an optical system, a 1 W laser has sufficient power to drive 1 Tb/s of information through a system. This is a very large data rate, and one that is difficult to achieve with purely electronic systems for a number of reasons (mostly related fundamentally to the skin effect and Maxwell's equations).

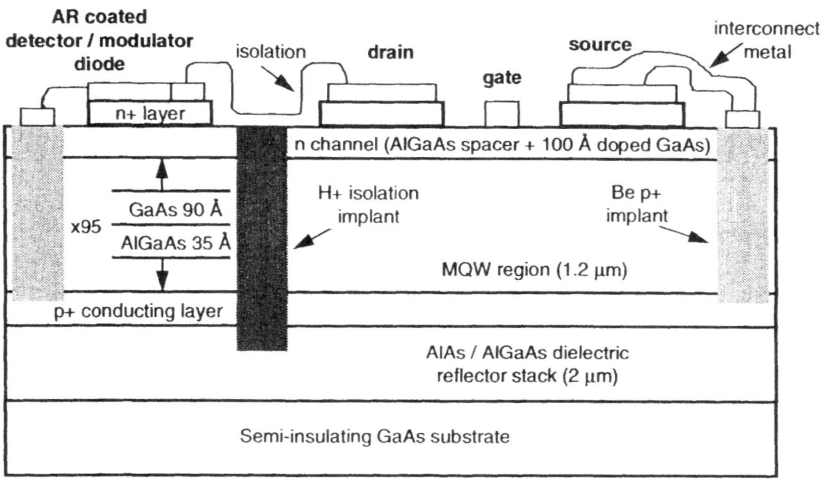

Fig. 21 FET-SEED layer structure and cross-section.

Another approach to such integration is to combine quantum well devices with silicon electronics. There has been success in monolithic integration of quantum well diodes onto silicon [29], and it seems likely that various other hybrid schemes might also be possible

7. Conclusions

The optical properties of quantum wells near to the optical band-gap energy have proved to be a fascinating laboratory for studying many novel physical mechanisms. In addition, several novel and practical devices have resulted that offer significantly new opportunities for optoelectronic systems. We can expect continued evolution of physics, devices, and novel information processing systems in the years to come.

Further Reading

For an introductory summary of quantum well optical physics and devices, see

D. A. B. Miller, "Optoelectronic applications of quantum wells", Optics and Photonics News **1**, No. 2, pp 7-15, February 1990.

For a longer treatment of the physics, see

D. A. B. Miller, D. S. Chemla, and S. Schmitt-Rink, "Electric Field Dependence of Optical Properties of Semiconductor Quantum Wells: Physics and Applications", and

D. S. Chemla, S. Schmitt-Rink, and D. A. B. Miller, "Nonlinear Optical Properties of Semiconductor Quantum Wells", both chapters in "Optical Nonlinearities and Instabilities in Semiconductors", ed. H. Haug, (Academic Press, Boston, 1988)

For an extensive discussion of quantum well optical physics see

S. Schmitt-Rink, D. S. Chemla, and D. A. B. Miller, "Linear and nonlinear optical properties of semiconductor quantum wells", Advances in Physics **38**, 89-188 (1989)

For an extended discussion of band structure and states in quantum wells, see

G. Bastard, "Wave mechanics applied to semiconductor heterostructures", (Les Editions de Physique, Les Ulis, France)

For extended treatments of quantum well optoelectronic devices see

D. A. B. Miller, "Quantum Well Optoelectronic Switching Devices", International Journal of High Speed Electronics **1**, 19-46 (1990)

D. A. B. Miller, "Quantum Well Self Electrooptic-Effect Devices", Optical and Quantum Electronics **22**, S61-S98 (1990).

References

[1] see, e.g., A. Y. Cho, "Advances in molecular beam epitaxy (MBE)", Journal of Crystal Growth **111**, 1-13 (1991)

[2] see, e.g., K. Furuya, and Y. Miyamoto, "GaInAsP/InP organometallic vapor phase epitaxy for research and fabrication of devices", Int. J. High Speed Electronics **1**, 347-67 (1990)

[3] see, e.g., W. T. Tsang, "Progress in chemical beam epitaxy", J. Crystal Growth **105**, 1-29 (1990)

[4] For a recent discussion of envelope function models, including discussion of the boundary conditions, see M. G. Burt, "The justification for applying the effective-mass approximation to microstructures", J. Phys: Condens. Matter **4**, 6651-6690 (1992)

[5] see, e.g., C. Weisbuch, "Fundamental properties of III-V semiconductor two-dimensional quantized structures: the basis for optical and electronic device applications", in "Semiconductors and Semimetals", vol. 24, ed. R. Dingle (Academic Press, New York, 1987), pp 1 - 117

[6] D. S. Chemla, D. A. B. Miller, P. W. Smith, A. C. Gossard, And W. Wiegmann, "Room temperature excitonic nonlinear absorption and refraction in GaAs/AlGaAs multiple quantum well structures", IEEE J. Quantum Electron. **20**, 265-275 (1984)

[7] S. Schmitt-Rink, D. S. Chemla, and D. A. B. Miller, "Theory of transient excitonic optical nonlinearities in semiconductor quantum-well structures", Phys. Rev. B **32** 6601-6609 (1985)

[8] P. W. Smith, Y. Silberberg, and D. A. B. Miller, "Mode locking of semiconductor diode lasers using saturable excitonic nonlinearities, J. Opt. Soc. Am. **B2**, 1228-1236 (1985); Y. K. Chen, M. C. Wu, T. Tanbun-Ek, R. A. Logan, and M. A. Chin, "Subpicosecond monolithic colliding-pulse mode-locked multiple quantum well lasers", Appl. Phys. Lett. **58**, 1253-1255 (1991); U. Keller, G. W. 't Hooft, W. H. Knox, and J. E. Cunningham, "Femtosecond pulses from a continuously self-starting passively mode-locked Ti:sapphire laser", Optics Lett. **16**, 1022-1024 (1991)

[9] D. A. B. Miller, D. S. Chemla, T. C. Damen, A. C. Gossard, W. Wiegmann, T. H. Wood, and C. A. Burrus, "Electric field dependence of optical absorption near the bandgap of quantum well structures", Phys. Rev. B **32**, 1043-1060 (1985)

[10] D. A. B. Miller, J. S. Weiner, and D. S. Chemla, "Electric-field dependence of linear optical properties of quantum well structures: waveguide electroabsorption and sum rules", IEEE J. Quantum Electron. **QE-22**, 1816-1830 (1986)

[11] G. D. Boyd, D. A. B. Miller, D. S. Chemla, S. L. McCall, A. C. Gossard, and J. H. English, "Multiple quantum well reflection modulator", Appl. Phys. Lett. **50**, 1119-1121 (1987)

[12] see, e.g., M. Whitehead, A. Rivers, G. Parry, and J. S. Roberts, "A very low voltage, normally-off asymmetric Fabry-Perot reflection modulator", Electronics Lett. **26**, 1588-1590 (1990)

[13] A. M. Fox, D. A. B. Miller, G. Livescu, J. E. Cunningham, and W. Y. Jan, "Quantum well carrier sweep out: relation to electroabsorption and exciton saturation", IEEE J. Quantum Electron. **27**, 2281-2295 (1991)

[14] J. A. Cavaillès, D. A. B. Miller, J. E. Cunningham, P. Li Kam Wa, and A. Miller, "Simultaneous measurements of electron and hole sweep-out from quantum wells and modeling of photoinduced field screening dynamics", IEEE J. Quantum Electron. **28**, 2486-2497 (1992)

[15] G. D. Boyd, J. A. Cavaillès, L. M. F. Chirovsky, and D. A. B. Miller, "Wavelength dependence of saturation and thermal effects in multiple quantum well modulators", Appl. Phys. Lett. **63**, 1715 - 1717 (1993)

[16] J. S. Weiner, D. A. B. Miller, and D. S. Chemla, "Quadratic electro-optic effect due to the quantum-confined Stark effect in quantum wells", Appl. Phys. Lett. **50**, 842-844 (1987)

[17] J. E. Zucker, K. L. Jones, M. G. Young, B. I. Miller, and U. Koren, "Compact directional coupler switches using quantum well electrorefraction", Appl. Phys. Lett. **55**, 2280-2282 (1989)

[18] D. A. B. Miller, D. S. Chemla, T. C. Damen, T. H. Wood, C. A. Burrus, A. C. Gossard, and W. Wiegmann, "The quantum well self-electrooptic effect device: optoelectronic bistability and oscillation, and self linearized modulation", IEEE J. Quantum Electron. **QE-21**, 1462-1476 (1985)

[19] D. A. B. Miller, "Quantum-well self-electrooptic effect devices", Optical and Quantum Electron. **22**, S61-S98 (1990)

[20] A. L. Lentine, H. S. Hinton, D. A. B. Miller, J. E. Henry, J. E. Cunningham, and L. M. F. Chirovsky, "Symmetric self-electrooptic effect device: optical set-reset latch, differential logic gate, and differential modulator/detector", IEEE J. Quantum Electron. **25**, 1928-1936 (1989)

[21] M. E. Prise, "Optical computing using self-electro-optic effect devices" in "Digital Optical Computing", ed. R. A. Athale, SPIE Critical Reviews of Optical Science and Technology, **CR35**, 3-27 (1990); M. E. Prise et al., "Optical digital processor using arrays of symmetric self-electro-optic-effect devices", Appl. Optics, **30**, 2287-2296 (1991); H. S. Hinton and D. A. B. Miller, "Free-Space Photonics in Switching", AT&T Technical Journal, **71**, No. 1 (Jan/Feb), 84-92 (1992); F. B. McCormick, T. J. Cloonan, F. A. P. Tooley, A. L. Lentine, J. M. Sasian, J. L. Brubaker, R. L. Morrison, S. L. Walker, R. J. Crisci, R. A. Novotny, S. J. Hinterlong, H. S. Hinton, and E.

Kerbis, "Six-stage digital free-space optical switching network using symmetric self-electro-optic-effect devices", Appl. Optics **32**, 5153-5171 (1993).

[22] A. L. Lentine, D. A. B. Miller, J. E. Henry, J. E. Cunningham, and L. M. F. Chirovsky, "Multistate self-electrooptic effect devices", IEEE J. Quantum Electron. **25**, 1921-1927 (1989)

[23] A. L. Lentine, D. A. B. Miller, J. E. Henry, J. E. Cunningham, L. M. F. Chirovsky, and L. A. D'Asaro, "Optical logic using electrically connected quantum well PIN diode modulators and detectors", Appl. Optics **29**, 2153-2163 (1990)

[24] D. A. B. Miller, "Novel analog self-electrooptic-effect devices", IEEE J. Quantum Electron. **29**, 678-698 (1993); E. A. de Souza, L. Carraresi, G. D. Boyd, and D. A. B. Miller, "Analog differential self-linearized quantum-well self-electro-optic-effect modulator", Optics Lett. **18**, 974-976 (1993)

[25] D. A. B. Miller, "Optics for low-energy communication inside digital processors: quantum detectors, sources, and modulators as efficient impedance converters", Optics Lett. **14**, 146-148 (1989)

[26] D. A. B. Miller, M. D. Feuer, T. Y. Chang, S. C. Shunk, J. E. Henry, D. J. Burrows, and D. S. Chemla, "Field-effect transistor self-electrooptic effect device: integrated photodiode, quantum well modulator and transistor", IEEE Phot. Tech. Lett. **1**, 61-64 (1989)

[27] L. A. D'Asaro, L. M. F. Chirovsky, E. J. Laskowski, S. S. Pei, T. K. Woodward, A. L. Lentine, R. E. Leibenguth, J. W. Focht, J. M. Freund, G. G. Guth, and L. E. Smith, "Batch fabrication and operation of GaAs-AlGaAs field-effect transistor-self-electrooptic effect device (FET-SEED) smart pixel arrays", IEEE J. Quantum Electron. **29**, 670-677 (1993)

[28] A. L. Lentine, F. B. McCormick, T. J. Cloonan, J. M. Sasian, R. L. Morrison, M. G. Beckman, S. L. Walker, M. J. Wojcik, S. J. Hinterlong, R. J. Crisci, R. A. Novotny, and H. S. Hinton, "A five stage free-space switching network using arrays of FET-SEED switching nodes", Conference on Lasers and Electrooptics, May 2-7, 1993, Baltimore, Maryland, Postdeadline Paper CPD24 (Optical Society of America, 1993)

[29] K. W. Goossen, G. D. Boyd, J. E. Cunningham, W. Y. Jan, D. A. B. Miller, D. S. Chemla, and R. M. Lum, "GaAs-AlGaAs multiquantum well reflection modulators grown on GaAs and silicon substrates", IEEE Photonics Tech. Lett. **1**, 304-306 (1989)

INTEGRATED OPTICS AND ALL-OPTICAL WAVEGUIDE SWITCHING

GEORGE STEGEMAN and PATRICK LIKAMWA
Center for Reseaerch and Education in Optics and Lasers
University of Central Florida
Orlando, Florida, 32826
USA

Introduction

Nonlinear optics to be efficient, and hence useful, requires a number of parameters to be optimized. All of nonlinear optics is driven by high intensities, i.e. large power per unit area. Any sample geometry which has a small cross-section (to minimize the required power) and which allows high intensities to be maintained over long distances offers a distinct advantage. An optical waveguide meets the above requirements. It can have a cylindrical geometry known as a fiber, and nonlinear optics in fibers is treated elsewhere in this book. Waveguides which derive from planar geometries belong to the field of integrated optics and will be discussed here.

The advantages of waveguides for nonlinear optics was recognized as early as the 1960s when both nonlinear and integrated optics were in their infancies.[1] The initial thrust was into efficient frequency doublers, and this area has seen a renaissance recently for applications in data storage.[1] Interest in waveguide third order phenomena originated with fibers and it is only in the last decade that this field has thrived in integrated optics.[1]

In this chapter we will first review the basic principles of integrated optics needed for a complete discussion of all-optical switching devices. We then show how nonlinear optics in waveguides differs from plane wave nonlinear optics. Instead of reviewing a number of all-optical switching devices, we will examine in depth switching in the nonlinear directional coupler (NLDC) which has proven to be the most versatile in this class of devices. Examples of the issues encountered will be based on NLDCs implemented in semiconductors near the band gap, and for photon energies below one half the band gap.

Integrated Optics: Basic Principles

The field of integrated optics traditionally deals with waveguides operated at power levels for which nonlinearly induced effects are negligible. In almost all cases, useful nonlinear effects are obtained without significant change in the guided mode field distributions. This is the case dealt with here. We note that the situation in which the waveguide modes are altered by high guided wave intensities has been discussed in the literature.[1]

PLANAR (SLAB) WAVEGUIDES

The prototypical integrated optics waveguide is shown in Fig. 1, along with sketches of its corresponding field distributions in Fig. 2.[2] For the classical slab waveguide, a thin film of

A. Miller et al. (eds.), Nonlinear Optical Materials and Devices for Applications in Information Technology, 285–320.

refractive index higher than the two semi-infinite media bounding it is required. The most common and simplest example is a thin film deposited on a substrate (s) with air as the cladding (c) medium. However, there are many other ways of making a planar waveguide. All that is required is a planar region of high index surrounded by media of lower index. Other examples are high index surface regions created by ion-implantation, ion exchange, chemical modification etc. From Fig. 2 it is clear that the field is a maximum inside the high index region, and decays, usually exponentially, with distance from the high index boundaries. Therefore the highest intensities are obtained in the high index regions, and it is therefore most advantageous to have the nonlinearities there.

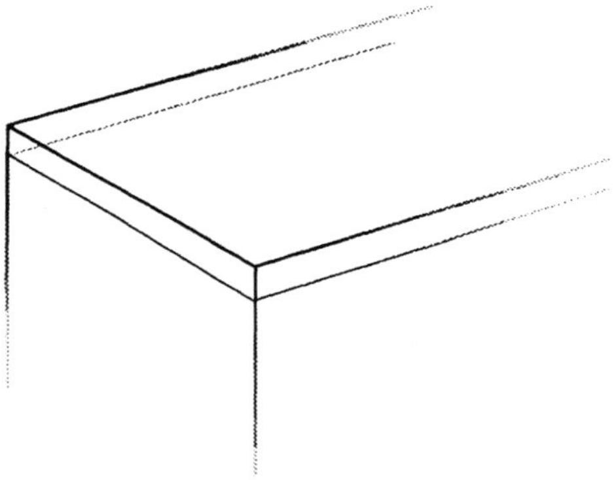

Fig. 1 Slab waveguide consisting of a thin, high refractive index film on a substrate of lower refractive index.

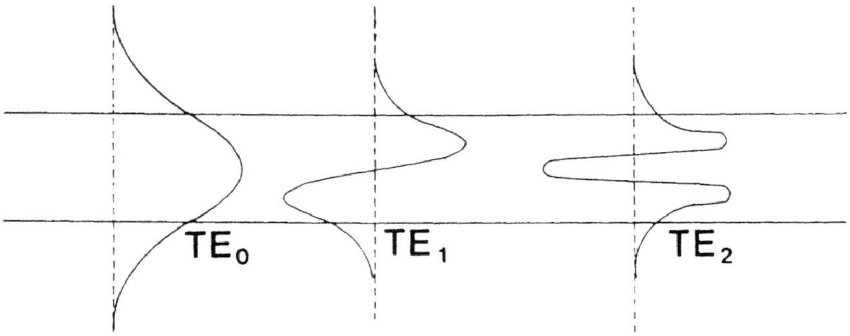

Fig. 2 The field distributions of the three lowest order TE modes of a thin film waveguide. The film boundaries are shown by the solid lines.

There are two orthogonally polarized normal mode fields, labelled TE and TM for the slab case.[2] Assuming for simplicity optically isotropic media, it is easy to show from Maxwell's equations that the field components E_y, H_x and H_z are associated with the TE mode and that the TM mode involves only E_x, E_z and H_y. For more complicated situations, there are still two orthogonal modes, each of which may include all of the field components. Sticking to the isotropic case, the convenient way to write the fields is

$$F_y(\mathbf{r}, t) = \frac{1}{2} F_y(x)\, a(z)\, e^{i(\omega t-\beta z)} + c.c. \tag{1}$$

where F_y denotes E_y for TE modes and H_y for TM modes, and the field amplitudes are chosen so that $|a(z)|^2$ is the guided wave power per unit distance along the wavefront (y-axis). The propagation wavevector β is bounded by $n_f\omega/c > \beta > \mathrm{Max}[n_c,n_s]\omega/c$. Dropping the subscript y to avoid confusion when the subscripts c, f and s are used to describe the cladding, waveguiding and substrate regions respectively, the transverse field distributions are given by

$$F(x) = F_c e^{\kappa_c x} \qquad x \le 0 \tag{2}$$

$$F(x) = F_f[\cos(\kappa_f x) + R_f \sin(\kappa_f x)] \qquad h \ge x \ge 0 \tag{3}$$

$$F(x) = F_s e^{-\kappa_s [x-h]} \qquad x \ge h \tag{4}$$

The propagation wavevector β is related to the media properties by substituting into the wave equation, i.e. $\beta^2 = n_{(s,c)}{}^2\omega^2/c^2 + \kappa_{(s,c)}{}^2$ and $\kappa_f{}^2 = n_f{}^2\omega^2/c^2 - \beta^2$ where the κ_i are the "transverse" wavevectors.

The relations between the field parameters (F_c, F_f, R_f and F_s) are obtained by satisfying the electromagnetic boundary conditions across each interface. That is, E_y, E_z, H_y and H_z are continuous across the film-cladding and film-substrate interfaces. For example, the y-component of the electric field is continuous across the x=0 interface so that $F_c = F_f$ etc. Solving yields the ratios F_f/F_c, $F_f R_f/F_c$ and F_s/F_c, and the dispersion relation $D(\beta,h) = 0$. Note that the values are different for TE versus TM modes. Furthermore, F_c is fixed by the normalization chosen for $|a(z)|^2$. For example, for TE modes ($F_y \rightarrow E_y$),

$$E_c = 2\kappa_f \left[\frac{\beta}{\omega\mu_0}\left(h + \frac{1}{\kappa_c} + \frac{1}{\kappa_s}\right)(\kappa_f^2 + \kappa_c^2)\right]^{-1/2} \tag{5}$$

$$\frac{E_f}{E_c} = 1 \qquad R_f = \frac{\kappa_c}{\kappa_f}$$
$$\frac{E_s}{E_c} = \cos[\kappa_f h] + R_f \sin[\kappa_f h] \tag{6}$$

and

$$D(\beta, h) = \tan[\kappa_f h - m\pi] - \frac{\kappa_f(\kappa_c + \kappa_s)}{\kappa_f^2 - \kappa_c\kappa_s} \tag{7}$$

where the angle argument of the tangent function is restricted via the $m\pi$ term to the range $-\pi/2$ to $\pi/2$. The results for TM modes are similar.

Solutions to the dispersion relation given by Eqn. 7 yield the normal modes. For a given thickness h, there are discrete values allowed for β, i.e. β_m with each corresponding to a unique field distribution $E^{(m)}$ (sketched previously in Fig. 2). The modes are orthogonal to each other so that the normalization condition becomes

$$\frac{1}{2}\int_{-\infty}^{\infty}[E_x^{(m)}(x)\,H_y^{(n)*}(x) - E_y^{(m)}(x)\,H_x^{(n)*}(x)]\,dx = \delta_{m,n} \tag{8}$$

in MKS units. For TE_m, there are m field maxima in E_y within the boundary of the film. Correspondingly for TM_m the field H_y has m field maxima. There is a "cut-off" thickness h_m below which the m'th mode will not propagate. The dispersion in β_m with normalized thickness $h\omega/c$ (hk_0) is shown in Fig. 3. In general, β_m (= $N_m k_0$ where N_m is the "effective index") varies from the larger of $n_c k_0$ and $n_s k_0$ at cut-off, to $n_f k_0$ as $h \rightarrow \infty$. Note that in this limit, the modes essentially become plane waves travelling along the z-direction. These features are also exhibited by TM_m modes, but with $\beta_m(TE) > \beta_m(TM)$ for optically isotropic media.

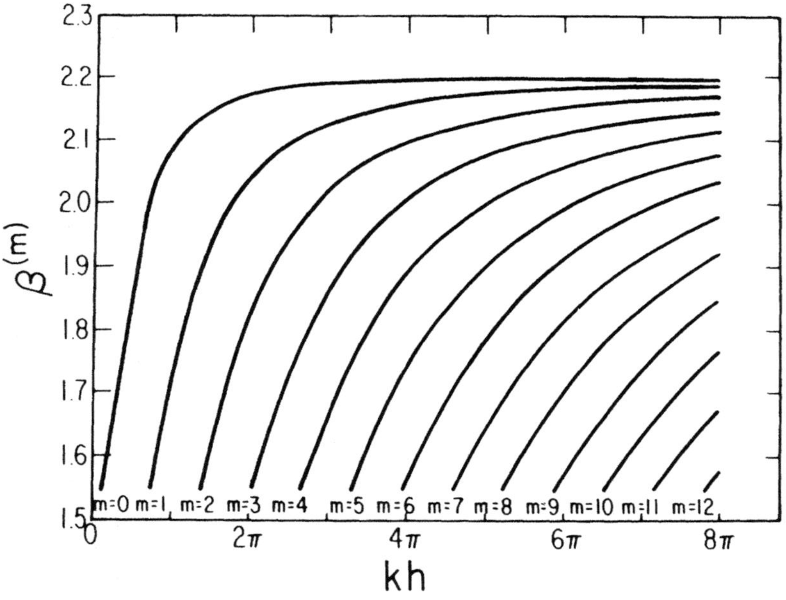

Fig. 3 Variation in the propagation wavevector for different TE_m modes with normalized film thickness (k_0h) for a slab waveguide with a substrate of index 1.54, a film of index 2.2 and air as the cladding.

Here we are primarily interested in nonlinear optical phenomena for which one would like to maximize the local intensity. The key question is: For a given input power in a beam W wide (along y), what is the film thickness which optimizes the intensity? Note that in Eqn. 6 an effective thickness $h + \kappa_c^{-1} + \kappa_s^{-1}$ appears and indeed this is a good approximation for the effective optical field E_y width along the x-direction. Note that the absolute minimum for h_{eff} occurs for m = 0, and most devices indeed do utilize the lowest order modes possible. For small h the field penetrates a long way into the bounding medium for which the index is the highest. For large h, the thickness h dominates and again the intensity is small. Between these two limits one finds the film thickness which optimizes the intensity. It depends on the details of the material optical constants. Note also that the optimum film thickness will depend on the frequency ω. This is evident from Fig. 3 in which the modes depend on the "normalized" thickness $k_0h = \omega h/c$.

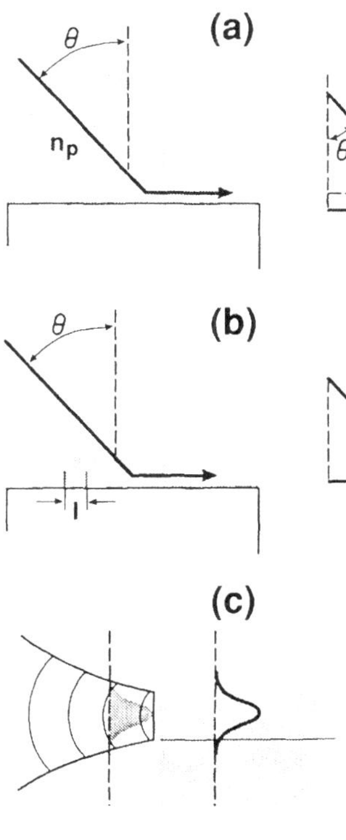

Fig. 4 Techniques for coupling radiation fields into a waveguide. (a) Prism coupler and the wavevector matching condition. (b) Grating coupler with the wavevector matching condition ($l \equiv L_p$). (c) End-fire coupling with the field profile matching condition.

Just how does one excite a waveguide mode? Clearly illuminating the film with a plane wave (or focussed beam) through one of the bounding media will not work since $\beta > n_{c,s}\omega/c$, i.e. it is impossible to match the incident to the guided wave wavevectors. Note however that if one of the interfaces is corrugated, or there is a periodic perturbation of the refractive indices along z, then $\beta = n_{c,s}k_0\sin\theta \pm K$ can be satisfied where θ is the angle of incidence from the normal to the film interfaces in medium n_c or n_s. Also $K = 2\pi/L_p$ where L_p is the period associated with the index or surface modulation. Such a scheme is shown in schematically Fig. 4. Coupling via wavevector matching can also be implemented using a high index prism separated from the film by a thin cladding layer, usually air. The idea is that light is incident onto the base of the prism at an angle θ so that $n_p k_0 \sin\theta = \beta$ where the prism index $n_p > \mathrm{Max}[n_c, n_s]$. Note that by adjusting the incidence angle it is possible to preferentially excite different modes (labelled by m). TE versus TM modes are discriminated by controlling the input polarization. Clearly the field in the gap must be evanescent because $n_p k_0 \sin\theta = \beta > n_c k_0$ and the coupling efficiency depends critically on the size of this gap. The dependence of the coupling efficiency, η, on waveguide parameters has been reported, and values of η in excess of 10% are easy to achieve.

The coupling techniques discussed above relied on matching the wavevector of the excitation field to that of the guided mode. Another approach is to match the field profiles of the incident to guided modes at the end-face of the sample as shown in Fig. 4c using a cylindrical lens. If the incident field is not well-matched in position and transverse profile to the guided wave mode, a superposition of modes is excited in multimode waveguides. This approach is best suited for the lowest order mode since higher modes require input fields with oscillating transverse profiles.

CHANNEL WAVEGUIDES

Although planar waveguides provide diffractionless propagation and a cross-section of the order of the wavelength of light along the direction normal to the interfaces, diffraction still occurs in the plane of the surface. To maintain high intensities, confinement in the second dimension, i.e. along the y-axis is also desirable. This requires the fabrication of structures in which a two-dimensional region of high index is surrounded on all sides by media of lower index, as shown in Fig. 5a.[2] Samples of other two-dimensional guiding structures are also given in Fig. 5. Representative field distributions are reproduced in Fig. 6, showing two-dimensional confinement.

The analysis of "channel" waveguides follows the same principles outlined in the preceding section for slab waveguides. That is, the fields are the solutions to the wave equation in each of the media, and the usual transverse field components must be continuous across each boundary. This problem cannot be solved analytically and requires numerical technique.[2,3] For example, the total field can be written as an expansion of analytical field functions especially chosen to reflect the symmetry of the channel waveguide. In the high index region, the functions usually have some oscillatory nature, and decay with distance from the boundaries in the bounding media. Therefore the functions will be different in the "core" and bounding regions. The relative weighing of the different terms is determined by imposing the usual boundary conditions at a discrete number of points along the interfaces. The larger the number of functions chosen, the denser the number of points along the boundaries at which the field components can be forced to be continuous, and the better the approximation to the actual field distribution. The net result of this procedure is to find the modal wavevectors β and their corresponding field distributions.

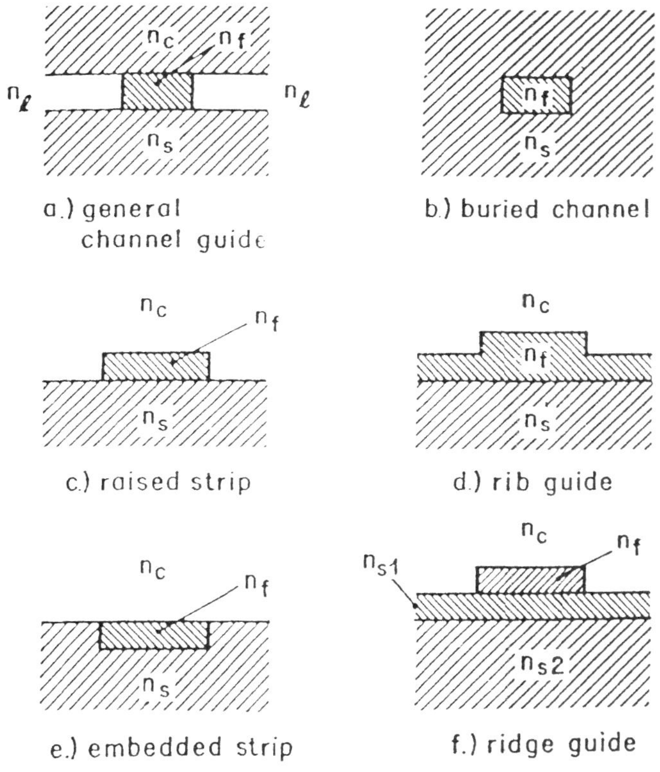

Fig. 5 Various channel waveguides and their designated names.[2]

There is an approximate technique known as the "effective index method" which can be used to estimate the propagation wavevector from slab waveguide formulae.[2] The procedure is shown schematically in Fig. 7. First, three planar waveguides are assumed with their normals along the x-axis. In the example shown, the effective index is calculated for film waveguides of thickness "t" and "t-h" respectively with the material refractive indices n_f, n_c and n_s, i.e. three slab waveguides (2 of which are the same) differing only in film thickness. This procedure yields effective indices N_f and N_t respectively. Next, one treats a "new" slab waveguide with cladding and substrate indices N_t, and a film of thickness w and index N_f. The effective index of this artificial guide should be a good approximation to N_{eff} for the channel.

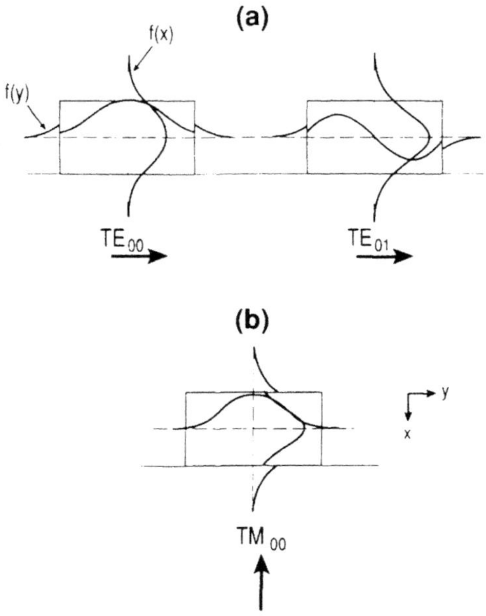

Fig. 6 Typical field distributions for a channel waveguide, and the connection between nomenclature and the dominant electric field.

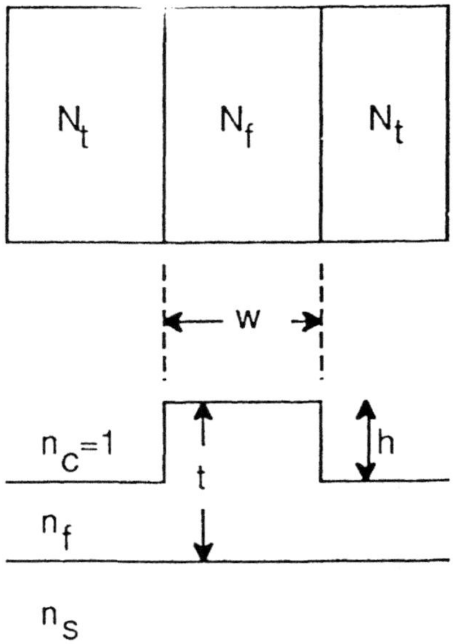

Fig. 7 Schematic representation of the effective index method. The bottom picture is the end facet view of the rib waveguide. The top picture is the effective slab waveguide.

The field distributions now exhibit oscillatory behavior in the core across both dimensions, i.e. along x and y. Just as in the planar case, the propagation wavevector β is identified by the number of zeros of the field inside the core across the two dimensions, i.e. β_{mn} where m and n are associated with the x and y coordinates respectively. β_{mn} disperses with both h_x and h_y for the case of rectangular channels with dimensions h_x and h_y in a fashion similar to the slab case of Fig. 3. Note that waveguides can be single mode in one dimension, e.g. m=0 and multimode in the other, e.g. n > 0.

The field distributions are more complicated than for the slab case. There is a dominant field component which identifies the mode as primarily TE_{mn} or TM_{mn}. For TE_{mn} modes, the dominant field component is E_y, where-as for TM_{mn} it is H_y. Because boundary conditions must now be satisfied in two dimensions, each mode contains all three field components. For example, for TE_{mn}, $E_y > E_x$ and E_z. The (m,n)th modal field is now written as

$$F_y^{(m,n)}(\mathbf{r},t) = \frac{1}{2} F_y^{(m,n)}(x,y)\, a_{m,n}(z)\, e^{i(\omega t - \beta_{m,n} z)} + c.c. \qquad (9)$$

where $F_y^{(m,n)}(x,y)$ denotes $E_y^{(m,n)}(x,y)$ for TE modes and $H_y^{(m,n)}(x,y)$ for TM modes, and the field amplitudes are chosen so that $|a_{m,n}(z)|^2$ is the guided wave power. This requires for each mode that

$$\frac{1}{2}\int_{-\infty}^{\infty}\int_{-\infty}^{\infty}[E_x^{(m,n)}(x,y)\, H_y^{(p,q)*}(x,y) - E_y^{(m,n)}(x,y)\, H_x^{(p,q)*}(x,y)]\, dx\, dy = \delta_{m,p}\delta_{n,q} \qquad (10)$$

in MKS units.

Channel waveguide modes can be excited with prisms or gratings with the incident radiation in the form of a line source along the z-axis. This proves to be inconvenient. Normally, end-fire coupling is used as sketched in Fig. 4c. With care, sufficiently good field overlap can be achieved to obtain > 80% coupling efficiency. For nonlinear optics, the end face must be very smooth and clean in order to avoid end-face damage at high powers.

COUPLED MODE THEORY

A formalism is needed to evaluate the effect on guided modes of a perturbation.[2] The equivalent is the "slowly varying phase and amplitude approximation" in plane wave optics. The primary difference between plane and guided waves is that the transverse distributions of guided wave fields are finite in extent where-as plane waves are usually considered to fill all space. First, any field near the waveguide is expanded into normal modes which include guided waves, as well as radiation fields which can enter and leave the waveguide volume. For example, we write

$$\boldsymbol{E}(\boldsymbol{r},t) = \sum \frac{1}{2}\boldsymbol{F}^{(m,n)}(x,y)\, a_{m,n}(z) \exp i(\omega t - \beta_{m,n} z) + \frac{1}{2}\int \boldsymbol{F}(x,y)\, b(z) \exp i(\omega t - \beta z)\, d\beta + c.c. \tag{11}$$

where the summation is over the guided modes and the integral is over the radiation modes. For our purposes we are going to ignore the radiation modes since they will at best be very weakly excited in the cases of interest here. Most nonlinear optical perturbations can be expressed in the form of a polarization source term, for example of the form

$$\boldsymbol{P}^{P}(\boldsymbol{r},t) = \frac{1}{2}\boldsymbol{P}^{P}(x,y,z) \exp[i(\omega t - \beta_p z)] + c.c. \tag{12}$$

This polarization generates fields (given by Eq. 11) via the polarization driven wave equation:

$$-\nabla^2 \boldsymbol{E}(\boldsymbol{r},t) + \frac{n^2}{c^2}\frac{\partial^2}{\partial t^2}\boldsymbol{E}(\boldsymbol{r},t) = -\mu_0 \frac{\partial^2}{\partial t^2}\boldsymbol{P}^{P}(\boldsymbol{r},t) \tag{13}$$

Noting that the individual fields in Eq. 11 each satisfy the wave equation, retaining only the waveguide terms, multiplying both sides by $\mathbf{F}^{(m,n)*}(x,y)$, integrating over x and y, and using mode orthogonality gives:

$$\frac{d}{dz} a_{m,n}(z) = i \frac{k_{vac}}{2N_{m,n}\varepsilon_0} \frac{\iint \boldsymbol{P}^{P}(x,y,z) \cdot \boldsymbol{F}^{(m,n)*}(x,y)\, dxdy}{\iint |\boldsymbol{F}^{(m,n)}(x,y)|^2 dxdy} \exp[i(\beta_{m,n} - \beta_p) \tag{14}$$

where we need to assume that the z-variation in $\mathbf{P}^{P}(x,y,z)$ is slow over one wavelength. Integrating over the z-coordinate gives the complex magnitude $a_{m,n}(z)$ of the (m,n)'th normal mode which is excited by the polarization source.

Nonlinear Refractive Index Effects in Waveguides

The perturbed polarization appropriate to an intensity-dependent refractive index can easily be written in terms of the Fourier components of the general third order susceptibility tensor $\chi^{(3)}(-\omega_4; \omega_3, \omega_2, \omega_1)$.[4] For this case $\omega_1 = \omega_3 = -\omega_2$ so that $\omega_4 = \omega_1 = \omega$. Assuming for simplicity isotropic media, the nonlinear polarization induced at the frequency ω_1 due to an intense beam at ω_2 (or at ω_1 for crossed polarized beams of equal frequency) can be written in the form

$$P_i^{P}(\boldsymbol{r},\omega_1) = \epsilon_0 \chi^{(3)}_{ijji}(-\omega_1;\omega_2,-\omega_2,\omega_1)\, E_j(\boldsymbol{r},\omega_2)\, E_j^{*}(\boldsymbol{r},\omega_2)\, E_i(\boldsymbol{r},\omega_1) \tag{15}$$

The total polarization induced, including the linear term associated with the refractive index, is

The rationale for a field-dependent refractive index stems from the terms in the square brackets which can be interpreted as $n^2 - 1$ where n is the refractive index. As a result we can write the index experienced by the ω_1 field as

$$P_i(\boldsymbol{r},\omega_1) = \epsilon_0[\chi_{ii}^{(1)}(-\omega_1;\omega_1) + \chi_{ijji}^{(3)}(-\omega_1;\omega_2,-\omega_2,\omega_1)|E_j(\boldsymbol{r},\omega_2)|^2]E_i(\boldsymbol{r},\omega_1) \tag{16}$$

$$n(\omega_1) = n_0(\omega_1) + n_2(\omega_1;\omega)I_2 + n_2(\omega_1;\omega_1)I_1 \tag{17}$$

where I_1 and I_2 are the intensities of the ω_1 and ω_2 beams respectively. Note that if beam 2 is at ω_1, but orthogonally polarized to beam 1, this case is retrieved with $\omega = \omega_1$. Given our definition of the susceptibility in Eqn. 15, the nonlinear index coefficient for plane waves is given by

$$n_2(\omega_1;\omega_1) = \frac{3\chi_{1111}^{(3)}(-\omega_1;\omega_1,-\omega_1,\omega_1)}{n_0^2(\omega_1)c\epsilon_0} \tag{18}$$

$$n_{2\parallel} = \frac{6\chi_{1111}^{(3)}(-\omega_1;\omega_2,-\omega_2,\omega_1)}{n_0^2(\omega_1)c\epsilon_0} \tag{19}$$

$$n_{2\perp} = \frac{6\chi_{1221}^{(3)}(-\omega_1;\omega_2,\omega_2,\omega_1)}{n_0^2(\omega_1)c\epsilon_0} \tag{20}$$

$$n_{2\perp} = \frac{6\chi_{1221}^{(3)}(-\omega_1;\omega_1,-\omega_1,\omega_1)}{n_0^2(\omega_1)c\epsilon_0} \tag{21}$$

where the $\parallel$ and $\perp$ subscripts refer to the I_2 beam being polarized either parallel or perpendicular to the ω_1 beam. For isotropic media and for ω_1 and ω_2 sufficiently close to each other so that there is no dispersion in the linear or nonlinear susceptibility, $n_{2\parallel}(\omega_1;\omega_2) = 2n_2(\omega_1;\omega_1) = 3n_{2\perp}(\omega_1;\omega_2) = 3n_{2\perp}(\omega_1;\omega_1)$.

In extending these concepts to guided waves, two differences between plane and guided waves need to be noted. First, the guided wave fields vary across the transverse (to the propagation wavevector) dimension. Second, field components exist along all three axes, although usually only one electric field component is dominant. Technically, one substitutes the expression for the nonlinear polarization, Eqn. 15, into Eqn. 14.[5] For beam 1 this gives

$$\beta = \beta_0 + \Delta\beta_{11}P_1 + \Delta\beta_{12}P_2 \tag{22}$$

where $P_1 = |a_{m,n}|^2$ and $P_2 = |a_{p,q}|^2$ are the powers of guided waves 1 and 2. Here

$$\Delta\beta_{ij} = \frac{\omega_1}{2cN_{m,n}} \frac{\iint \chi^{(3)}_{ijji}(-\omega_i;\omega_j,-\omega_j,\omega_i)\,|E^{(m,n)}(\boldsymbol{r},\omega_i)|^2 |E^{(p,q)}(\boldsymbol{r},\omega_j)|^2}{\iint |E^{(m,n)}(\boldsymbol{r},\omega_i)|^2 dxdy} \tag{23}$$

To briefly summarize, an intensity-dependent refractive index leads to a power-dependent propagation constant. The details of this constant depends on the mode polarizations and mode numbers involved.

Theory of the Nonlinear Directional Coupler

A number of all-optical switching devices have now been implemented in waveguides.[1] There has, however, been one device which has by far been the most frequently studied. That is the nonlinear directional coupler shown in Fig. 8. (The electro-optic version of this device has proven very popular in integrated optics for switching.) This is a very versatile four port device in which there are two input and output ports.

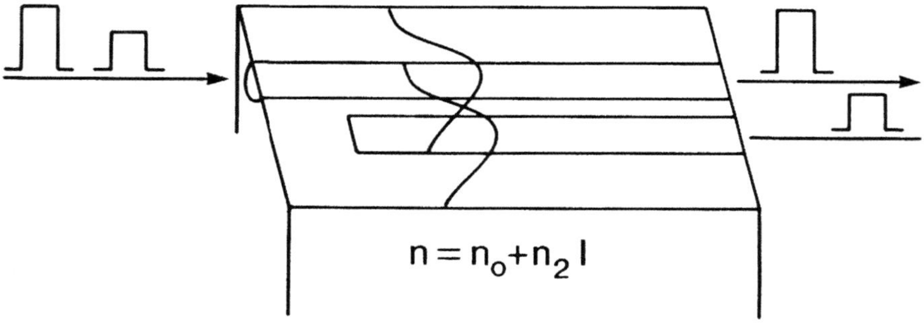

Fig. 8 Schematic picture of a nonlinear directional coupler showing the guide mode field overlap.

There are two separate phenomena occurring simultaneously in the nonlinear directional coupler. The perturbation polarization is written as $P^p = P^p$(waveguide coupling) + P^P(nonlinear effects).[6] Waveguide coupling is a result of field overlap between waveguide modes guided by the two parallel channel waveguides. This coupling can described as follows. The unperturbed system consists of two **isolated** waveguides. Therefore the perturbation polarization induced in the vicinity of waveguide 2 by a field in waveguide 1 is given by

$$P_i^P(\mathbf{r},t) = \frac{1}{2}\epsilon_0 n^2(\mathbf{r}) E_i^{(m,n)}(\mathbf{r}) a_{m,n}(z) e^{i(\omega t - \beta_{m,n} z)} + cc \qquad (24)$$

As far as the second waveguide is concerned, this is an additional polarization that would not be present if the first waveguide was far away. Therefore this polarization drives a field in waveguide 2 which can be quantified by substituting into Eq. 14. This gives

$$\frac{d}{dz} a_{p,q}(z) = i\kappa_{21} a_{m,n}(z) \exp[i\Delta\beta z] \qquad (25)$$

where

$$\kappa_{21} = \frac{\omega}{2cN_{p,q}} \frac{\iint n^2(\mathbf{r}) E_i^{(m,n)}(\mathbf{r}) E_i^{(p,q)*}(\mathbf{r})\, dxdy}{\iint |E^{(p,q)}(\mathbf{r})|^2 dxdy} \qquad (26)$$

and $\Delta\beta = \beta_{p,q} - \beta_{m,n}$. Similarly the field in waveguide 1 is coupled via

$$\frac{d}{dz} a_{m,n}(z) = i\kappa_{12} a_{p,q}(z) \exp[-i\Delta\beta z] \qquad (27)$$

where

$$\kappa_{12} = \frac{\omega}{2cN_{m,n}} \frac{\iint n^2(\mathbf{r}) E_i^{(p,q)}(\mathbf{r}) E_i^{(m,n)*}(\mathbf{r})\, dxdy}{\iint |E^{(m,n)}(\mathbf{r})|^2 dxdy} \qquad (28)$$

to the field in waveguide 2. Note that in the usual case the waveguides are fabricated to be as similar as possible, and are usually single mode in both transverse dimensions, i.e. (m,n) = (p,q) = (0,0). Unfortunately small wavevector mismatches $\Delta\beta$ sometimes occur and will be kept in the analysis. In such a case the field distributions are effectively the same, simplifying the calculation of the coupling coefficients so that $\kappa_{12} = \kappa_{21}$. These equations now describe the linear directional coupler.

There is also a nonlinear polarization induced when the fields inside the coupler are large.[6] We assume that the fields are co-polarized which is the usual case, and label the fields in waveguides 1 and 2 by subscripts 1 and 2. For this particular case the subscripts do not identify polarization components. We need to examine the field product $[E_1(r,t) + E_2(r,t)]^3$ and isolate the terms which oscillate like $\exp[i\omega t]$. They are of the form $|E_1|^2E_1$, $|E_2|^2E_2$, $|E_1|^2E_2$, $|E_2|^2E_1$, $E_1^2E_2^*$ and $E_2^2E_1^*$. It is possible to include all of these terms in the analysis. To simplify the problem we assume that the two waveguides are sufficiently separated so that the overlap of the fields is weak and hence only the self phase modulations terms $|E_1|^2E_1$ and $|E_2|^2E_2$ are retained. In actual fact this is an excellent approximation for couplers of length greater than 1 mm, the usual case, and a useful approximation for couplers >0.1 mm long. This simplifies the nonlinear polarization to

$$P^{NL}(\mathbf{r},t) = \frac{3}{8}\epsilon_0\chi^{(3)}[|E_1(\mathbf{r})|^2E_1(\mathbf{r})\exp(-i\beta_1 z) + |E_2(\mathbf{r})|^2E_2(\mathbf{r})\exp(-i\beta_2 z)]\exp(i\omega t) + c.c. \quad (29)$$

Substituting into Eqn. 14 now gives:

$$\frac{d}{dz}a_1(z) = i\Delta\beta_{11}|a_1(z)|^2a_1(z)$$
$$\frac{d}{dz}a_2(z) = i\Delta\beta_{22}|a_2(z)|^2a_2(z) \quad (30)$$

We now note that both perturbations are small and essentially decoupled from one another. Therefore they can be combined using Eqns. 25, 27 and 30 to give the coupled mode equations

$$-i\frac{d}{dz}a_1(z) = \kappa a_2(z)\exp[i\Delta\beta z] + \Delta\beta_{11}|a_1(z)|^2a_1(z)$$
$$-i\frac{d}{dz}a_2(z) = \kappa a_1(z)\exp[-i\Delta\beta z] + \Delta\beta_{22}|a_2(z)|^2a_2(z) \quad (31)$$

which describe the operation of a nonlinear directional coupler. Note that for the usual case where the two channels are almost identical, $\kappa_{12} = \kappa_{21}$ and $\Delta\beta_{11} = \Delta\beta_{22}$.

These equations have analytical solutions for certain boundary conditions, i.e. input conditions at z = 0.[6] Note that for signals input into both channels, the relative phase between them is important. For the simplest case of an input into one channel only, say channel 1 with $[P_1(0)]$, the field in channel 1 is given by

$$P_1(Z) = P_1(0)\frac{[1 + cn(2Z|m)]}{2} \qquad m = \frac{P_1^2(0)}{P_c^2} \quad (32)$$

where $Z = \kappa z$ and the critical power $P_c = 4\kappa/\Delta\beta_{11}$. Here $cn(\phi\,|\,m)$ is a Jacobi elliptic function. Because power is clearly conserved in this system, $P_2(z) = P_1(0) - P_1(z)$. The detailed dependence of the power evolution in channel 1 with distance is shown in Fig. 9.

The detailed behavior of this device depends on the input power. At low powers, the effects of the nonlinearity are minimal and power is periodically (with distance) exchanged between the two channels. The length for complete transfer from one waveguide to the other is given by $L_c = \pi/2\kappa$. This is called the "half beat length coupler". As the input power is increased, the period of oscillation increases and eventually goes to infinity at the critical power. At that power (and just exactly that power), the output at infinity is split between the two channels. For P_c - ΔP, the output eventually returns completely to channel 1 and oscillates slowly with propagation distance between 0 and approximately one half. For $P_c + \Delta P$, the power remaining in channel 2 oscillates between 1 and about 0.5, with the oscillation period increasing as $\Delta P \rightarrow 0$. For $P_1(0) > P_c$, the power stays primarily in channel 1 (the input channel) and the oscillation period is halved. Therefore in a half beat length coupler, at low input powers the output appears in channel 2, and at high powers stays in channel 1. The complete power evolution versus incident power is shown in Fig. 10.

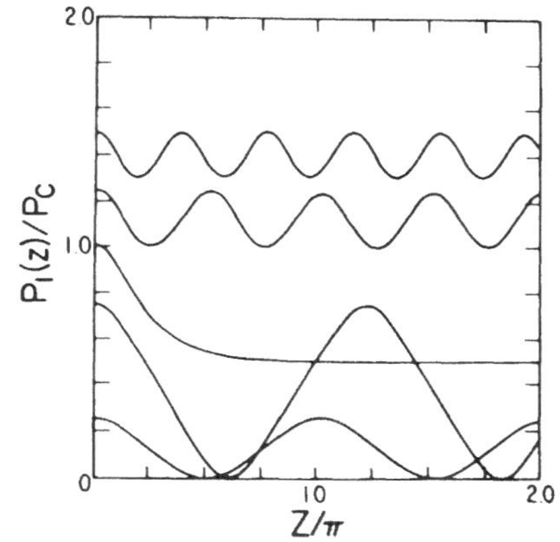

Fig. 9 Power remaining in the input waveguide, normalized to the critical power, as a function of normalized distance down the waveguide.

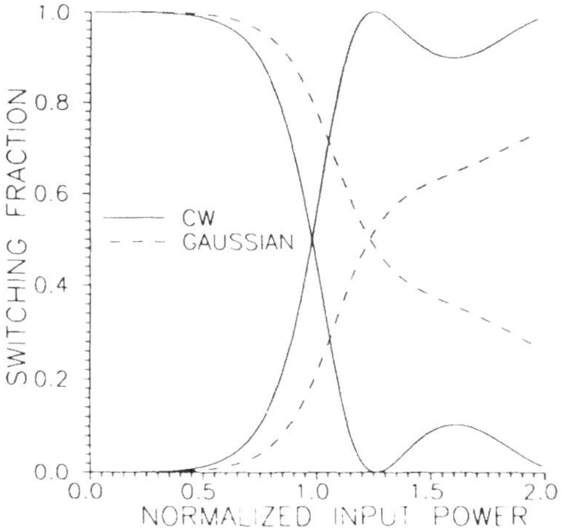

Fig. 10 Switching characteristics of a half-beat length NLDC as a function of input power (normalized to the critical power). The curves starting at 0 and 1 identify the bar and cross states respectively. The response for a cw input (solid line) and a gaussian pulse input (dashed line) are shown.

Complete switching is obtained for $P_s = 1.25\ P_c$. Note that this response is for a cw input signal. In actual fact, when pulses are used, different parts of the pulse are switched with different efficiencies (from Fig. 10), leading to distortion in the output pulse.[7] Shown as the dashed line in Fig. 10 is the case for a gaussian pulse in time where the integral over the pulse temporal cross-section has been carried out, i.e. the dashed line gives the switching response for the pulse energy. Clearly complete switching is not easily obtained since the low power part of the pulse tends to stay in the cross channel. The corresponding pulse distortion is shown in Fig. 11 for different power levels. Note that accompanying the temporal pulse distortion is a distortion of the frequency spectrum of the pulse.

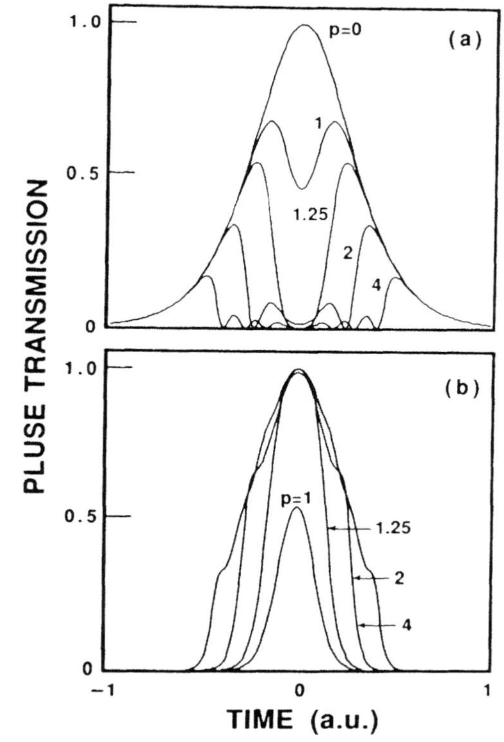

Fig. 11 Temporal profile of pulses output from the bar (a) and cross (b) states of a half-beat length NLDC with a Kerr nonlinear response and with input pulses of variable power $p = P(0)/P_c$.

Ultrafast Kerr Materials

The key question now is the choice of material to implement such an all-optical switch.[1] We can obtain considerable insight into how to choose the right material by examining the minimum nonlinear phase shift required from the input channel of a nonlinear directional coupler, i.e.

$\Delta\phi^{NL} = \Delta\beta_0 P_c L_c = 2\pi$. That is, at least 2π of nonlinear phase shift must be available. (For a material with a saturating index change, i.e. $\Delta n \to \Delta n_{max}$ as $I \to \infty$, the minimum nonlinear phase shift required is 4π.)

We now inquire as to what mechanisms can limit the nonlinear phase shift in a material. The input intensity required for switching a given device length L_c may surpass the damage threshold, i.e. there is a Δn_{max} just before material breakdown. Alternatively, the nonlinear phase shift may saturate because of the physics of the nonlinear mechanism involved, again giving a Δn_{max}. This could happen, for example, in a two level atom nonlinearity in which photon absorption equalizes the population of the two levels, totally bleaching out the dispersion in the refractive index. But, in principle it is possible to increase the device length so that the required phase shift is acquired at input intensities less than those required for damage or saturation. However, the absorption α, both linear (α_1) and nonlinear ($\alpha_2 I + \alpha_3 I^2 + ..$) can limit the effective device length to α^{-1}.

It proves convenient to use the concepts discussed above to define figures of merit (FOM) for comparing materials. In the first case we assume 1), that there is a maximum index change Δn_{max}, 2) that the effective device length is limited by linear absorption, i.e. $\alpha_1^{-1} > L$, and 3) that a phase shift of 2π is required for switching (as discussed previously). This gives

$$W = \frac{\Delta\phi^{NL}}{2\pi} = \frac{\Delta n_{max}}{\alpha_1 \lambda} > 1 \tag{33}$$

$W > 8$ is required for >90% device throughput. Note that if $L > \alpha_1^{-1}$, switching can be still obtained, but with large throughput losses and incomplete switching. That is the light can never fully return to the input channel in a nonlinear directional coupler, even for the cw case.

In the second case, two photon absorption dominates the loss, i.e. $\alpha_2 I >> \alpha_1$. Assuming again that $L < [\alpha_2 I]^{-1}$ in order to obtain reasonable throughput, then

$$T^{-1} = \frac{\Delta\phi^{NL}}{2\pi} = \frac{n_2}{\alpha_2 \lambda} > 1 \tag{34}$$

The variation in the nonlinear phase shift $\Delta\phi^{NL}$ and the transmission versus the parameter $\alpha_2 I(0)L$ is shown for representative T in Fig. 12. The trade-offs are clear. When there is both linear and two photon loss, the combined figure of merit becomes

$$\frac{W}{1 + WT} > 1 \tag{35}$$

One can continue to generate additional FOM by considering higher order multiphoton absorption processes. Described in some detail in reference 8 are experimental examples of how switching in NLDCs is affected when materials are used which do not satisfy the figures of merit, or which have nonlinearities dominated by thermal effects.

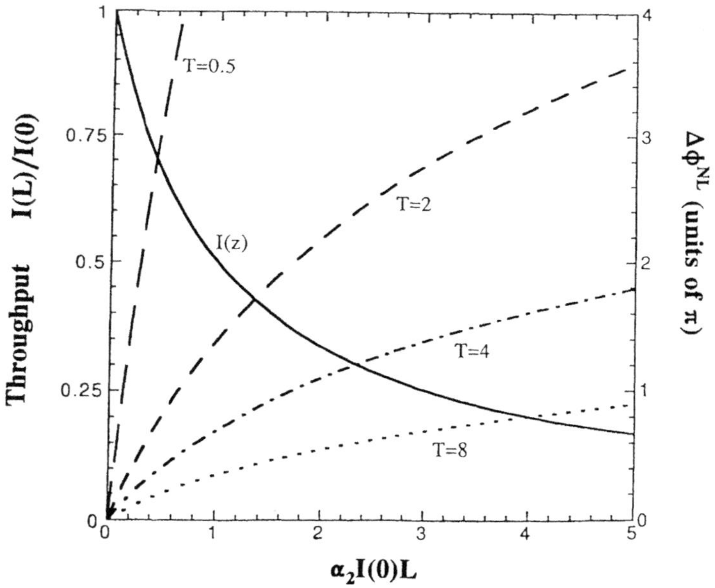

Fig. 12 Variation in intensity and nonlinear phase shift with normalized propagation distance $\alpha_2 I(0)L$.

One of the problems in choosing appropriate Kerr materials is that measurements of n_2 (ultrafast only), α_1, and α_2 over broad spectral ranges are rare. A listing of representative materials is shown in Table 1. Clearly semiconductors used with photon energies just below one half of their bandgap, some organic materials over selected spectral ranges, and glasses for wavelengths beyond 1000 nm are all interesting prospects. Of these, the nonlinearities for glasses are too small for use in integrated optics waveguides just a few cms long, and the information on organics is too recent for devices to have been implemented.

An ideal medium is $Al_{0.18}Ga_{0.82}As$ whose band gap was chosen so that the photon energy corresponding to one half of the band gap falls at wavelengths just shorter than the communications band around 1550 nm. The details of the physics are discussed in detail in Chapter xx. The expected variation in both n_2 and α_2 with wavelength are shown in Fig. 13, based on a two parabolic band model for the semiconductor.[9] (The linear losses are usually small, typically $\alpha_1 < 0.1$ cm^{-1} giving $W > 8$ for $I = 1$ GW/cm^2.) The corresponding experimental values of n_2 and α_2 are shown in Fig. 14.[10] Note that the two photon coefficient is non-zero for wavelengths longer than those corresponding to half the gap, apparently in contradiction to theoretical predictions. This effect is similar to the Urbach tail seen in linear absorption at wavelengths longer than the bandgap wavelength.[11] These results indicate that $T < 0.5$ and $n_2 = 1 \rightarrow 2\text{x}10^{-13}$cm^2/W at 1555 nm, ideal for switching with a nonlinear directional coupler.

TABLE 1

MATERIAL	n_2 cm^2/W	α cm^{-1}	W	T	λ microns
Semiconductors					
AlGaAs (.79μm)	-4 x 10^{-12}	18	2.5	0.9	0.81
AlGaAs (.75μm)	2 x 10^{-13}	0.1	8	<0.3	1.56
Organics					
PTS (crystal)	-2 x 10^{-12}	0.8	40	4	1.06
poly-4BCMU	5 x 10^{-14}	<0.2	>2.5	<1.0	1.3
DANS (polymer)	8 x 10^{-14} 8 x 10^{-14}	<0.2 <0.2	>4.0 >5.0	≈1 ≈0.2	1.06 1.31
DEANST (20%solution)	6 x 10^{-14}	<10^{-2}	>40	<1	1.06
DAN (crystal)	5 x 10^{-12} -10^{-12}	3	26		0.63 1.32
DAN2 (polymer)	2 x 10^{-13}	<1	>2	<1	1.06
Glass					
SiO_2	2 x 10^{-16}	10^{-6}	>10^3	<< 1	>1.06
RN (Corning)	1.3 x 10^{-14}	0.01	13	<0.1	1.06

Intensity (assumed) = 1 GW/cm^2 → 100 W Power
W = $n_2 I/\alpha\lambda$ (Goal: W>1) T = $2\lambda\beta/n_2$ (Goal: T<1)

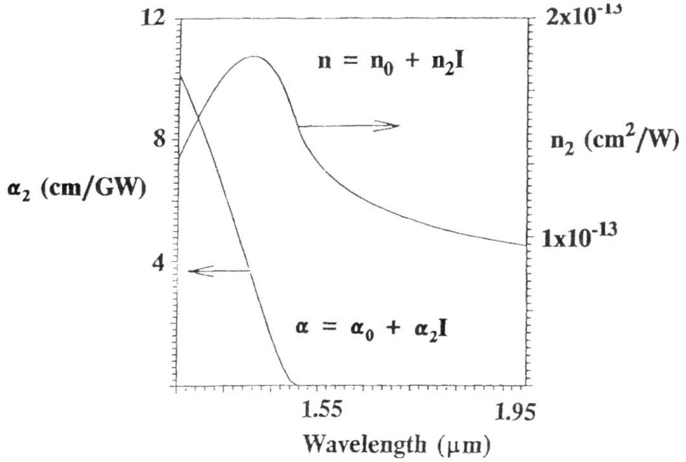

Fig. 13 The calculated nonlinear refractive index n_2 and two photon absorption coefficient α_2 versus wavelength near one half the band gap (1.54 μm) for $Al_{0.18}Ga_{0.82}As$.

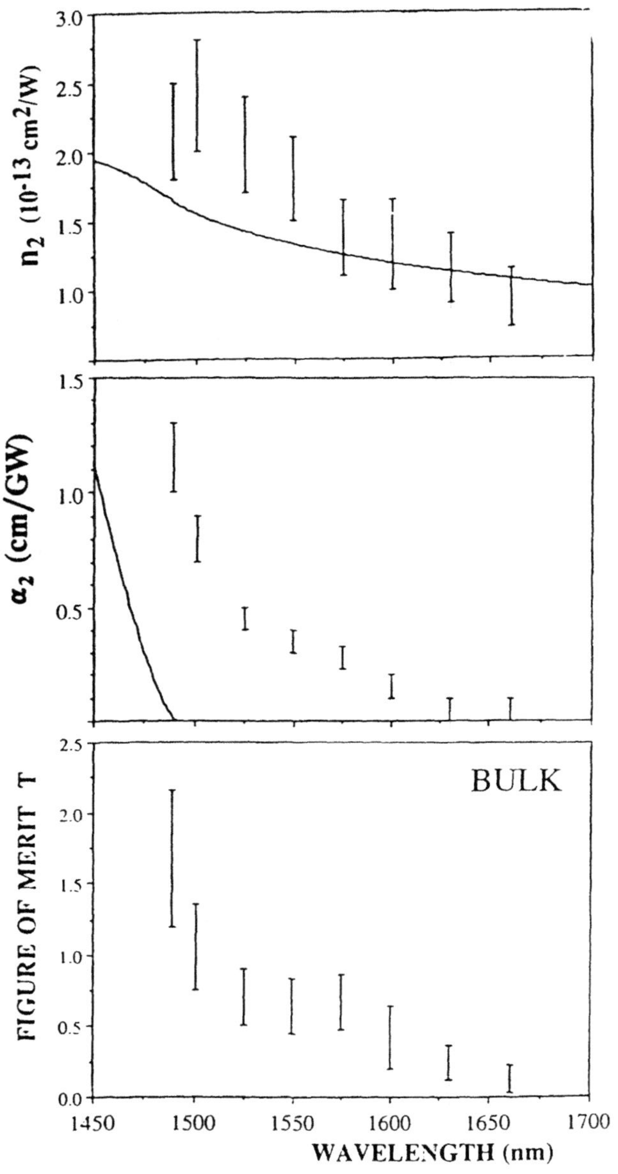

Fig. 14 Experimental values of the n_2 and α_2 versus wavelength in $Al_{0.18}Ga_{0.82}As$ channel waveguides.

Nonlinear Directional Coupler in AlGaAs

To date both bulk AlGaAs and Multiple Quantum Well (MQW) have been used for switching.[12-17] The best results have been obtained in bulk AlGaAs samples.

The nonlinear directional coupler studied is shown in Fig. 15. The rib width and etching depth are chosen to produce a single mode waveguide, and the rib separation is chosen to produce a half-beat length of about 6 mms. Because approximate techniques are almost always used to calculate the field profiles, usually a set of devices are fabricated with a range of guide widths and separations centered about the nominal calculated values. Then trial and error testing is used to find the particular device whose response is as close as possible to that of a half beat length directional coupler.

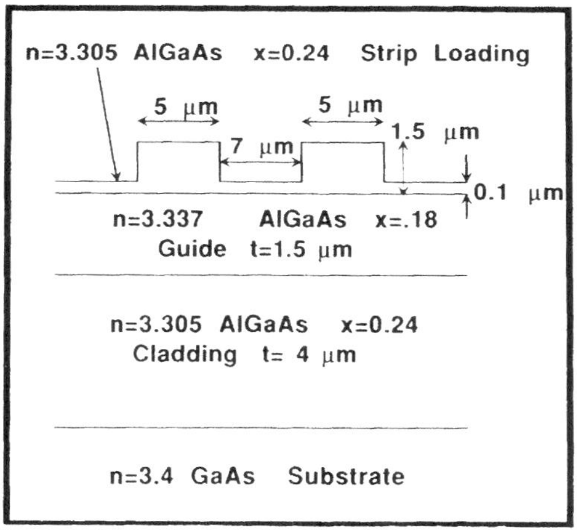

Fig. 15 AlGaAs NLDC structure, viewed end-on.

This device was studied using a color center laser, operating at 1555 nm with 400 fsec pulse widths.[12] The results for the "cross" ($P_1(L)/[P_1(L)+P_2(L)]$) and "bar" ($P_2(L)/[P_1(L)+P_2(L)]$) channels are shown in Fig. 16. The dashed line is the expected response, including the effects of two photon absorption. The difference is due to three photon absorption (3PA), i.e. $\alpha = \alpha_3 I^2$. If we now assume that the loss is dominated by 3PA, we can define a three photon figure of merit V

$$V^{-1} = \frac{n_2}{\lambda_0 \alpha_3 I} \tag{36}$$

and V^{-1} greater than about 0.7 is needed for complete switching.[18] Noting that $V^{-1} \propto I_s \propto L_c$, then V can be decreased by increasing the L_c for the half beat length coupler. As a result we increased the device length to 2 cms by increasing the channel spacing and repeated the experiments.[17] In Fig. 17(a) both the experimental and theoretical curves are shown to agree very well for 10 psec pulses for this longer NLDC. There is no nonlinear absorption, as evidenced by the linear variation of the output with input. With 400 fsec pulses, much higher peak intensities are available and nonlinear absorption is evident at high input intensities. The switching energy in that case is 60 picojoules.

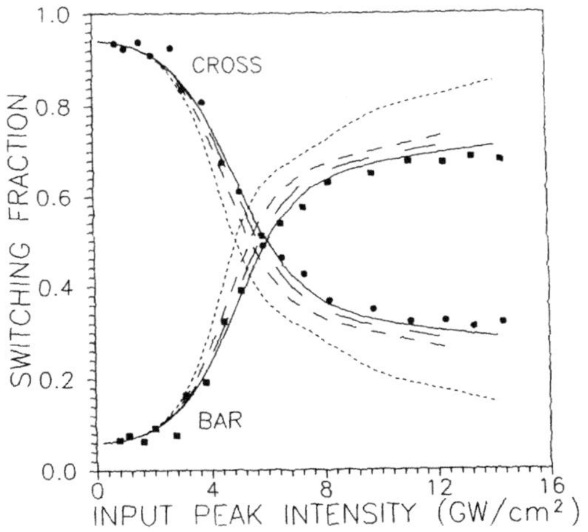

Fig. 16 Switching in a AlGaAs NLDC 6 mm long. The fit includes two and three photon absorption. The short-dash line corresponds to setting $\alpha_3 = 0$ and shows that the three photon absorption plays a very important role at high intensities. The long-dash line for $\alpha_2 = 0$ shows that two photon absorption is not important.

Fig. 17 Switching in a 2.1 cm NLDC with (a) 4 psec and (b) 400 fsec pulses.

For such fast pulses there is no way to measure the output pulses directly in real time. An autocorrelation technique was developed in the early days of nonlinear optics to gain some useful information about pulse shapes.[19] The idea, based on second harmonic generation

(SHG) with orthogonal input beams is shown schematically in Fig. 18. By changing the delay (τ) between the two beamsplit parts of the signal [$E_s(t-\tau)$, $E_s(\tau)$] and detecting the SHG with a "slow" (relative to the pulse width) detector, the output signal [$S(\tau)$] is given by

$$S(\tau) \propto \int_{-\infty}^{\infty} \chi^{(2)} E_s(t-\tau) E_s(\tau)\, dt \tag{37}$$

Shown in the collage in Fig. 19 are experimental autocorrelator outputs, along with the calculated ones based on gaussian input pulses and an instantaneous nonlinearity.[12] The good agreement is an indication that the nonlinearity response time is much faster than the pulse width, i.e. less than 100 femtoseconds.

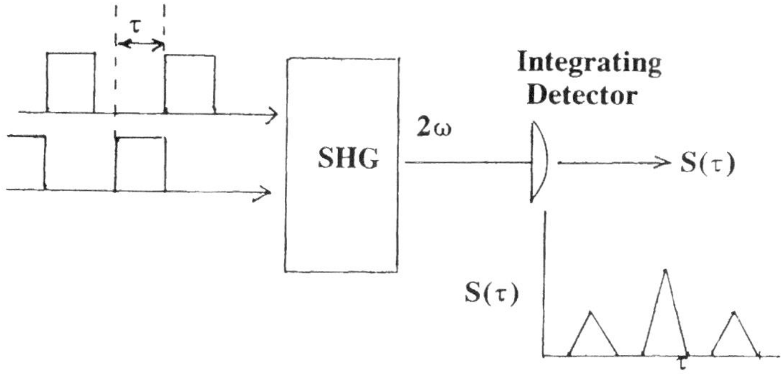

Fig. 18 Autocorrelation technique for measuring a pulse structure using second harmonic generation.

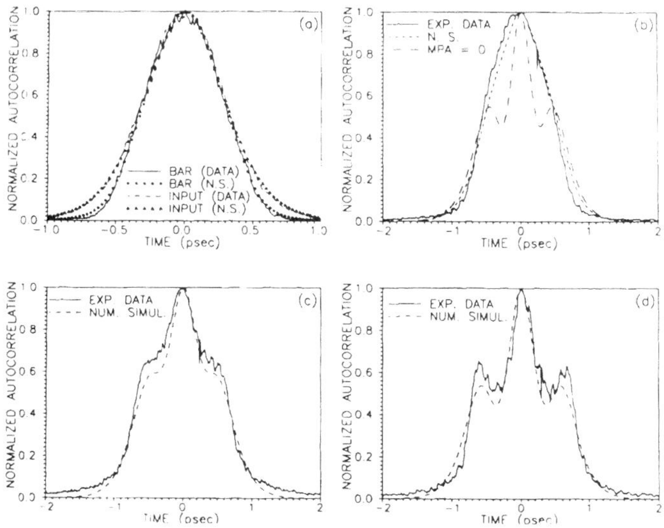

Fig. 19 Autocorrelation traces of pulses emerging from the bar channel of a half-beat length NLDC as a function of increasing power [(a) →(d)]. The dashed curve is the numerical simulation. Shown in (b) is the change in the autocorrelation trace when all multi-photon absorption effects are set to zero (long-dash).

There are other issues which need to be addressed in such an all-optical switch. An important one is group velocity dispersion. Short pulses have a large frequency spectrum, and the different frequency components travel with different velocities, thus dispersing the pulse. This effect is very important in regions of strong spectral dispersion, for example near the band gap. Just below half the band gap the effect is minimal for half beat length devices, but is important for full beat length switches.

The NLDC has also been used as the basic geometry for demultiplexing signals in a data stream. The concept is shown in Fig. 20. The control and signal pulses are orthogonally polarized and $P_c >> P_s$ where "c" and "s" refer to the control and signal respectively. The response of the NLDC is linear for the signal pulses which get switched to the cross channel in the absence of the control pulse. When a control pulse overlaps a signal pulse, the channels are wavevector mis-matched and hence the signal pulse stays in the input (bar) channel. Therefore the control pulse determines the output channel for the signal, i.e. a demultiplexer.

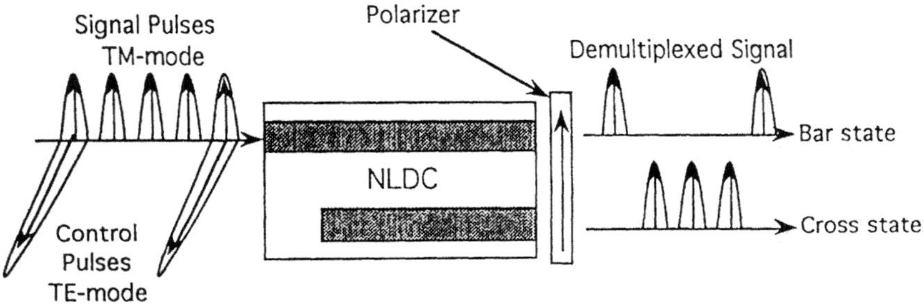

Fig. 20 The NLDC geometry used for demonstrating demultiplexing.

The analysis follows the sequence discussed before for self-switching in a NLDC. Noting that the strong control pulse stays in the incidence channel, the coupled mode equations are

$$-i\frac{d}{dz}a_1(z) = \kappa a_2(z) + \Gamma a_1(z)$$
$$-i\frac{d}{dz}a_2(z) = \kappa a_1(z) \tag{38}$$

where $\Gamma = \Delta\beta_{1c}P_c$. The solution is

$$a_2(z) = \frac{\kappa}{[4\kappa^2+\Gamma^2]^{1/2}}a_1(0)\,[\exp(i\Delta_+z)-\exp(i\Delta_-z)]$$
$$\Delta_\pm = \frac{\Gamma}{2} \pm \frac{1}{2}[\Gamma^2 + 4\kappa^2]^{\frac{1}{2}} \tag{39}$$

In the demultiplexer limit, the absence and presence of the control pulse is given by the limits $\Gamma/\kappa \rightarrow 0$ and $\kappa/\Gamma \rightarrow 0$ respectively. For signal input in channel 1 ($a_2(0) = 0$), the results are

$$\lim\frac{\Gamma}{\kappa} \rightarrow 0 \qquad a_2(z) \rightarrow i a_1(0)\sin(\kappa z)$$
$$\lim\frac{\kappa}{\Gamma} \rightarrow 0 \qquad a_2(z) \rightarrow i\frac{2\kappa}{\Gamma} a_1(0)\sin(\Gamma z) < a_1(0) \tag{40}$$

For a half-beat length coupler $z = L_c = \pi/2\kappa$ so that when $P_c \neq 0$ the signal remains in the input channel and for $P_c = 0$ the signal output appears in the parallel channel.

This concept was implemented with TM signal pulses for which the NLDC is $1.4L_c$ long. As a result, there is always some output in the incidence channel. The results are shown in Fig. 21, and as expected there is always signal output in channel 1. However, the demultiplexing concept clearly works.

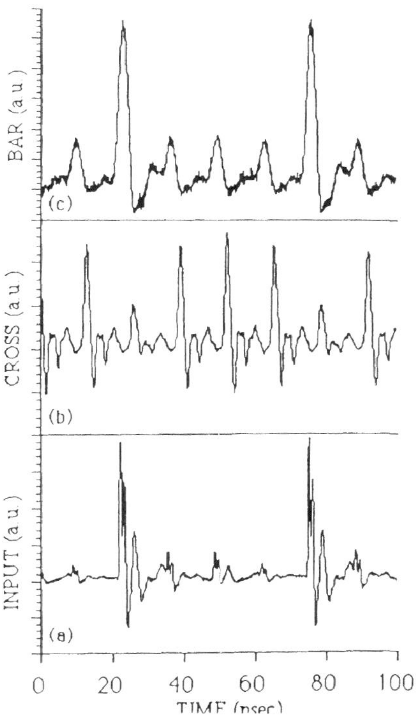

Fig. 21 The signals emerging from the bar (c) and cross(b) states for the input control pulses (c) shown.

Other all-optical switching devices with classical switching characteristics have also been reported in AlGaAs operated below one half its bandgap. They include an X-switch and a Mach-Zehnder interferometer.[13-15] Some polarization switching has also been observed.[16]

In all of these cases, the responses observed are in agreement with calculations based on classical Kerr-law nonlinearities, i.e. ultrafast n_2.

Zero Gap Nonlinear Coupler: Theory

To this point only devices based on "instantaneous" nonlinearities have been discussed. Many switching devices based on nonlinearities due to carrier generation and exciton bleaching in semiconductors near the bandgap have also been reported.[1,20] They are attractive because an amplifier can in principle be fabricated on the same substrate as the switching device, giving switching and routing with enough gain for fan-out.

The particular device of interest here consists of a single waveguide which can support two modes, (TE_{00} and TE_{10} in a channel waveguide).[21,23] The fields are written as

$$E(\mathbf{r}, t = \frac{1}{2}[E^{(0,0)}(x,y)\,a_{0,0}(z)\exp[-i\beta_{0,0}z] + E^{(0,1)}(x,y)\,a_{0,1}(z)\exp[-i\beta_{0,1}z] + c.c.] \quad (41)$$

To simplify notation, in this discussion we now contract (0,0) to (0) and (0,1) to (1). The normal mode field distributions are shown in Fig. 22. Since the two modes are orthogonal and propagate with different wavevectors β_0 and β_1, the relative phase between the two fields changes with distance as $[\beta_1 - \beta_0]z = \Delta\beta z$ at low input powers. As indicated in Fig. 22, after some distance $L_s = \pi/\Delta\beta$, the two fields which were initially in phase fall 180^0 out of phase. As a result the position of the maximum of the total field shifts from one side of the waveguide to the other, i.e. it switches in space. This is shown in Fig. 22.

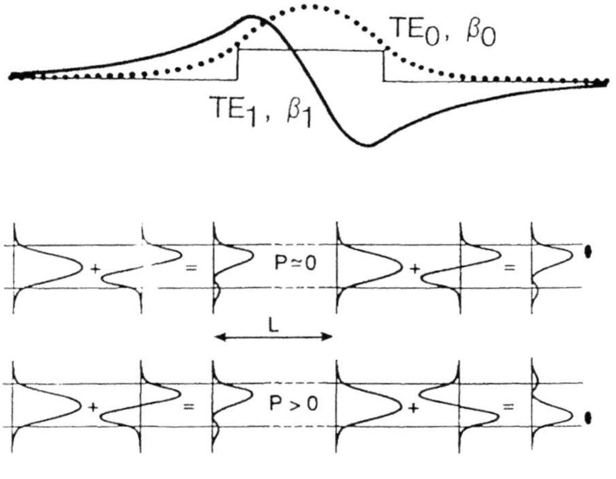

Fig. 22 Top: The field distributions of the two lowest order modes of a single channel waveguide used as a zero gap coupler. Bottom: The decomposition of the input field into the lowest order modes and the output field a distance L down the waveguide at low powers (upper) and high powers (lower).

If the waveguide material is nonlinear ($|n_2| > 0$), self-modulation and cross-phase modulation of the two fields produce a nonlinear polarization.[21] It takes the form

$$
\begin{aligned}
P^{NL}(\mathbf{r},t) &= \frac{1}{2}\epsilon_0\chi^{(3)}_{1111}(-\omega)\{[E_0^3(x,y)|a_0(z)|^2 a_0(z) \\
&+ 2E_0(x,y)E_1^2(x,y)|a_1(z)|^2 a_0(z)]\exp[-i\beta_0 z] \\
&+ E_0(x,y)E_1^2(x,y)a_1^2(z)a_0^*(z)\exp[-i(2\beta_1-\beta_0)z] \\
&+ [E_1^3(x,y)|a_1(z)|^2 a_1(z) \\
&+ 2E_0^2(x,y)E_1(x,y)|a_0(z)|^2 a_1(z)]\exp(-i\beta_1 z) \\
&+ E_0^2(x,y)E_1(x,y)a_0^2(z)a_1^*(z)\exp[-i(2\beta_0-\beta_1)z]\} + c.c.
\end{aligned}
\tag{42}
$$

Substituting into Eqn. 14, we search for terms which oscillate slowly in space. Terms which oscillate quickly, for example as β_0, β_1 etc. result in rapidly oscillating changes in the amplitudes $a_0(z)$ and $a_1(z)$ and are therefore neglected. In Eqn. 42, the first three terms contribute to the distance variation in $a_0(z)$, and the last three to the z-dependence of $a_1(z)$. These equations contain self-phase modulation terms, as well as terms that lead to power exchange between the two modes. Writing $a_i(z) = P_i^{1/2}\exp(i\phi_i)$, $P = P_0(z) + P_1(z)$ where P is clearly independent of z due to power conservation, $U(z) = P_0(z) - P_1(z)$ and $\theta = \phi_0 - \phi_1 + \Delta\beta z$ leads to

$$
\frac{d}{dz}U(z) = \Delta\beta_{10}[P^2 - U^2(z)]\sin(2\theta) \tag{43}
$$

$$
\begin{aligned}
\frac{d}{dz}\theta &= \Delta\beta + \frac{\Delta\beta_{00}-\Delta\beta_{11}}{2}P \\
&+ \frac{\Delta\beta_{00}+\Delta\beta_{11}}{2}U(z) - [2+\cos(2\theta)]\Delta\beta_{01}U(z)
\end{aligned}
\tag{44}
$$

These equations can be solved in terms of elliptical integrals. We do not show the detailed solutions here due to their complexity.

Just like in the NLDC, there is also a critical power for this interaction. Although there is no simple analytical expression for P_c, its significance can be deduced from Fig. 23. For $P \sim P_c$, there is a strong power exchange between the two modes. For a self focusing nonlinearity, the TE_{01} power eventually converts totally into the TE_{00} mode at $P = P_c$. Conversely, for $n_2 < 0$, all of the power evolves into the TE_{01} mode at that power. Note also that the phases of the two modes are locked together for $P \geq P_c$, where-as for $P_c >> P$ the phase difference between the modes increases linearly with propagation distance. The results shown in Fig. 23 are actually only valid for $P < P_c$ since coupled mode theory breaks down for larger P. In that limit the nonlinearly induced wavevector change $\beta_0 - \beta_1$ is comparable to the total dispersion in propagation wavevector when the waveguide dimensions are varied from cut-off to the plane wave limit.

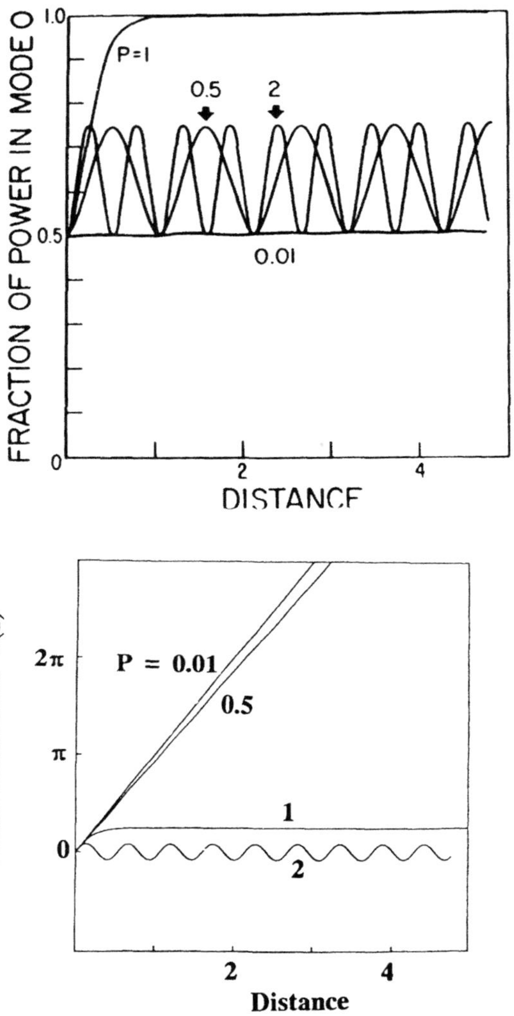

Fig. 23 The evolution of power in the TE_0 mode, and the phase difference between the two modes as a function of distance down the waveguide, for different normalized input powers p (= $P_0(0)/P_c$).

It is this last mode of operation, a linearly increasing phase differential with distance, that has been utilized for all-optical switching devices. In this limit the power in the two modes does not exchange with distance and the phase difference between the modes is approximately $\theta(z) = \Delta\beta z + [\Delta\beta_{00}P_0(0) - \Delta\beta_{11}P_1(0)]z \approx \Delta\beta z + (\Delta\beta_{00}-\Delta\beta_{11})P_0$. As a result the interference pattern some distance z down the waveguide can be changed by increasing the input power, i.e. all-optical switching occurs.

Zero Gap Nonlinear Coupler: Experiment

Very important and large nonlinearities in semiconductors are associated with the absorption of photons.[11,23] (This is in contrast to the preceding case in which the nonlinearity was not accompanied by photon absorption.) The origin and details of these nonlinearities have been discussed previously in Chapter xx. The net effect is summarized for GaAs in Fig. 24.[11,23] High intensity radiation inputs change the absorption coefficient due to a number of mechanisms. By Kramers-Kronig, this leads to a change in the refractive index. Initially the index change is linear in the input intensity, and in this limit the nonlinearity resembles a self-defocusing n_2. At higher intensities, the index change becomes sublinear in intensity, effectively saturating. The goal is to use the index change in the regime where it is linear in input intensity.

Whether the index change can be described by an n_2 or not is determined by the ratio of the pulse width to the nonlinearity relaxation time (usually limited by carrier recombination), i.e. $\Delta t/\tau$. For $\Delta t/\tau >> 1$, the carrier concentration follows the input intensity in a quasi-equilibrium fashion and n_2 is a useful concept for the carrier-related nonlinearity. However for $\Delta t/\tau << 1$, the carrier concentration (and therefore Δn) increases over the duration of the pulse leading to an "integrating" nonlinearity. This makes the details of the switching process complex, potentially leading to asymmetric pulse distortion.

The zero gap nonlinear coupler has been studied with pulses of different duration, corresponding to the limits $\Delta t/\tau >> 1$ and $\Delta t/\tau << 1$.[22,24,25] Experiments with cw, microsecond, picosecond and femtosecond pulses have all been reported. The sample geometry for the cw and 100 nsec pulses is shown in Fig. 25. This waveguide is overmoded when excited at 865 nm with most of the field localized in the MQW layer. It was studied with an argon-ion pumped Styryl-9 dye laser which was carefully focussed onto one half the waveguide in order to excite a linear combination of the TE_{00} and TE_{01} modes. For a 400 μm long sample, the output as a function of input power, as detected by a vidicon camera, is shown in Fig. 26. The spatial switching of the output is clearly seen.

The switching power for this device is not given by a constant in the limit $1 > \Delta t/\tau_r$. In this case, the index change accumulates over the duration of the pulse, i.e.

$$\Delta n(z) = \frac{\xi \Delta t \alpha}{h\nu} I(0) \exp[-\alpha z] \qquad (45)$$

where ξ is the proportionality constant linking the absorption induced index change ($\Delta n = \xi \Delta N$) to the carrier change ΔN, and $h\nu$ is the photon energy. The nonlinear phase change after a distance L_c is given by

$$\Delta\phi^{NL}(L_c) = \frac{\xi \Delta t k_0}{h\nu} I(0) \exp[1 - \exp(-\alpha L_c)] \qquad (46)$$

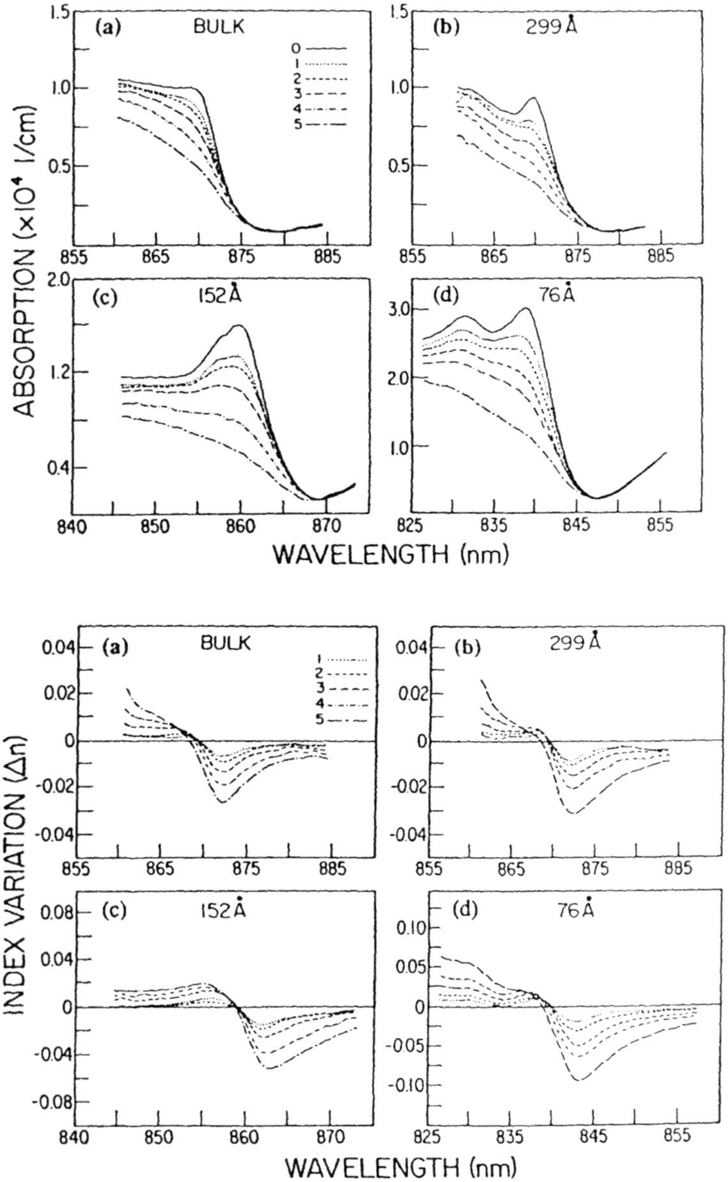

Fig. 24 Top: The change in the absorption coefficient as a function of wavelength and increasing incident power (0→5) for GaAs multi-quantum well samples with different well widths. Bottom: The calculated change (via Kramers-Kroenig) in refractive index based on the upper absorption data.

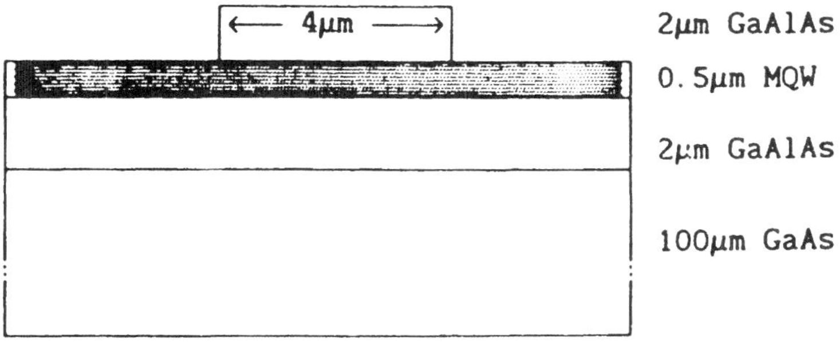

Fig. 25 The zero gap coupler geometry implemented in the GaAs materials system.

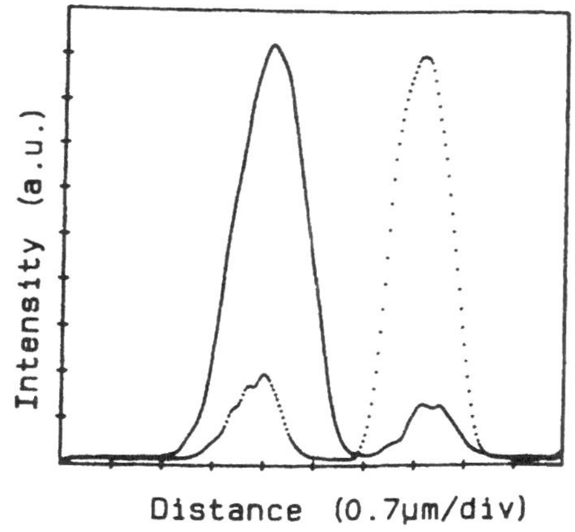

Fig. 26 Output response of a nonlinear zero gap coupler at different input powers. As shown, the two channels are separated by about 2μm. The solid and dotted lines correspond to the outputs for low and high power inputs.

The maximum phase change is obtained for $\alpha L_c >> 1$. However for this limit there is very little output from the device. A comprise is a 1/e throughput which requires a pulse energy $E_p > 1.56hc/\xi$ which depends on the absorption coefficient only through $\alpha L_c = 1$. The switching pulse energy is independent of pulse width.

The response of a zero gap device containing GaAs quantum wells with AlGaAs barriers and claddings has been studied using psec pulses.[24,25] In the first device lateral confinement was achieved through the photo-elastic effect by a 0.2μm thick SiN_x stripe 6 μm wide on the surface of the waveguide layer. The total length of the device used in the experiment was 1 mm, although modelling predicted that the half-beat length was about 0.25 mm. The second device was a strip-loaded (ridge) waveguide geometry shown schematically in Fig. 26.

A pump-probe (demultiplexer) geometry was used to study the switching characteristics of this waveguide. Prior to input, the beam was split into a strong (pump) and weak (probe, signal), and the polarization of the strong beam was rotated 90^0 to make the two beams orthogonal. The pump beam excites the carriers which leads to the index change encountered by the signal beam. Typical results obtained by scanning an aperture across the waveguide output face are shown in Fig. 27. Note the good discrimination found by varying the input power. Here the usual ratios of the third order susceptibility do not apply and the index change is the same for both the pump and the probe. It was found that 150 picojoules (before the sample) was needed for switching. By delaying the probe relative to the pump beam and by monitoring the switching, the carrier decay time was found to be 1.5 ns. In terms of a device, this means that another signal pulse will also be switched up to nanoseconds after the passage of a pump pulse.

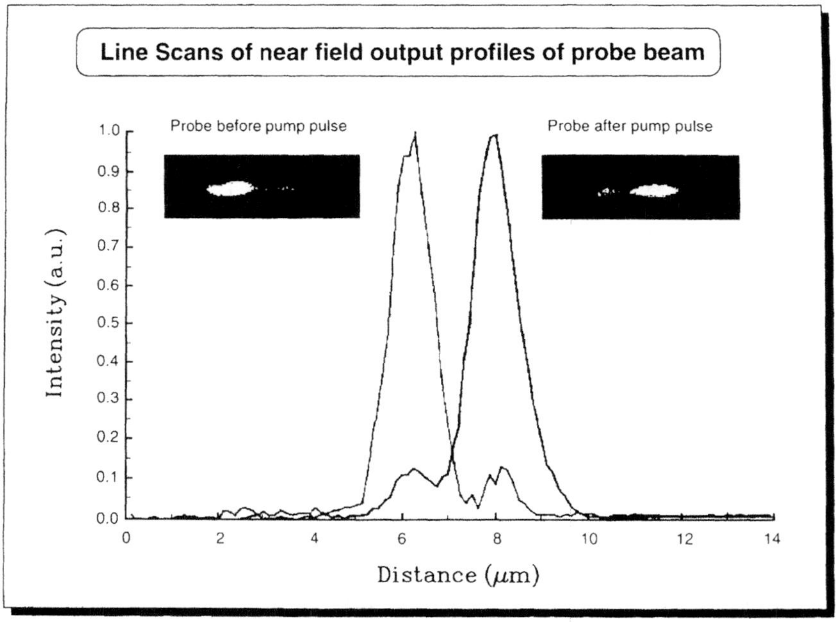

Fig. 27 The output of a zero gap coupler for a signal before and after a pump pulse.

Other devices based on carrier nonlinearities have been reported. The most popular has proven to be the nonlinear directional coupler.[1] Recently, a nonlinear Mach-Zehnder interferometer was demonstrated.[20] In general, a common problem with all of these devices has been the low throughput.

Subsequent work with this device has concentrated on reducing the carrier decay time in the waveguide volume.[25] The shorter the decay time, the sooner a second pulse pair can enter the waveguide without its switching characteristics being influenced by the passage of the preceding pulse pair, i.e. "cross-talk". One approach which has proven successful is to apply an electric field to sweep the carriers out of the waveguide region. To date, sweep out times of 130 psecs have been demonstrated. An example of the effect of the decay on the switching characteristics is shown in Fig. 28.[25] The ultimate limit could be as low as a few picoseconds.

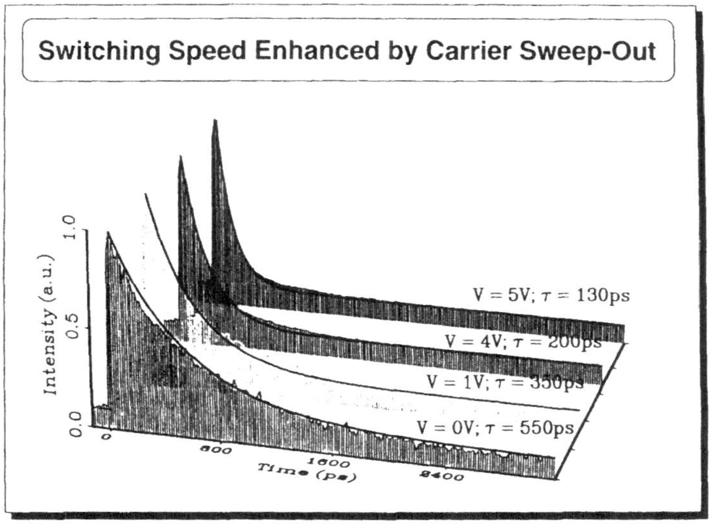

Fig. 28 The switching response (turn-off time) of a zero gap coupler for different sweep-out voltages.

Summary

Waveguides are great for nonlinear optics. They do provide the high intensities and propagation distances needed to make practical devices with small input powers. When used in semiconductors, various all-optical switching devices have been demonstrated with switching energies of order 100 picojoules.

References

1. reviewed with extensive references in: G.I. Stegeman, G.I. and Miller, A. (1993) 'Physics of all-optical switching devices', in J. Midwinter (ed.), Photonic Switching, Academic Press, Orlando; Stegeman, G.I. and Wright, E.M. (1990) 'All-Optical Waveguide Switching', J. Optical and Quant. Electron. 22, 95-122 ; Stegeman, G.I. (1992) 'Nonlinear Guided Wave Optics', in G. Agarwal and R. Boyd (eds.) Nonlinear Optics, Academic Press, New York, pp. 1-40;

2. Marcuse, D. (1974) Theory of Dielectric Optical Waveguides, Academic Press, New York; numerous articles in (1990) T. Tamir (ed.) Guided-Wave Optoelectronics, 2nd ed., Vol 26 in Electronics and Photonics, Springer-Verlag, Berlin.

3. Saad, S. M. (1985) 'Review of numerical methods for the analysis of arbitrary shaped microwave and optical dielectric waveguides', IEEE Trans. on Microwave Theory and Technique, MTT-33, pp. 894-899.

4. Hopf, F. and Stegeman, G.I. (1986) Advanced Classical Electrodynamics Vol. II: Nonlinear Optics, John Wiley and Sons, New York; Boyd, R. W. (1992) Nonlinear Optics, Academic Press, San Diego.

5. Stegeman, G.I. (1982) 'Guided wave approaches to optical bistability', J. Quant. Electron., QE-18, 1610-1619

6. Jensen, S.M. (1982) IEEE J. Quant. Electron. 18, 1580-3; Maier, A. (1982) 'Optical transistors and bistable devices utilizing nonlinear transmission of light in systems with unidirectional coupled waves',Sov. J. Quant. Electron. 12, 1490-4.

7. Hoffe R. and Chrostowski, J. (1986), 'Optical pulse compression and breaking in nonlinear fiber couplers', Opt. Commun. 57, 34-8.

8. Stegeman, G.I., (1993) 'Material figures of merit and implications to all-optical switching', SPIE Proceedings on Nonlinear Optical Properties of Advanced Materials, in press

9. Sheik-Bahae, M., Hutchings, D.C., Hagan, D.J. and VanStryland, E.W. (1991),'Dispersion of bound electronic nonlinear refraction in solids', IEEE J. Quantum Electronics 27, 1296-1309.

10. Villeneuve, A., Yang, C.C., Stegeman, G.I., Lin, C.-H. and Lin, H.-H. (1993) 'Nonlinear Refractive-Index Near Half the Band Gap in AlGaAs', Appl. Phys. Lett., 62, 2465-7.

11. Haug, H. and Koch, S. W. (1990) Quantum Theory of the Optical and Electronic Properties of Semiconductors, World Scientific, Singapore.

12. Villeneuve, A., Aitchison, J.S., Yang, C.C., Wigley, P.G.J., Ironside, C.N. and Stegeman,

G.I, (1992) 'Ultrafast all-optical switching in semiconductor nonlinear directional couplers at half the bandgap', Appl. Phys. Lett. 61, 147-9.

13. Al-hemyari, K., Aitchison, J.S., Ironside, C.N., Kennedy, G.T., Grant, R.S. and Sibbett, W. (1992) 'Ultrafast all-optical switching in GaAlAs integrated interferometer in 1.55 μm spectral region', Electronics Lett. 28, 1090-3.

14. Aitchison, J.S., Villeneuve, A. and Stegeman, G.I. (1993) 'All-optical switching in a nonlinear AlGaAs X-junction', Appl Phys. Lett., in press

15. Villeneuve, A., Al-hemyari, K., Kang, J.U., Ironside, C.N., Aitchison, J.S. and Stegeman, G.I. (1993) 'Demonstration of all-optical demultiplexing at 1555 nm with an AlGaAs directional coupler', Electron. Lett., 29, 721-2.

16. Yang, C.C., Villeneuve, A., Stegeman, G.I., Lin, C.-H. and Lin, H.-H. (1993) 'Nonlinear Polarization Switching Near Half the Band Gap in Semiconductors', Opt. Lett., in press

17. Al-hemyari, K., Villeneuve, A., Kang, J.U., Aitchison, J.S., Ironside, C.N., and Stegeman, G.I., 'Ultrafast All-Optical Switching in GaAlAs Directional Couplers at 1.55 μm Without Multi-Photon Absorption', Appl. Phys. Lett., submitted

[18] Yang, C.C., Villeneuve, A., Stegeman, G.I. and Aitchison, J.S., (1992), 'Effects of three-photon absorption on nonlinear directional coupling', Opt. Lett., 17, 710-2.

19. Ippen, E. and Shank, C.V. (1977) 'Ultrafast Light Pulses', in vol 18 of Topics in Applied Physics, Springer-Verlag, Berlin.

20. Nakamura, S., Tajima, K., Hamao, N. and Sugimoto, Y. (1993) 'High speed all-optical switching experiment in Mach-Zehnder configuration using GaAs waveguide' Appl. Phys. Lett. 62, 925-7.

21. Silberberg, Y. and Stegeman, G.I. (1987) 'Nonlinear coupling of waveguide modes, Appl. Phys. Lett. 50, 801-3.

22. LiKamWa, P., Burke, S.V., Pate, M.A., Roberts, J.S. and Roberts, P.N. (1989) 'All-optical cross switching of modes of a strip loaded GaAs/GaAlAs multiple quantum well waveguide' Technical Digest of Nonlinear Guided Wave Phenomena: Physics and Applications, Optical Society of America, Washington DC, pp. 214-217; LiKamWa, P., Robson, P.N., Roberts, J.S., Pate, M.A. and David, J.P.R. (1988) 'All-optical switching between modes of a GaAs/GaAlAs multiple quantum well waveguide' Appl. Phys. Lett. 52, 2013-4.

23. reviewed in Peyghamberian, N. and Koch, S.W. (1990) 'Semiconductor Nonlinear Materials', in H.M. Gibbs, G. Khitrova and N. Peyghamberian (eds.), Nonlinear Photonics,

Springer-Verlag, New York, pp. 7-60

24. LiKamWa, P., Miller, A., Park, C.B., Roberts, J.S. and Robson, P.N. (1990) 'All-optical switching of picosecond pulses in a GaAs quantum well waveguide coupler', Appl. Phys. Lett., 57, 1846-8.

25. LiKamWa, P., Miller, A., Roberts, J.S. and Robson, J.S. (1991), '130 psec recovery of all-optical switching in GaAs multiquantum well directional coupler', Appl. Phys. Lett., 58, 2055-7.

DIGITAL OPTICAL COMPUTING

B. S. WHERRETT
Department of Physics
Heriot-Watt University
Edinburgh, EH14 4AS,
Scotland, UK

ABSTRACT. Digital optical computing architectures are surveyed, from the use of optics within electronic machines to potential all-optical systems. Many of the schemes that have been implemented optically are described, including brief comments on the active logic devices that they exploit. Certain of these active devices are based on the nonlinear mechanisms described elsewhere in this text. Improvements in their specifications rely on new developments in nonlinear optical materials and device fabrication.

1. Contents

A. Miller et al. (eds.), Nonlinear Optical Materials and Devices for Applications in Information Technology, 321–361.

2. Optics in Computing

2.1 THE MOTIVATION FOR OPTICS

Photons do not interact with each other in free space whereas electrons interact strongly through their Coulomb repulsion.

This simple fundamental statement is key to the roles of optics and electronics within information communication and processing. The interaction between electrons facilitates the nonlinear responses required for decision making, that is, for switching and logic. In contrast, the absence of such an interaction between photons allows for high-bandwidth communication. This bandwidth may be exploited in time or in space. The former leads to possibilities of high repetition rates for sequentially-encoded data; the latter leads to high spatial parallelism in the form of 2-D data images.

In order to get a feel for the time-bandwidths concerned, note that present telecommunications standards range from $2Mbs^{-1}$ (two million bits of information per second) for 30 voice channels, through to $600Mbs^{-1}$ with a capacity to handle up to 8 TV channels. All of this is achievable using electrical signals, employing coax cable for $>8Mbs^{-1}$, and signal restoration at repeater stations placed every 1-3 km. The electrical limit, set by crosstalk between data pulses is of the order of a few Gbs^{-1}. By comparison the shortest laser-pulse duration measured to date is a mere 6fs (6×10^{-15} seconds) and whilst it is not proposed to use such short pulses in information systems, picosecond pulses with comparable repetition rates can and have been propagated through optical fibres as soliton pulses, over thousands of kilometres without the need for signal restoration. Single-mode fibres are in place for long-distance telecoms and $2Gbs^{-1}$ systems are already in use; repeaters need be placed only every 5-50 km, and in 1995 the $5Gbs^{-1}$, TPCS transpacific link will be in operation. The demand is for application to local-area-networks for computers, cable TV, multi-media, virtual-reality simulators and interactive TV. A Tbs^{-1} optical system is a recognised possibility, so there is almost three orders-of-magnitude bandwidth capability in optics that is not yet being exploited.

Turning to space-bandwidth. VLSI feature sizes of 0.25 microns can now be achieved, and 10^4 logic gates (or even 10^6 memory elements) can be fabricated on a single chip of area less than $1cm^2$. However, wire bonding-pad dimensions tend to be of order 100 microns. Therefore there is room for less than 400 input/output connections from a single chip. This pin-out restriction is a serious drawback. In the optical case, the diffraction-limited spot-size for visible radiation is approximately $1\mu m^2$. Hence a fundamental space-bandwidth limit is 10^8 per cm^2. In practise, however, active device sizes of optoelectronic components presently under consideration are of order $(10\mu m)^2$ and it may be necessary to maintain $100\mu m$ separations between components. The achievable on-chip parallelism is therefore unlikely to exceed by much, that attainable in electronics. Where optics does gain in space-bandwidth though, is in the interconnect capability. The freedom provided by propagation through the third dimension in connecting 2-D arrays of active devices means that in excess of 10^4 communication channels between chips is a possibility. Furthermore, the interconnections need not be simple one-to-one direct imaging but could affect either a complicated reordering of data on the 2-D planes or a one-to-many (or many-to-one) connection scheme. Again the non-interacting nature of photons is being used; light beams do not exhibit mutual crosstalk when they are close to one another or even if they pass through each other. The same can certainly not be said for electrical signals in wires.

In attempting to push to the limits of information processing performance there are three hardware options to be considered, each of which leads presently to a bottleneck:

1. All-Electronics
2. Hybrid Electronics-Optics
3. All-Optics

In the all-electronics option, favoured until recently, the bottleneck lies in the interconnections. In the hybrid, optoelectronic option, where electronic devices are connected optically, then the bottleneck is the absence of suitable interface technologies. Finally, in the all-optical option the same lack of a photon-photon interaction that facilitates communication means that it is difficult to carry out processing using optics. Indeed, the so-called optical switches and logic devices, discussed later in this article and elsewhere in this text, all rely on the presence of electronic charge that in effect catalyses the photon interaction. It is the movement of charge, albeit maybe only in the local distribution as described by temporary changes in the electronic-wavefunctions, that provides the nonlinearity manifested in the optical response of an optical switching device. The search for materials of high optical nonlinearity encompasses the desire to provide low energy optical switches that could overcome this processing bottleneck to the all-optical-computing option.

2.2. CHRONOLOGICAL PERSPECTIVE

The history of optical information processing can be traced back as far as the work of Abbe in the 19th century [1]. Aspects of analogue processing were implemented in the 1950's, with work on synthetic aperture radar processing as an example [2]. Here unconventional optical configurations are employed to deconvolute complex images obtained from radar reflections to moving aircraft (the tilted plane processor), with processing of maybe 10^9 bits of data per second but with relatively low accuracy. With the advent of lasers in 1960 the analogue techniques expanded into Fourier-based pattern-recognition schemes exploiting the 2-D spatial coherence of laser-based images [3], and into Bragg-cell 1-D Fourier processing (spectrum analysis of sequential signal streams) [4]. More recently the Fourier schemes, in conjunction with thresholding (logic) devices have been used for optical neural network studies, where the inaccuracy of analogue methods may be an asset rather than a hindrance [5].

A number of proposals for digital optical components and for digital optical processor architectures also appeared in the early 1960's [6]. In the present context digital is taken to refer to binary data representation, with calculations carried out to high accuracy. The proposals were prompted by the concept of binary number representation by laser pulses of high and low energy (as was soon to be exploited in telecoms.), combined with the recognition that a range of nonlinear optical phenomena could be attained at the power levels achieved in short laser-pulses. The nonlinearities as understood at that time were far too small to offer the required low energy switches and progress in optical computing during the 1970's was confined primarily to further applications of Bragg-cell acousto-optic devices for numeric processing such as matrix-matrix multiplication [7]. The demonstration of optical bistability, at low laser power and in small semiconductor samples, in 1979 [8], heralded new consideration of true digital optical processing, which has in turn led to the development of the symmetric-self-electrooptic-effect-device [9] and to one range of so-called smart-pixel devices. These components, and their underlying physical mechanisms, are described in this and other articles in this text. A further type of active component, that has been employed in both digital and analogue processor demonstrations, is the 2-D spatial-light-modulator (SLM). There are many versions of this component, several of which exploit the high electro-optic coefficients of liquid crystal materials. SLMs can be either electrically or optically addressed, their functionality can be very similar to that of bistable device arrays in the latter case. It is also possible to fabricate smart-pixel SLMs.

The possibility of an all-optical commercially viable computer has always been a distant one. Whilst there is still work aimed at such a system, it is established that optics has a less general but distinct role to play within information processing. This may be in special purpose processors, or as modules within a hybrid machine. In the short term, however, optics is encroaching into electronic machines in the form of optical disc storage and, recently, erasable memory. Optics is also appearing as a means of communication between and within electronic computers. It is this latter role that is addressed in the next subsections. For some ten years or so now, optical interconnection has featured, along with digital optical processing , optical neural networks, and novel optical memory mechanisms, in the annually held international optical computing conferences.

Meetings on optical information processing can be traced backed to 1964. A list is given in the bibliography of section 6, along with early journal special issues that were dedicated to the subject, and a selection of texts that have appeared covering digital optical computing. It will be apparent that the subject expanded considerably in the 1980's; only one book was available prior to 1984.

Computational tasks that might involve optics, in order of increasing complexity, may be considered as:

1. Clocking
2. Data storage
3. Communication
4. Routing
5. Logic processing
6. Programming

The clocking tasks requires only the communication of pulses that have no information content other than to enable the operation of the components by which they are received. Data storage requires static encoding, whereas communication obviously requires the movement of data from place to place. If this movement is not along fixed pathways then the routing task is to determine those pathways that are taken. That data will be altered by logic processing and finally the particular processing is determined by the computer programming. This article concentrates on the logic processing task (sections 3,4), but communication (interconnection) is discussed in subsections 2.3,4 and routing is mentioned briefly in section 2.5.

2.3 INTERCONNECTIONS

There are several potential advantages to the use of optics for interconnections within computers.

1. Freedom from crosstalk
2. Low-loss fan-in
3. Reduction in clock-skew
4. Reduction of the power budget
5. Immunity to electromagnetic interference
6. Compatibility with optical storage.
7. Freedom from planar constraints

The time and space bandwidth advantages of optics have been quantified above. The attainment of high space-bandwidth is partly possible because of the use of propagation off-chip in the third dimension. Systems have been constructed using free-space propagation along with mirrors, lenses and/or holograms as described below. In particular, optics offers the possibility of non-local connection patterns that are not achievable using wires. It may also eventually prove easier to achieve interconnect reconfigurability using optical schemes. For high time-bandwidth both planar waveguides and fibres have been employed.

If information initially in a single channel (a stream of electrical pulses in a wire or optical pulses in a fibre, say) is split such that the same information passes along two channels then clearly the pulse energies must be lowered. The information is said to be fanned-out from one to two channels; along with a fan-out of 1-to-N there is an obvious power reduction by N^{-1} for each bit of information in a given channel. For single-mode fibres, or in the electrical case, there is also an N^{-1} loss for fan-in. In the electrical case the output and (N-1) inputs form loads for the N-th input. The optical case is exampled by the constant brightness theorem: a practical manifestation is that a beam splitter cannot be used to combine two optical beams into one, without loss (unless the initial beams have orthogonal polarisations). This loss can, however, be avoided if the product of the beam area and angular convergence of the inputs is less than that of the output. A simple example is input to a detector. The use of multi-mode operation, eventually with detectors, may alleviate the fan-in loss problem. Alternatively it may prove convenient to use fibre amplifiers within optical systems.

In order to maintain control of computational processing it is usually essential that the operation at the gate-level is synchronised. This requires that information arrives at appropriate gates at closely the same time. However, distances between gates on a single chip or on different chips must differ. Because the information propagation rate is finite this means that signals become out of synchronism. Known as clock-skew, this problem is not alleviated by reduction in feature size. This is because the propagation and switching time-delays are determined by R-C time constants and whereas the Capacitances scale according to area/length the Resistances scale as length /area; the RC-product is scale-invariant. Various optical clocking schemes have been proposed whereby optical signals are broadcast from a single, pulsed source to selected gates on the chips. In effect the third dimension is used in order to equate the optical paths to these gates so as to achieve simultaneous arrival of the pulses, which must then enable the electronic logic operation of the gates.

There are circumstances when optical interconnect preference is made on power grounds alone. Typically electrical chip-to-chip connections require sufficient power *(P)* to drive the gates to approximately 1 volt (V) over a period of 1 nanosecond (τ). The capacitance of the wire bonding pad (C_b) will dominate this power; giving $P = C_b V^2 / 2\tau$ of order 100μW. If the connection length exceeds the pulse length, however, it is necessary to employ impedance-matching in order to avoid reflection problems. If this is the case then the coaxial connectors will have a characteristic impedance $R_0 = 50\Omega$ and the dissipation at the terminating load will give $P = V^2 / R_0 = 20\text{mW}$. Given a signal propagation velocity of around c/2 the critical length is a few centimetres, although coax is also used for shorter distances in order to minimise crosstalk by pickup. By comparison the total power dissipation in an optical link is of order 250μW. This is determined by the source (laser) efficiency and the detector requirements. An example is an $\eta = 25\%$ efficient laser, combined with a detector of responsivity $\Re$=0.5AW^{-1}. The input power requirement is then $P = CV(\eta\Re\tau)^{-1}$, where *C* is the detector capacitance (~ 5fF). A given laser must operate above its lasing threshold, which might be around 1 mW. Nevertheless, there is a power advantage in using optics for connections beyond a few cms, regardless of the bandwidth considerations.

Regarding electromagnetic interference immunity, optical schemes are useful in hostile environments where radiation can cause information corruption. Equally it is far more difficult to tap into an optical system, which therefore provides information security.

Finally, with the increased use of high density optical disc storage within otherwise conventional computers, there is automatically some transfer of information by optical means. At present this is almost exclusively sequential transfer of a few bits of information at a time. Parallel readout of disc storage is a future prospect.

2.4 EXAMPLE PROTOTYPE OPTICAL SCHEMES

In order to demonstrate the freedom from planar constraints offered by optics, it is useful to consider a hierarchy of interconnects between components of a computer, figure 1.

At the *machine-to-machine* level optical fibre connections are well established. Fibre links of lengths from metres to kilometres carry signals at data rates of 500 Mbs^{-1} to 1 Gbs $^{-1}$.

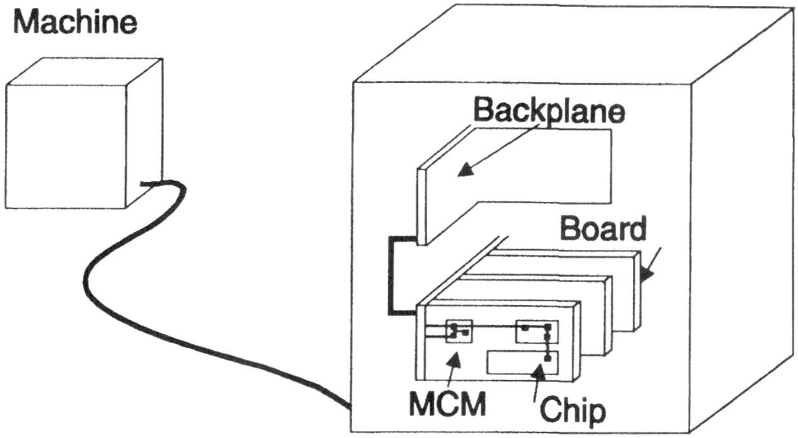

Figure 1. Computer module hierarchy, from machine to chip. (MCM ≡ Multichip module).

Within a single machine there are several modules. Each of these consists of a backplane onto which a number of boards are attached. *Backplane-to-backplane* separations may be 10cm to 10m. Fibre connects are employed in the commercially available Cray supercomputers and an example of a system with free-space connections is Dialog.H, an experimental machine put together at the Electrotechnical Laboratory in Tsukuba, Japan [10]. An optical bus that allows the different processing units of the machine to communicate with one another is achieved by reflecting the signals emitted from one unit off of a cylindrical mirror onto the adjacent units. The units are placed in a circle around the central reflecting cylinder.

At the *board-to-board* level, data must be bused via a backplane at rates exceeding $10Gbs^{-1}$, over distances of less than 1m. Several optical schemes are currently under development. One intriguing scheme is used in the NTT prototype called COSINE III [11]. This machine consists of four boards on each of which is a 4×4 array of T-800 transputers (individually these are powerful processors). Each transputer is connected electrically to its four nearest neighbours on any one board; but between boards there is a 4×4, two-way optical interconnect array. In this way a 4×4× 4 nearest-neighbour architecture is established. In the first system the optical links are achieved using highly inefficient Light Emitting Diodes, focussing lenses, and Si-PIN detectors. The current $20Mbs^{-1}$ data rate used by transputers is easily achieved and registration is such that error rates are not optically limited. As a bonus the scheme allows removal of the boards from the backplane without physically disconnecting the links as would be required in a hard-wired version. Other backplane optical buses include a fibre ribbon version with $32Gbs^{-1}$ capacity, in which the fibres feed into a star-coupler switch where the precise interconnections are determined, constructed under the European ESPRIT programme [12]. In the same project a system in which each board supports one emitter and a set of detectors has been constructed. Using reflective holograms on the backplane, each emitted signal is passed to a defined detector on each of the other boards; GHz operation at Bit Error Rates of less than 5×10^{-11} has been achieved [13].

At the *chip-to-chip* level a number of companies have demonstrated the combination of optical fibre coupling, optoelectronic interfacing and GaAs electronic circuitry. An example of such an OEIC (optoelectronic integrated circuit) contained four detectors each followed by signal amplification and demultiplexing to a data rate handleable by the electronics, multiplexing back after processing and signal output via integrated lasers. A 1GHz rate per channel was achieved back in 1989 [14].

None of the above schemes have high 2-D parallelism and whilst this at first sight should be an area in which optics can contribute uniquely, there is one electronics technology that does have such potential. This is the flip-chip solder-bump technology in which typically 20μm-dimension solder droplets are formed on arrays of solder-wettable pads. In combination with narrow vias through the chips it is possible to make perhaps 10,000 one-to-one connections between chips stacked on top of one another. On each chip local metallic connections complete an architecture with processor element dimensions of ($100 \mu m^2$) and 3-D connections to nearest neighbour elements. An example of a similar architecture is a stack of 16 4-inch CMOS wafers forming a 128x128-processor neural computer [15].

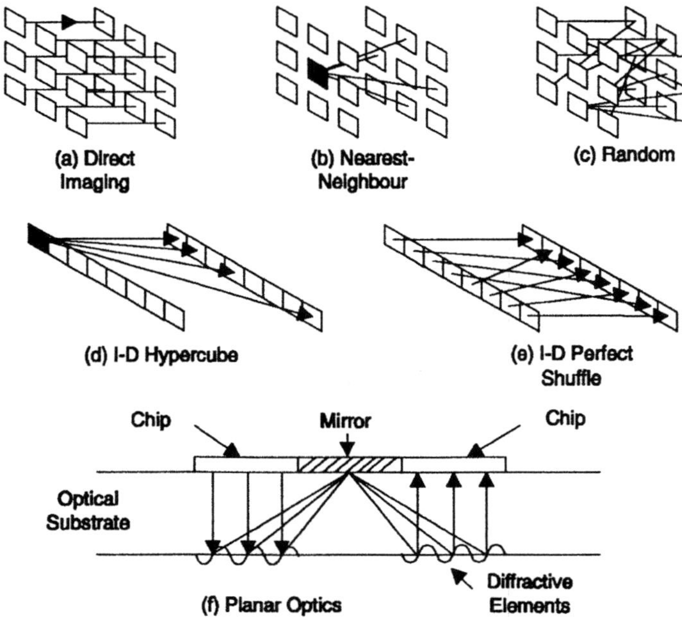

Figure 2. Example free-space optical interconnect architectures. In (b) and (d) all elements have the same output pattern. 2-D versions of (d) and (e) also exist. (f) indicates a planar-optics implementation.

The use of the third dimension for multiple connections as offered by optics is varied. Figure 2 shows a hierarchy of architectures, from the one-to-one imaging that corresponds to the flip-chip scheme, through a nearest-neighbour (locally connected) space-invariant scheme in which every processing element has an identical interconnection environment, to non-local schemes. In the hypercube case (shown in one dimension) space invariance exists but each element is connected to several others, some of which are distant (non-local). The perfect-shuffle architecture is both space-variant and non-local; this particular scheme is the basis for a number of fast algorithms in information processing. A totally random interconnect is also shown, with both space-invariance, non-locality and various numbers of connections from or to a given element. Electronics has great difficulty in achieving these nonlocal schemes in a single processing cycle. The space-bandwidth product (the number of resolvable pixels) for an arbitrary space-variant connection scheme is considerably less than for direct imaging. The limit is set by the large grating period required of a hologram used to produce a small deflection angle. For a space-invariant scheme a single hologram performs the interconnection for the entire beam array. Hence the hologram area may be comparable to the array area and a space-bandwidth product of order 10^7 cm^{-2} is achievable. In the opposite limit, where a separate subhologram is required for each array element, only $10^{7/2}$ cm^{-2} is possible. Fortunately some variant schemes, including the perfect shuffle, can be achieved using space-invariant techniques.

Note that all of the topologies in figure 2 refer to interconnections between different (optoelectronic) chips; the processing elements might contain one or many gates. There have been almost no suggestions that optical connections locally *gate-to-gate* on a single planar array would be sensible, although distant regions of a chip might be usefully connected via the third dimension. As with chip-to-chip connections the free-space propagation could be replaced by propagation within transparent layers, as indicated in figure 2f. This packaging scheme is known as planar optics.

In summary, there are three regimes in which optical interconnections offer a potential beyond that of hard-wired electronics:

1. High sequential bit rate
2. Very high parallelism
3. High complexity, medium parallelism.

2.5 SWITCHING NETWORKS

Just as in a telecoms system there is a requirement in many computers for a switching network. The role is to direct the flow of information between those modules that must communicate with each other. As this procedure is the next step towards optical computing a few comments are included here for completeness.

If we accept that the data is communicated in the optical domain (as it is in long distance telecoms), then there are three technology options for routing devices:

1. Optics-to-electronics conversion, electronic routing, electronics-to-optics reconversion
2. Electronic control of the optical signals.
3. Optical control of optical signals.

In the first option the data must be demultiplexed to a rate at which the electronics can cope. The potential for data transparency, using the second option, has led to intense research into electro-optic directional couplers. For example in 1990 a 2-input, 2-output, $LiNbO_3$ waveguide system operated at $72Gbs^{-1}$ [16]. The relatively large area required per coupler is likely to limit the switching-fabric size to less than 128-to-128. A similar statement must be made concerning the possibilities of all-optical waveguide directional couplers. These devices are described in the article by Stegeman in this volume. In order that a directional coupler can be used as a cross-bar switch the following scheme is possible. Consider information flowing in through channel A (figure 3a) and, in the absence of a control signal, exiting through channel B'. Now suppose that the control is applied of sufficient power to produce the refractive index change required to switch the information into output A'. Whilst the control pulses are present along with each data bit then the redirection continues. When the control is removed then the data reverts to channel B'. The control pulse must however be stripped off of the output data stream. Otherwise the contrast ratio of the output would be far lower than that of the input. This can be accomplished by using a control that is polarised orthogonally to the data, combined with polarisation beamsplitters at the input and output. As in the electro-optic case there is no gain in such a coupler and therefore a network of such devices would need to be followed by amplification. Further, the coupler response must be such as to maintain the contrast ratio of the incident signals.

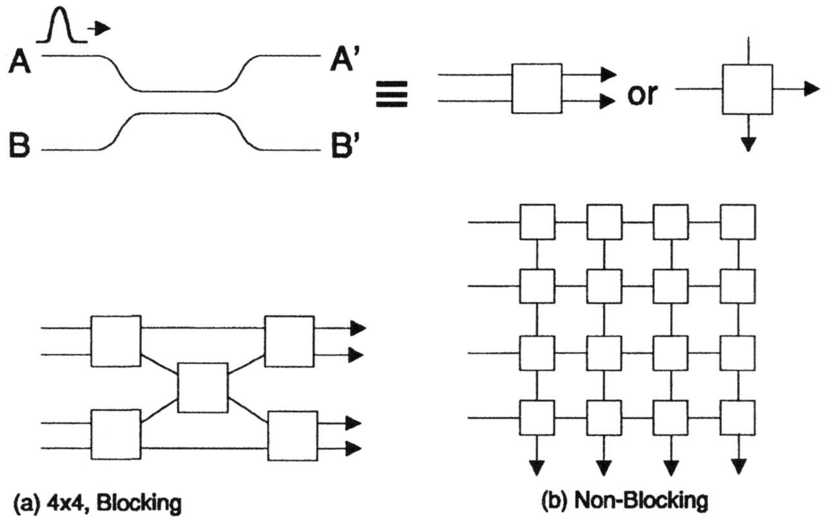

Figure 3. Schematic of a 2×2 directional coupler and its use to construct 4×4 networks.

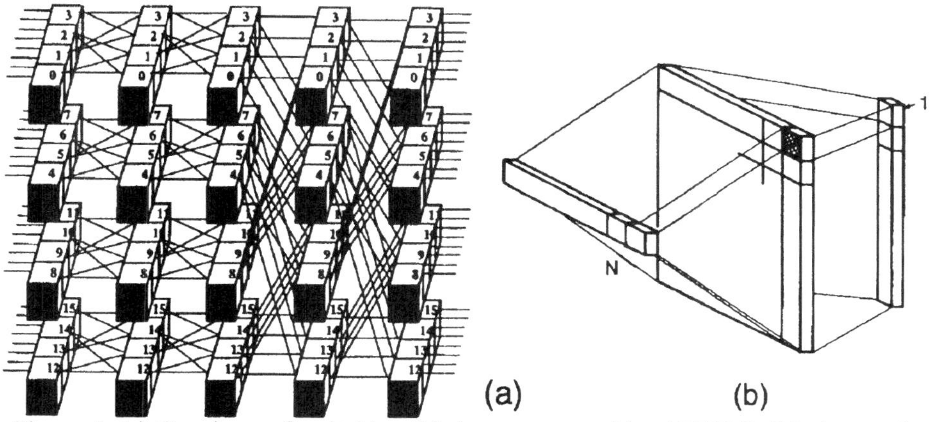

Figure 4. (a) Topology of switching fabrics constructed by AT&T Bell Laboratories using S-SEED nodes.
(b) The generalised crossbar, vector/matrix/vector switch architecture.

The architecture of switching networks forms a subject by itself. Some example networks are presented in figure 3. A blocking network is one in which, having established one interconnection, then others may be impossible to achieve. A non-blocking scheme, at the cost of greater coupler numbers, allows arbitrary interconnection. Partial blocking designs are used as a compromise. Figure 4a shows a system demonstrator constructed by AT&T Bell Laboratories, using the S-SEED technology described by D.A.B. Miller in this volume. The use of nonlocal interconnection through the third dimension between 2-D arrays of coupling nodes is seen. In the SEED scheme the data is restored at each node stage, but at no time is communicated as electronic signals. The capability of the SEEDs currently allows 50MHz operation of 32×32 input-output fabrics, for example. A yet more powerful architecture is enabled in principle in optics - the generalised crossbar architecture shown in figure 4b. The key to the scheme is a device called a Spatial Light Modulator (SLM). The SLM consists of a 2-D array of elements, each of which can be configured either to pass on, or to block, incident information. From an array (1-D) of N optical inputs, each data stream is fanned out to N 'windows' on the SLM. Each output receives data that has been passed by N windows. An orthogonal configuration such as that pictured allows any one (or group of) input elements to be connected to any one (or group of) outputs. The cost of this highly flexible, non-blocking architecture is that the SLM contrast ratio must be very high, whereas a ratio close to 2:1 is acceptable in the SEED scheme above regardless of the magnitude of N. In the vector-matrix architecture a contrast exceeding N:1 is essential if N low power signals at a detector are to be distinguished from a single high power signal. A current project is targeted at a 64×64 size operating at 640 MHz, although with a 10μs reconfiguration time [17]. The SLM is based on electro-optic liquid crystal material with Si-circuitry control of the element reflectivities. A low voltage applied across the liquid crystal determines the polarisation change imposed upon light that is transmitted (and then reflected back) through the material. A polariser in the output plane determines the irradiance of the output. Electrical address to individual cells, through the Si circuitry, therefore controls the SLM switching configuration. In electronics the limiting size of a vector-matrix switch is of order 16×16.

3. Optical Memory and Logic

3.1 LOGIC AND ALGORITHMS

Before describing the devices themselves it is useful to introduce the logic basis upon which various optical processing implementations have been constructed. Device principles will be discussed in this section; the devices that have been exploited in processing circuits are described along with the circuits in section 4.

At the heart of all digital processing is Boolean logic. Suppose that a given hardware element receives two signals, each of a level that we denote either as logic-1 or as logic-0, and transmits one output signal, at one of these levels. There are in principle 16 combinatorial logic functions that determine the output with respect to the inputs. It is sufficient, however, to work with a complete set of functions from which all others can be derived. The NOR function on its own forms one such complete set and a general purpose sequential computer can in principle be built using only nor-gates. Figure 5 shows the response function required for nor operation (assuming that inputs add arithmetically).

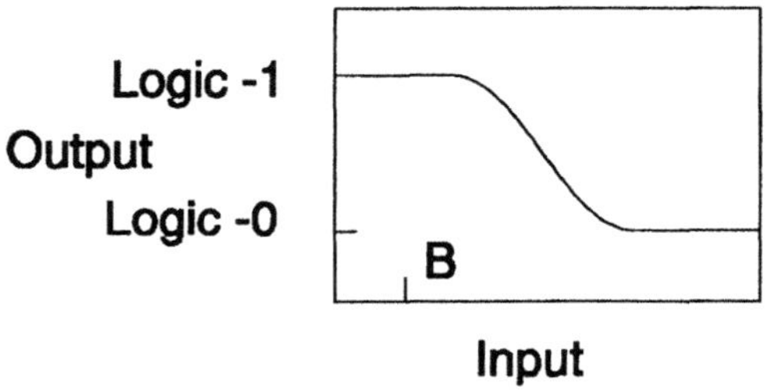

Figure 5. Output-input response required for a NOR-gate, also showing a possible bias position, B.

To example the implementation of a more complicated Boolean function, consider the operation of full-addition, where two binary numbers A and B must be combined with the carry from a lesser significant bit calculation (C) to provide a sum and a new carry. Figure 6 shows part of the Truth Table for full-addition and the Karnaugh Map representation of the table. Thus the Sum (S) must be logic-0 and the Carry (C) logic-1 if one of A,B,C is 0 and the other two are 1's etc. Note that in the Karnaugh maps $\overline{A}$ is used to denote the complement of A (i.e. if A =0 then $\overline{A}=1$ and vice versa), and AB is used to indicate A-and-B. The Sum function can be extracted from the map as

$$S = \overline{A}\,\overline{B}C + ABC + A\overline{B}\,\overline{C} + \overline{A}B\overline{C}.$$

This form is known as a sum-of-products; (A+B) means A-or-B. The same function can, if need be, be written entirely in nor form using DeMorgan's theorem:

$$AB = \overline{\overline{A} + \overline{B}}.$$

ABC	SC'
0 0 0	0 0
1 0 0	1 0
1 1 0	0 1
1 1 1	1 1

(a)

S	AB	A$\bar{B}$	$\bar{A}\bar{B}$	$\bar{A}$B
C	1	0	1	0
$\bar{C}$	0	1	0	1

C'	AB	A$\bar{B}$	$\bar{A}\bar{B}$	$\bar{A}$B
C	1	1	0	0
$\bar{C}$	1	0	0	0

(b)

Figure 6. Full-adder representations.

The resulting complexity of what is one of the most commonly required functions has, in part, led to alternative approaches. Figure 7 shows schematically a Programmable Logic Array, for a pair of inputs. The eight AND (minterm) combinations of the inputs and their complements are formed, following which various of the products are OR-ed together. In this way any sum-of-products can be constructed . The flexibility gained is at the cost of gate-number. For M distinct inputs the PLA must be 2^M layers wide and 2^2^M layers deep.

Another alternative, shown in figure 8, uses a technique called Symbolic Substitution (SS). Here the logic process is replaced by a pattern recognition task followed by replacement by a second pattern. For half-addition, four substitution rules are required.

Symbolic substitution is one example of a Binary Image Algebra, an algebra that can be used to describe the processing of 2-D images. In similar manner to the use of nor-gates for general purpose sequential computing, it can be shown that various discrete sets of functions can be combined to carry out arbitrary algorithms on 2-D data sets. One choice of such a set is given in figure 9; the functions are:

1. The formation of the complement of an image (the NOT function).
2. The union of two images, one of which is termed the reference image (OR function).
3. The dilation of two images (replacement of each logic-1 pixel of one image by the complete reference image). If the reference consists of just one, off-centre element then the dilation produces a simple shift of the original image.

Other image algebras that are subsets of BIA are called Cellular Logic, Cellular Automaton processing and Morphological processing [18].

Whilst bit-parallel techniques such as the SS half-addition can be employed in 2-D, it is often more efficient, given suitable tasks, to use a bit-slice representation of the data. As in figure 10, an array of numbers is represented by one plane for each significant bit Thus when the 2^0-plane of an array A is combined with the 2^0 plane of B the 2^0 sums and the 2^1 carries of all elements might be calculated simultaneously using the same processing function across the entire array. This architecture is known as Single Instruction Single Datastream (SIMD) and is ideally suited for optical implementation.

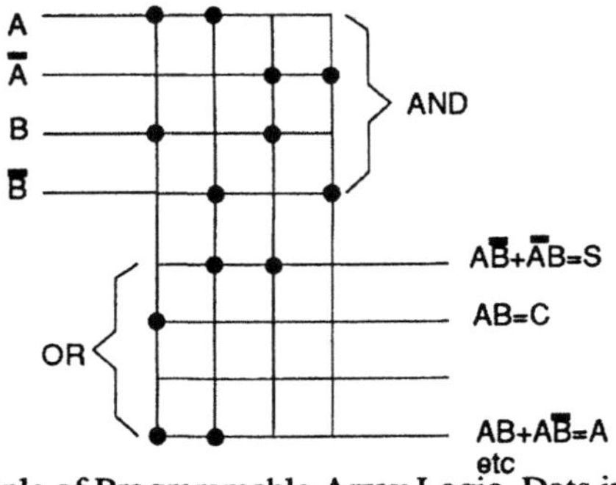

Figure 7. The principle of Programmable Array Logic. Dots indicate the inputs that are ANDed (vertically) and ORed (horizontally) to produce Half-addition outputs, or other combinatorial functions.

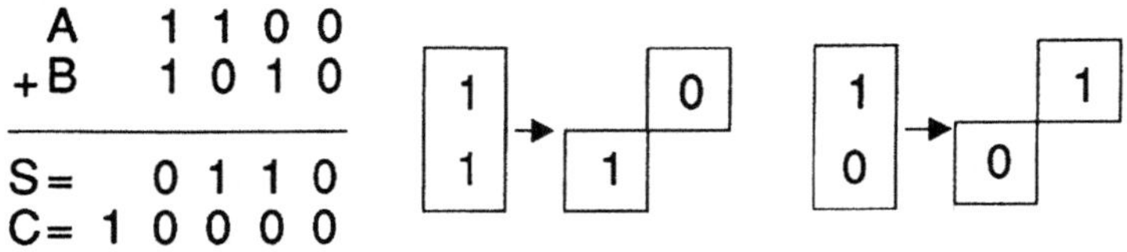

Figure 8. Explanation of two of the four symbolic substitution rules for half-addition.

Figure 9. The primitive operations of Binary Image Algebra.

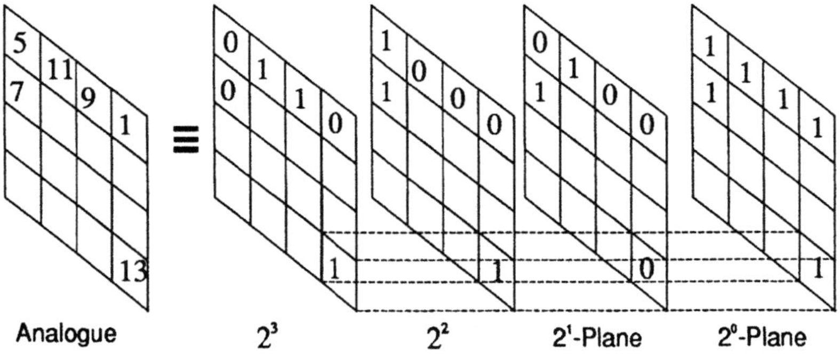

Figure 10. Bit-slice data representation.

Other situations where single instruction processing can map directly algorithms that are to be solved lie in the area of partial differential equations. The Poisson equation:

$$\nabla^2 u = f \quad ,$$

can be rewritten in iterative form:

$$u_{ij}^{k+1} = u_{ij}^{k} + \in \left(h^2 f_{ij} + 4u_{ij}^{k} - u_{i+1,j}^{k} - u_{i,j+1}^{k} - u_{i-1,j}^{k} - u_{i,j-1}^{k}\right)$$

This indicates that, at each element site, the new value to be calculated u_{ij}^{k+1} depends on the previous value u_{ij}^{k} at that site plus the values at the four nearest-neighbour sites $u_{i+1,j}^{k}$, etc. Hence 2-D single-instruction processing, combined with neighbour interconnects directly maps the problem. Similar results apply to any physical problem with local diffusion. More complicated tasks in multi-dimensions, such as weather forecasting, oceanography, chromodynamics, can also all be handled more efficiently in 2-D than by sequential (0-D) processing.

There are circumstances, however, where non-local rather than neighbourhood interconnects are used to speed the solution to a problem. Classic cases are the use of the perfect-shuffle in the Discrete Fast Fourier Transform and in the fast sort algorithm. These two tasks are widely used in information processing; they represent cases where a non-local interconnect architecture can produce a speed up essentially by a factor of log(N) to N, where N is the array dimension, provided that the connection requires only one computational cycle.

3.2 OPTICAL DATA REPRESENTATIONS

Binary data can, in principle, be encoded optically using any of the parameters of light:-
pulse energy, polarisation, wavelength , phase, position, or timing.

1. As in current telecommunications, the most common encoding for information processing is to represent logic-0 by a low pulse energy (perhaps zero energy) and logic-1 by a high pulse energy. The pulses should be of standard duration and spot-size (on the processors); hence the peak power or irradiance might alternatively be used to describe the logic levels. Logic-level restoration, which is key to computing, requires that the energies are returned to their standard levels following any processing and fan-out.

2. Orthogonal polarisations might be used to describe the two logic levels. Care has to be taken though that logic functions are polarisation sensitive rather than sensitive to the pulse energies, which should be well defined, and equal for the two logic-levels in this case.

3. Separate wavelengths λ_0 and λ_1 can in principle be used. Here the difficulty is to design devices that perform the wavelength conversion that Boolean logic would imply.

4. Standard phases would produce great demands on the coherence of the various light beams and, although interference effects have been employed in digital optics, the logic levels have to-date corresponded to photon number rather than beam phase.

5. There are several ways of encoding spatially. The most common is Dual Rail encoding in which two pulses arrive simultaneously, side-by-side (figure 11a). If pulse A is of higher energy than $\overline{A}$ then the pair is said to represent logic-1 and vice versa. This coding has the property that the total energy per data bit is the same for both binary values. Figure 11b shows a more complicated coding, which is termed here Karnaugh-map encoding, because two-input minterms are seen immediately from the overlay of a pair of inputs. Simple logic functions can be selected directly from the combined inputs. Unfortunately the output must be re-encoded in A or B form for further processing.

6. Time position coding, or time-shift keying, is achieved for clocked sequences of pulses that are detected during time windows with which they are meant to be synchronised. If a pulse is shifted in time outside of its detection window, due to the influence of other data pulses at some processing stage, then the detected signal energy becomes low (logic-0). Otherwise a high energy signal (logic-1) is detected.

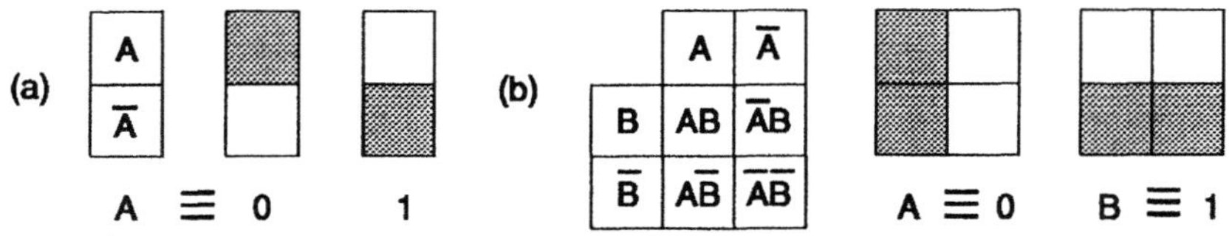

Figure 11. (a) Dual-Rail and (b) Karnaugh Map, Spatial Encoding.

3.3 CASCADABILITY, OPTICAL GAIN AND RESTORING LOGIC

The nor-function response of figure 5 can be implemented optically by a number methods. To be useful the resulting device response must have the following characteristics.

1. The output must be *cascadable*. This means that it must be able to act as the input to at least two further devices. For pulse-energy coding the logic-1 output energy must be at least twice as large as the energy of each information signal at the input. This implies that there must be optical gain in the information response. Such gain may be achieved either by providing an optical bias - a holding or clocking pulse that has no information content, or by using electronic amplification in a hybrid device. A simple all-optical scheme is to add a bias level into the total input (figure 5).

2. The second characteristic, which is made more restrictive by the use of a bias, is that logic-0 and logic-1 signals must be distinguishable. Further, to avoid the build-up of errors over a sequence of devices, the two levels should be consistent from input to output; this is known as *restoring logic*. This means that the contrast ratio, in the presence of the bias, must be sufficient to span the sloped region of figure 5. Also the extreme regions of the response should be flat. A

sharp threshold is advantageous here and there is a trade-off between the achievable gain and the acceptable contrast-ratio.

3. A third requirement concerns the *energy* involved in the device operation. This must be such that a high bit-rate can be achieved. For each watt of power available at the device a 1pJ element will allow a 1THz processing rate. This might be a sequence of fast devices each requiring high peak power or a highly parallel array of low power (but slow) devices. Local heat sinking limits are of order $10Wcm^{-2}$, which would allow the above rate to be achieved on a $1cm^2$ 'optical-chip'.

In order to produce a sharp energy threshold the power (or irradiance) response of the device should ideally exhibit an hysteresis loop, figure 12. The output power changes gradually as the input is increased, until the switch-down position Pd. The output then jumps to the lower level where it changes gradually with input. On reducing the input, however, the output remains at the lower level until the switch-up point Pu, and then changes gradually again. If this is the case then for pulses of low energy, corresponding to low peak power, the entire pulse experiences the higher response, whereas if the peak power exceeds Pd a certain fraction of the pulse produces a low power output; this fraction begins when P = Pd on the rising edge of the pulse and, because of the hysteresis, ends when P = Pu in the trailing edge. The sharpness of the energy-response threshold, and the contrast ratio, can therefore be tailored by choice of the duration of the pulse. In the region of the hysteresis loop there are two possible, stable outputs at each input, hence the expression optical bistability. Not only is it easier to satisfy contrast demands for a bistable case but also it helps achieve restoring logic. Consider the total input power consisting of a bias power plus the signals. This determines the position on the ordinate of figure 12b. If the bias is set within the hysteresis loop then, when the signal is removed the response will remain in that branch which the total input determined. That is, the device will latch. The output power level is therefore set by the two stable responses to the bias, and not by the detailed level of the signals. It is necessary only that the device recognises the logic levels of the two inputs. The bias response forms the standardised (restored) logic level. In practise it is also possible to accommodate variations in the input power by suitable device configurations that make the response slope within the loop close to zero; in this way the component becomes equivalent to an electronic three-port device.

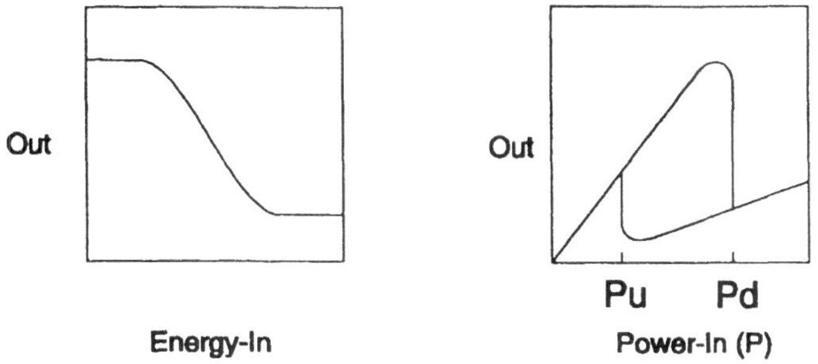

Figure 12. NOR-gate energy response and corresponding power response for the bistable situation.

In order to take advantage of this standardisation it is necessary to phase the flow of information between devices such that the output of one device is used as the input to others when only the bias is present, and that there is a previous period when the signals are also present.

3.4 OPTICAL VOLATILE MEMORY AND SYNCHRONISED DATA FLOW

The bistable response is usable not only for latching logic, but also as a memory. Suppose that there is only one signal present, plus the bias. The bias response will be determined by the combined input and remain for as long as the bias power is held on the device. This memorised signal can later be used as an input to further devices that it addresses by introducing bias beams onto them. The memory is volatile in that, when the bias is removed the device response returns to the zero-input value, from where the device can be reset by reintroducing the bias. The flow of information in a circuit can be controlled in this manner by a sequence of bistable memories that are clocked by phased bias beams, this forms a shift register either for a single channel of data or, if arrays of devices are used and addressed by arrays of bias beams, for 2-D images. A scheme in which three or more bistable arrays are linked in an iterative circuit has come to be called the lock-and-clock circuit; it enables both the synchronised control of data flow and logic level restoration [19].

4. Digital Optical Computing Architectures

4.1 AN ARCHITECTURE HIERARCHY

Figure 13 shows schematics of computer architectures relevant to optical implementation. In each diagram arrows indicate the communication of information (interconnects) and boxes indicate processing elements or memory components. The processors may be elementary Boolean (bit) processors as described above, which is known as the fine-grain limit, or they may be highly sophisticated modules such as transputer chips (themselves essentially computers), which is the course-grain limit.

The principle shown in figure 13a is that of a factory production line. Information flows into the first processor where it is transformed in some way and passed to the second, to be replaced by the next set of information in the first, etc. Each processor carries out one (perhaps complicated) function on a single set of data. This architecture is known as Single Instruction Single Datastream (SISD) Pipeline processing.

1-D parallelism is introduced in figure 13b, along with the alternative either of quite independent SISD pipelines or of the transfer of information between the pipelines. At a given processing stage each element of the 1-D array might carry out the same or different functions. The architecture is vector-parallel with either single instruction Multiple datastream (SIMD) or Multiple instruction multiple datastream (MIMD).

A folded vector parallel scheme is depicted in figure 13c. The sequence of 1-D arrays of 13b is simply rearranged physically so that the processing elements form 2-D arrays. In principle only one array is required in such an architecture.

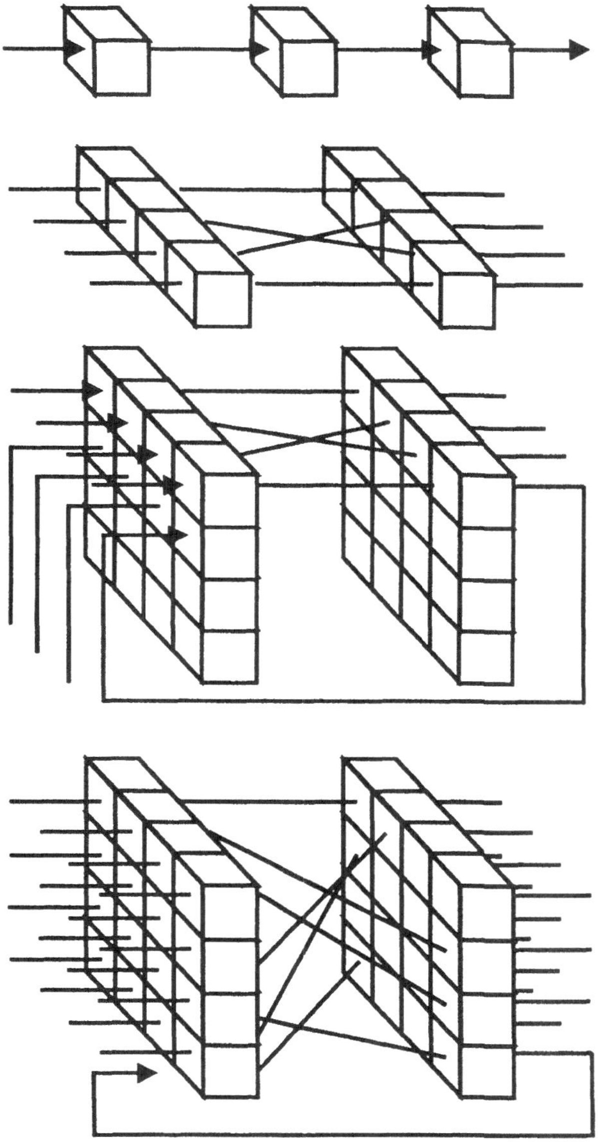

(a) (b) (c) (d)

Figure 13. An architecture hierarchy for digital optical computing (low fan-out) see text.

True 2-D parallelism is shown in figure 13d. Here the interconnects may rearrange the data in either dimension and the input and output are of image format. Again only one processing array is needed in this iterative architecture, but there must be a time delay between the output of this array and its input at the next processing cycle. Otherwise there would be contamination of the initial processing. Whilst in principle such a delay could be achieved by time-of-flight, in practise the device response times are such that synchronisation, as provided by the lock-and-clock scheme, is needed. In an SIMD mode the 2-D parallel architecture has a performance proportional to the instruction rate times the parallelism.

In any of the parallel architectures several further options exist. The interconnects may be local or nonlocal, space-invariant or variant. The processor logic could range from nor-gate only, through a set of primitive (2-input) logic functions, to more course grained possibilities. Also, either the processor functions and/or the interconnects could be reconfigurable on successive cycles as a means of programming the system. Different geometries of the picture elements (pixels) and of the processor arrays are also possible; triangular, square, or hexagonal geometries allow different numbers of nearest neighbour interconnects, for example.

The parallel architectures described above avoid a major problem that is present in conventional computers. In order to fetch information from memory a conventional machine uses an address module. This determines the memory location from which the stored information is to be acquired. The address method was introduced by Von Neumann in order to avoid the difficulties of multiple connections between processors and memory. The resulting slow access time for large sets of data has become known as the Von Neumann bottleneck. The parallel architectures, with volatile memories, correspond closely to the classic finite state architecture, which has no such bottleneck.

The use of parallelism in electronic computers has led to huge increases in performance. Since the CDC-6600, developed in the early 60's , increasing parallelism has been responsible for most of the improvement from around 10^5 floating point calculations per second to beyond 10^{10}. One of the most parallel machines is the Connection Machine sold by Thinking Machines; CM-1 had 65,536 processing elements. In effect this was a 256x256 bit-processor array with neighbour interconnects and routers that formed a hypercube network [20]. This degree of parallelism is comparable to that being discussed as viable for optical interconnections but has limited single-cycle configuration flexibility. Furthermore, the processors in distributed array machines are loaded and down-loaded from the array perimeter. These two features are the current performance bottlenecks. Indeed, recent Thinking Machine supercomputers have reverted from the nonlocal hypercube connectivity to a tree-like topology, because of the difficulty of achieving reliability at the huge pin-out numbers required.

4.2 OPTICALLY IMPLEMENTED COMPUTATIONAL DEMONSTRATORS

In this section a number of existing demonstrators are discussed. Essentially, the ordering corresponds to the increasing parallelism of figure 13. In each case the active processing devices should be thought of as a means to a proof-of-principle machine and not as viable competitors to electronic active components. The devices are not presently competitive either in power, speed, parallelism or tolerance terms in almost all cases.

4.2.1. *SPOC.* There has been just one demonstrator directed towards sequential optical computing. Constructed at the University of Colorado at Boulder and announced only in 1993 , the Stored Programme Optical Computer (SPOC) uses hybrid directional coupler switches and time-of-flight fibre delay lines [21]. Figure 14 depicts the functional structure of the machine and the nature of the active devices. In the example of the data format, logic-1 optical pulse energies are shown, logic-0 levels are in the noise.

The directional couplers are $LiNbO_3$ *electro-optic switches*. If a pulse is incident in the write-fibre then it is detected and the signal amplified in the electronics in order to produce a voltage across the coupler. As a result the incident data pulse, which must be synchronised with the write pulse, takes one waveguide channel. If the write pulse is absent then the data takes the other channel. The memory register, pictured in 14b, contains two such devices and many metres of fibre delay line. Signal regeneration (restoring) is achieved by switching the clock pulses using the weaker signals within the memory loops.

The first SPOC demonstrator contained 200 separate fibres and 70 switches. A 50 MHz clock rate was achieved but with 20μs memory access time. Careful tailoring of the lengths of all fibres was key to the success of the processor. Waveguide switches cannot hope to achieve a high density; some 70 switches per cm^2, with 40GHz operation and 1024 16-bit words stored per 164 m of fibre is a current target. The SPOC machine is unique in one aspect- the use of a sequence of optical pulses as the stored programme. This feature is, however, only significant for single-datastream processing. In the parallel, SIMD architecture the instruction rate may be relatively low and easily achieved by electronics.

4.2.2. *DOCIP.* The Digital Optical Cellular Image Processor, constructed at the University of Southern California L.A., is designed as one processing element of a Binary Image Algebra processor [22]. Figure 15 shows the optical circuit, the logic implemented and example data.

The active device in this case is an optically-addressed 2-D spatial light modulator. Known as a *liquid crystal light valve* (LCLV) the device responds to weak input signals on one surface [23]. Using a photo-conducting material the signal pattern creates a photocurrent with the same spatial pattern. Where initially a uniform voltage was set up across the device area the voltage will now also have the same pattern. The voltage is set up between the outer face of the photoconductor and a transparent electrode on the other device surface. A uniform light beam or an array of beams of identical power level incident on the other side of the device, passes through a layer of nematic liquid crystal, is reflected from a dielectric mirror, and passes back out through the liquid crystal. The liquid crystal experiences just the voltage pattern set up by the weak information image. For a suitable voltage change the liquid crystal has the property that it rotates the polarisation of the emerging light by 90 degrees. Hence, with the use of a polarising beamsplitter at the input and output of the uniform (read) beam, the polarisation image is converted back into an irradiance image. The read-beam may be of considerably higher power than the write beams so that the

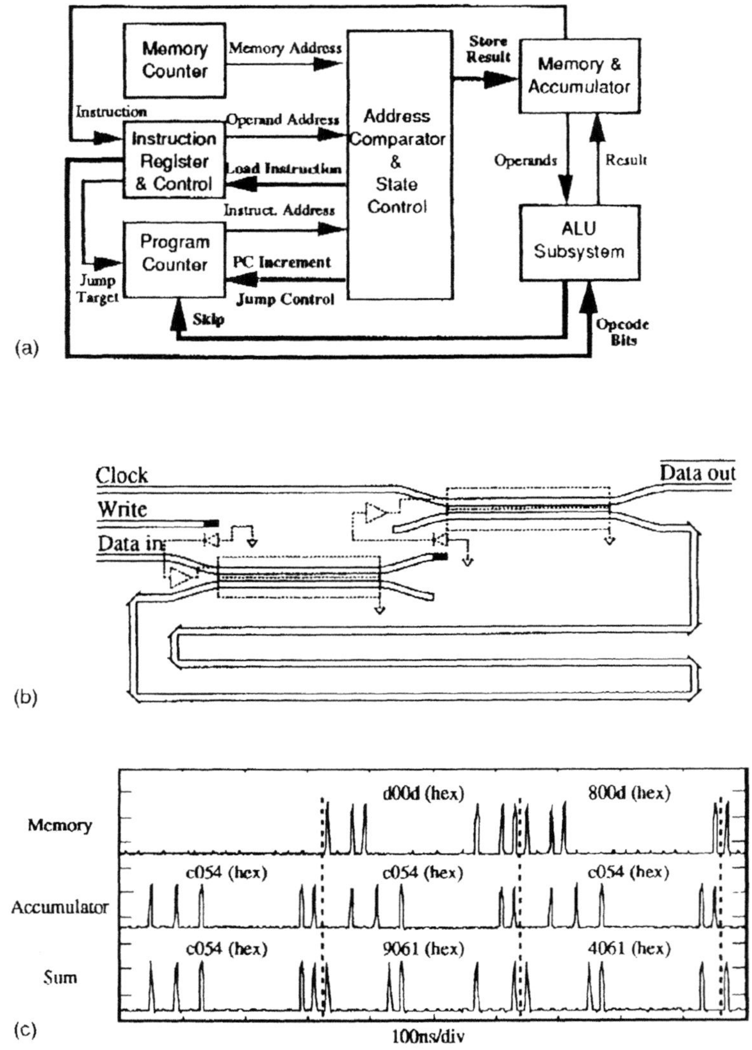

Figure 14. (a) Block diagram, (b) the active device, and (c) example data for the stored programme optical computer.

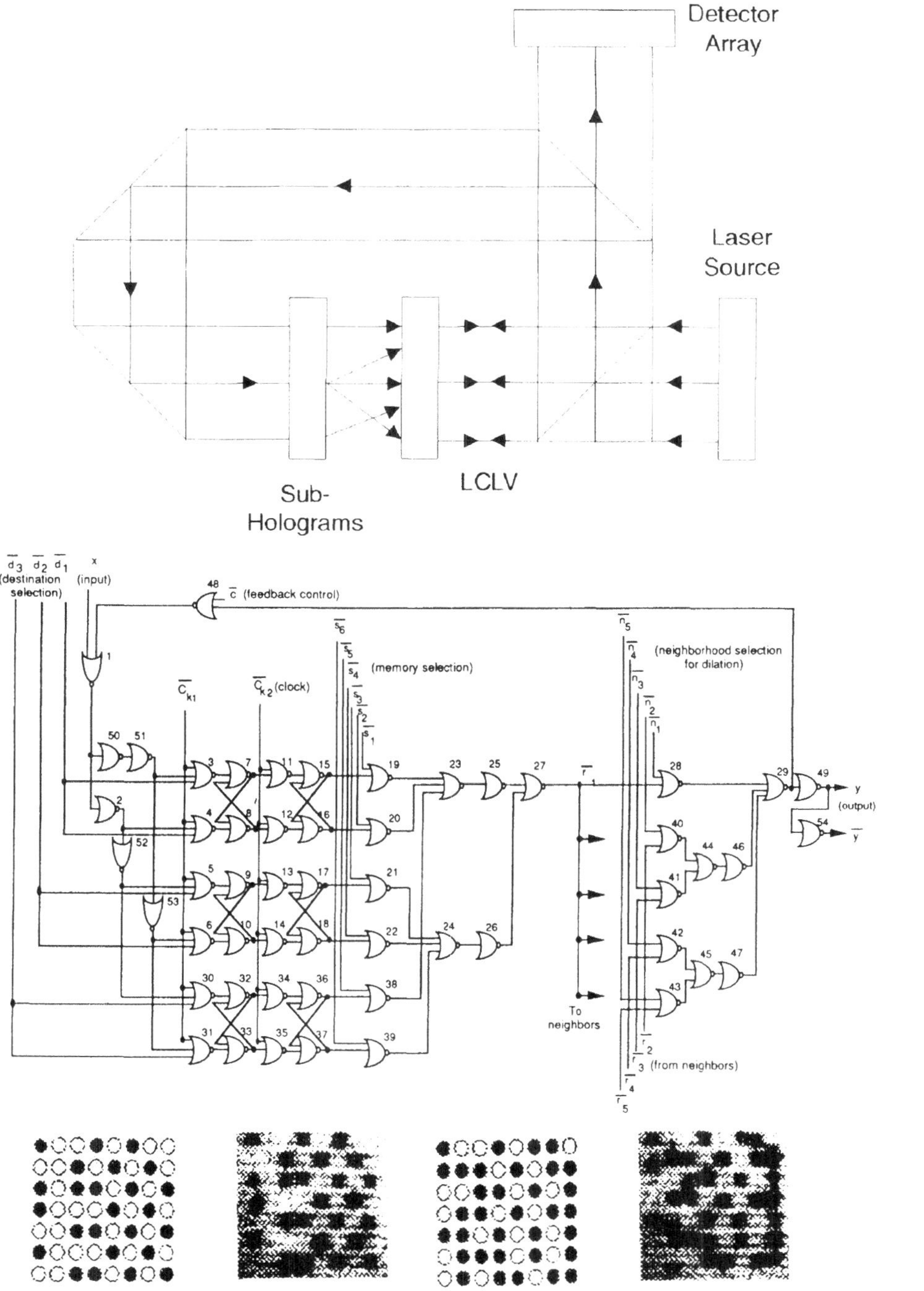

Figure 15. (a) Optical layout, (b) implemented circuit, and (c) example data for the digital optical cellular logic image processor.

device can have a high gain. In addition, if the polariser is set to give a high signal for the higher voltage and vice versa, then the LCLV operates precisely as a nor-gate. Whilst faster SLMs are now appearing on the market, the LCLV nematic device has a slow response time, of order milliseconds.

The LCLV is not a natural bistable device and cannot therefore act as a memory. As a consequence circuit clocking in DOCIP must be achieved by the combination of nor-gates into flip-flops, as in electronics. The DOCIP circuit is in fact a direct copy of the type of circuit implemented in electronic parallel machines, figure 15b. The circuit wiring is achieved by use of an array of sub-holograms. The output from one element of the LCLV is imaged on to a subhologram, from which it diffracts such as to fall on the write side of the LCLV at one or more positions. This can be thought of as optical wiring between the nor-gate position of the reflected read-beam and the nor-gates at these write positions. 54 separate positions on the LCLV are wired up in this way to achieve the pictured circuit. This particular circuit would correspond to just one pixel position of a BIA processor, although as it stands it is an SISD circuit. There is just one output that corresponds to the pixel information output. This will of course put high demands on the total achievable device parallelism if a large array of pixels is to be processed. The prototype DOCIP operates only at around 1Hz, but has complement and dilation functionality, where the reference image consists of the four nearest neighbour inputs, as well as clocking.

4.2.3. *ODP.* A vector parallel scheme, one of a range of architectures proposed by Murdocca [24], is shown in figure 16a. This is in effect a programmable logic array scheme in that data flows through a series of AND-gates followed by a series of OR-gates. The interconnection scheme is fixed and non-local, such as to allow selection of the minterms that can be set by stage-5. The figure indicates the construction of a full-adder by this means. Whilst the logic functions and the connections are fixed, a programmable mask is required after each stage in order to select particular pathways for the information. An electrically addressed SLM would satisfy this requirement. For any one function, such as full-addition there is considerable redundancy in this scheme.

The Optical Digital Processor, constructed at AT&T Bell Laboratories, implemented a folded vector-parallel scheme, using four 8×4 arrays of nor-gates in an iterative architecture [25]. A 1-D shift register and a two-to-four decoder (effectively constructing minterms) were demonstrated by using a split-and-shift interconnect along with fixed masks for pathway selection.

The nor-gate logic devices employed in this scheme were the *symmetric-self electro-optic effect devices* described herein by Miller. The dual-rail response of an s-SEED is presented in figure 16b. To understand what this response plot means it is necessary to consider the time dependence of s-SEED use. Note that the response is differential; the ratio of the output beams is related to the ratio of the inputs. The device has two windows (termed R and S) onto which it receives dual rail signals. Logic functionality can be established by an initial (preset) beam on just one window. If the preset falls on the S-window the system is driven to position A on the figure. The preset is now replaced by the two, dual-rail data signals; suppose that these are both logic-1, with higher power falling on S. The dual-rail contrast should be such that the system now goes to position B. Finally

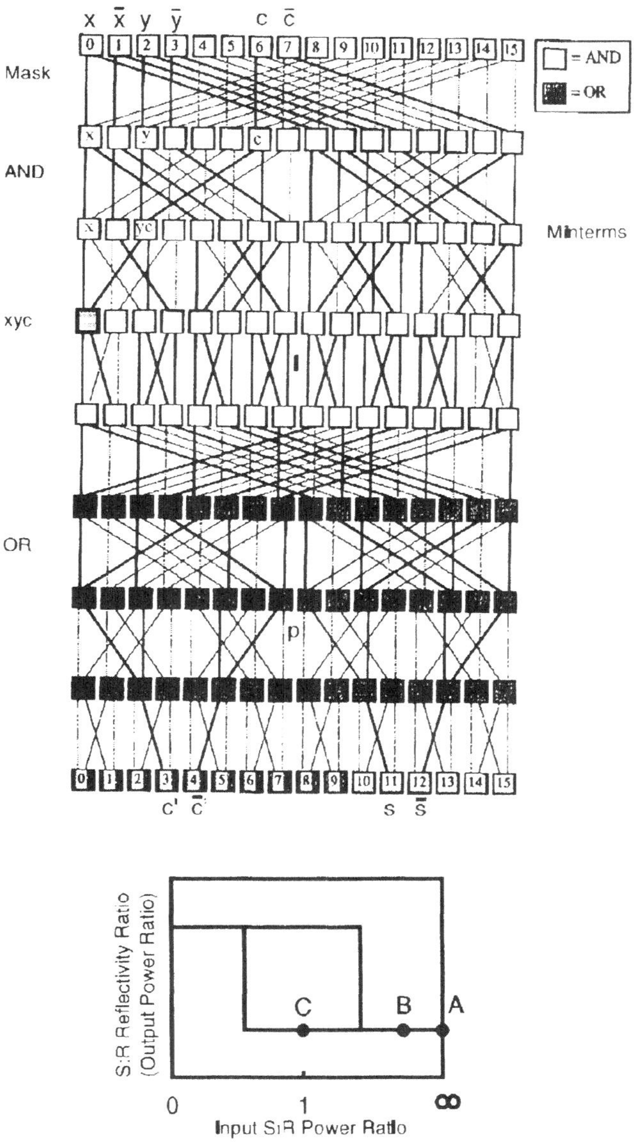

Figure 16. (a) 1-D pipeline architecture in PLA form for full-addition. (b) The S-SEED optical response.

replace the signals by two read-beams of equal power so that the power ratio falls to unity and the system goes to position C on the figure. The reflectivity ratio plotted now translates to a power ratio (because the read beam powers are equal), and the output can be used as an input, dual-rail signal to further devices. The s-SEED has gain because the read-beam powers can be far higher than those of the data; this is called time-sequential gain in view of the dynamics of the device operation. As described, two logic-1 inputs have produced one logic-0 output, as would be required of a nor-gate or nand-gate. Equally two logic-0s would produce a logic-1. The role of the preset has been to determine the result of a 0+1 input. The point is that the S:R power ratio is unity for such an input, so that there are in principle two stable output ratios. By starting from one or the other end of the response curve the particular ratio is predetermined, in this case as logic-1. The preset has therefore produced a nand functionality for the device. By presetting on the R-window instead then nor logic is obtained. The s-SEED can therefore be used with real time programmability. If necessary, and/or logic can be achieved by a fixed reversal of the R and S beams at the output of the device array.

4.2.4. *Symbolic Substitution.* Figure 8 showed two of the substitution rules required for half-addition. Given a 2-D image representing pairs of numbers to be added together then all of the pairs may be added simultaneously if each occurrence of the patterns in 8 is replaced by its corresponding substitution. There is a simple way to recognise every occurrence of a pattern optically. For example, to recognise two low power pixels in a vertical line: beam-split the image into two and then recombine, having shifted one image vertically by one pixel compared to the other. Every occurrence of a low power output corresponds to the presence of the vertical pairing in the original image. Multi-pixel patterns may be recognised by overlaying several shifted copies of the original. The identified positions of a pattern now form a new image. In order to substitute a new pattern one merely duplicates this new image and combines versions that have been shifted in the manner of the new pattern. The implementation of half-addition would require the four rules of figure 8 to be treated quite separately (one must start with four versions of the image even before tackling the rules and then finally recombine the four substituted patterns). There is also ambiguity in these rules because the presence of one low power beam would be recognised also when two are present. This problem is avoided by the use of dual-rail coding, because every bit-pair becomes a unique pattern of two low and two high-power beams. A symbolic substitution adder, in what is effectively a folded vector-parallel architecture, was constructed in Erlangen in 1992 [26]. For demonstration purposes the active gates were inverting threshold amplifiers between optical detection and LED sources, in a 16×16 array.

An entertaining version of symbolic substitution was devised as a means of demonstrating the complete computational ability of the scheme. It can be proven that general purpose computing requires the use of just one substitution rule [27]. In order to carry out logic functions dedicated

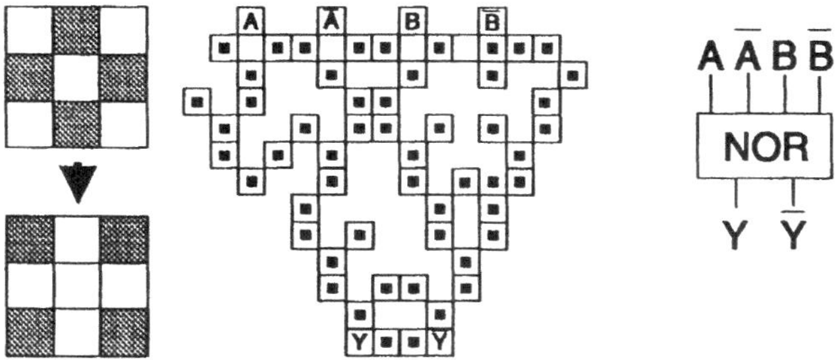

A	$\bar{A}$	B	$\bar{B}$	Y	$\bar{Y}$
0	1	0	1	1	0
0	1	1	0	0	1
1	0	0	1	0	1
1	0	1	0	0	1

Figure 17. Substitution rule for a one-rule cellular automaton; mask for NOR-gate implementation, and the corresponding truth table.

masks are used. The NOR-mask is shown in figure 17 along with the rule. Consider illumination from behind the mask, plus an input on the top row. Two out of the four input positions will form cross patterns. The resulting two square patterns cycle back onto the mask where they can now contribute to new crosses lower down the image. In this way the data flows down the picture to emerge as a single, dual-rail output. The generality achieved is clearly at a huge cost in the number of device active elements required for any given function.

4.2.5. *OPALS*. The Optical Parallel Array Logic System introduced by Ichioka [28] is most easily described in terms of its first implementation method- shadow casting. Figure 18 shows four LEDs that can illuminate one pixel of the input from images A and B; the Karnaugh map encoding of section 3.2 is used to determine the spatial form of the transmission at the input plane. The shadow of the 2x2 element pixel forms a 3x3-element image which is masked to allow only the light in the central element to transmit. As pictured, if only LED-4 is on then light is transmitted only if A is logic-1 and B is logic-1. Thus the four LEDs can be used to construct each of the 2-input minterms, or any combination of them. That is, all 16 combinatorial logic functions are programmable through the LEDs. There is no gain in this system; the light must be detected and recoded if further processing is to take place. An iterative architecture can however be implemented using LCLVs as the coding and gain elements [29].

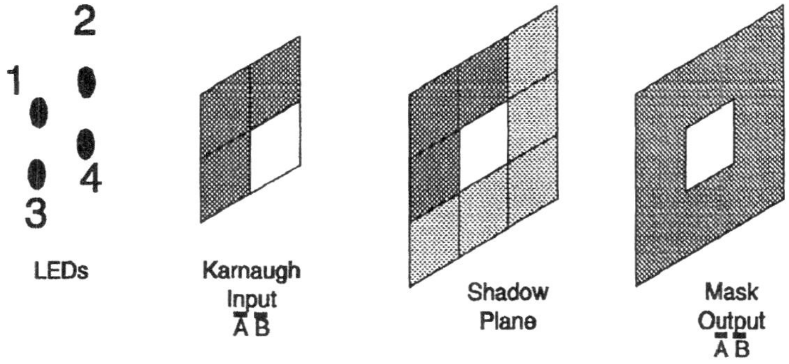

Figure 18. The shadow casting principle.

The basic OPALS architecture is SIMD; an MIMD version has been proposed by Yatagai [30]. Instead of transmitting only at the central element in the shadow plane, if the mask allows different elements to transmit then the logic functionality for each LED will alter. Hence a programmable, space-variant mask enables different logic across the data plane.

4.2.6. *O-CLIP*. The only computational circuit that has used true all-optical components to-date is the Optical-Cellular Logic Image Processor constructed at Heriot-Watt University in Edinburgh. The functionality of the circuit components is shown in figure 19, along with example data [31]. O-CLIP is an SIMD machine. A host, electronic computer controls just six laser beams. In turn the laser pulses are used to clock and to programme the machine. Input comes from a spatial light modulator that is read using a seventh laser. To achieve 2-D parallelism each laser beam is fanned out to an array of equal power sub-beams. There are four optically-bistable plates in the iterative circuit. Each receives a read-beam array. Two of the plates form a programmable

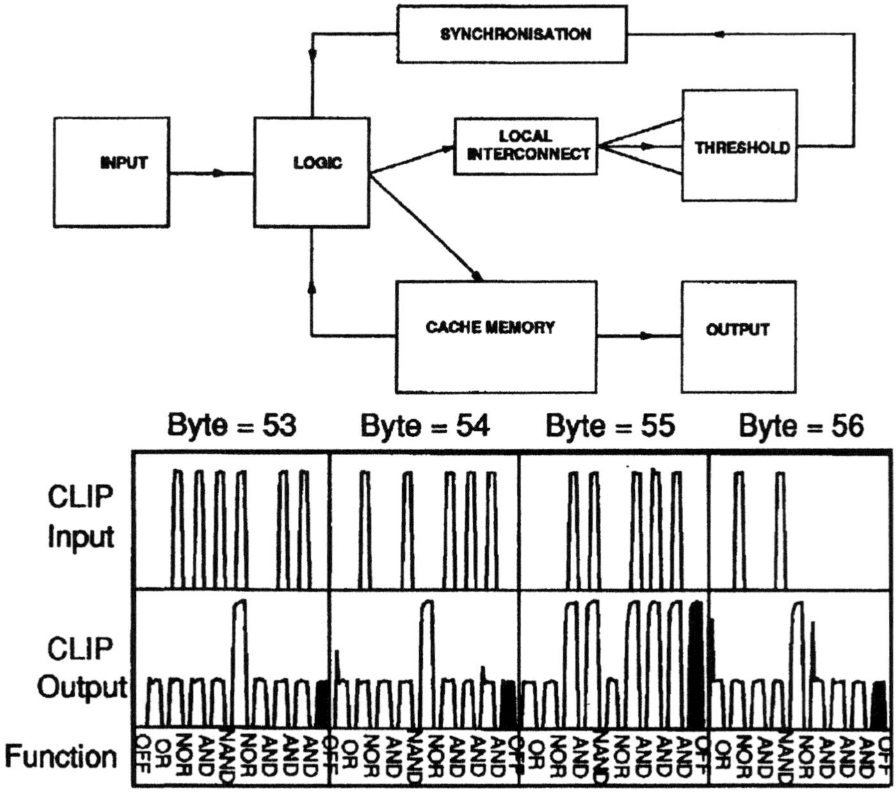

Figure 19. (a) O-CLIP block diagram and (b) single-channel data format.

logic unit. These two have additional control-beam arrays that can be varied in power level such as to produce eight of the combinatorial logic functions. These plates act on the input image at a given cycle, combined with the output image from the previous cycle. Following the logic unit is an interconnect. In the first O-CLIPs a nearest-neighbour scheme is used (figure 2b). Simultaneously, over the whole image the information from one pixel is connected to the four nearest pixel positions on the next active device. Because each element of that device receives four bits of information (from its nearest neighbours) thresholding must be carried out in order to return the data to binary representation. The third bistable device performs this role. In principle three bistables should then be used to achieve lock-and-clock synchronisation and a delay, before recycling. In practise, because of the use of bistables for thresholding and logic just one further component is necessary.

The data of figure 19b was taken for a single channel version of O-CLIP (no fanning and no interconnect element). This version was used to implement recognition, comparison, full-addition and subtraction tasks [32]. In the figure, recognition of a sequence of bits (55) is shown. The input data is a sequence of laser pulses (8 bits followed by a zero), repeated for each input number. The control lasers are used to programme the sequence of logic functions OFF, NOR, AND, etc., which form the algorithm for recognising the binary representation of 55. The output after each cycle is combined with the next input bit according to these functions. To check the algorithm note that if a given bit is recognised correctly, then the CLIP output is designed to be the next bit in the sequence and vice versa. Immediately a bit is incorrect then all subsequent bits will be. In the penultimate stage of the algorithm the selected output is reversed so that a complete recognition generates a logic-1 in the final stage, and vice versa.

In a 2-D neighbour-interconnected version of O-CLIP algorithms such as noise-removal, contour detection, target recognition labelling and maze solution can be implemented. The simplicity of parallel noise removal is seen in figure 20. It requires only two cycles, corresponding to nine computational steps (including the data synchronisation). The computer code is shown in different forms in the figure. Bear in mind that a conventional computer would take at least this many steps to consider just one pixel. O-CLIP processing can be thought of as the recognition stage of symbolic substitution. Instead of pattern substitution, however, the recognition of one pixel's environment determines the next cycle value for just that pixel. For more general processing the O-CLIP circuit must be supplemented by a two further loops, one of which plays the role of an image inverter (complement) and the other of which acts as an image shift register [33].

The active device that is key to the success of O-CLIP as a demonstrator is the *bistable, nonlinear interference filter* [34] in the configuration known as a BEAT (bistable etalon with absorbed transmission). The structure and device response are shown in figure 21. The structure is a multilayer, dielectric narrow-band-pass interference device (a Fabry-Perot cavity). The irradiance level inside the device (I) is determined by the read-beam irradiance (Io) and the cavity optical path-length (nD). With constants A and F depending on the cavity absorption and reflection:

$$I = AI_0\left(1 + F\sin^2[2\pi nD/\lambda]\right)^{-1}$$

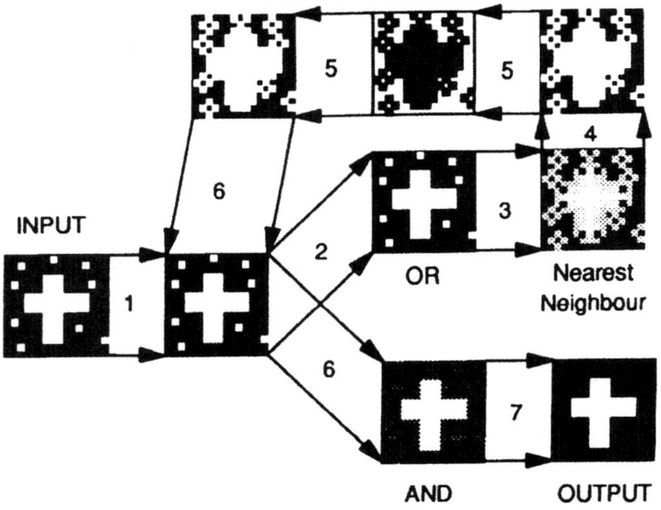

B.I.A. formalism

$\overline{\overline{X}\ U\ \overline{(X \oplus R)}}$ R = 

CL^2 language

```
FETCH(IMAGE)
ENABLE(N,OR)
SET THRESH(1)
STORE(N)
FETCH(IMAGE,N)
ENABLE(D,AND)
STORE(OUT)
```

object code

0	1	2	3
1	0 0	0 00	0
1	1 0	0 00	0
0	1 0	0 00	0
0	1 0	1 00	0
0	0 0	1 00	1
1	0 0	0 00	1
1	1 1	0 00	1
0	1 1	0 00	0

Figure 20. (a) Simulation of noise-removal; using O-CLIP, and (b) corresponding codes.

The cavity material is chosen to be nonlinear; it has an irradiance-dependent refractive index. In the O-CLIP case ZnSe based filters were used and the nonlinearity was due to material heating induced by the radiation.

$$n = n_0 + n_2 I \ .$$

These two equations must be solved self-consistently for the steady-state solutions for I and n, as illustrated in figure 21. For low or high Io there is just one solution; for a range of intermediate values three steady states exist, the central one of which can be shown to be unstable. The dramatic nonlinearity at the switching irradiances is produced by the positive feedback in the cavity. Note that n_0 lies to one side of the cavity resonance in the absence of radiation. Switching is produced when the refractive index change moves the cavity closer to resonance. This causes an increase in the internal irradiances (which can be considerably greater than I_0) and thus a further change in n. Switching is due to an avalanching through feedback.

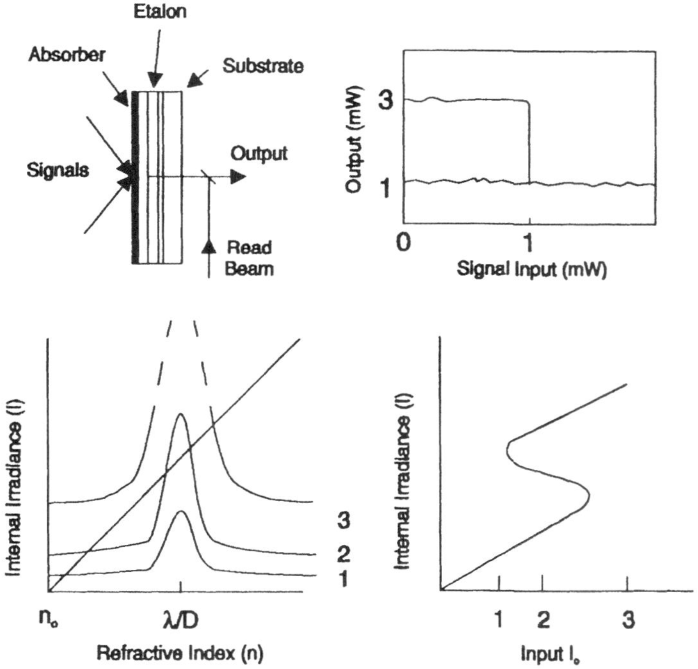

Figure 21. (a) The BEAT structure and its optical response. (b) Optical bistability responses (see text).

The device output in the BEAT system is the reflected bias (read) beam, figure 21. To operate as a logic device, the bias is set within the bistable region. The signal beams are then introduced onto the opposite side of the structure where they are incident on an absorbing coating and, by conduction, produce a fixed temperature rise and corresponding index change inside of the cavity. The bias beam effectively sees a shift in the cavity resonance frequency, a horizontal movement on fig 21b, and may switch from one bistable branch to the other, where it will remain after the signals have been removed. The resulting output response to the signal input is a latching logic response. Programmable logic is achieved by adding a control beam along with the signals. This determines the switching threshold and enables nor/nand programming. By suitable coupling of two such devices, eight functions were achieved in the single channel O-CLIP. Whilst the contrast ratio of the signal is not high (figure 19b) the output results of the Filter-based O-CLIP are unambiguous and the circuit operated for tens of thousands of cycles before failure. Significantly O-CLIP proved that all of the functions of a computer could be demonstrated all-optically.

The thermal nature of the above devices restricts the processing rate of O-CLIP (the millisecond scales could in principle be reduced towards microseconds but no further), it also prevents operation of much beyond the 16x16 level (due to thermal sinking limits) and places high power demands. All-optical switches of electronic-nonlinearity basis should in principle reduce the switching energies from the nanojoule to the picojoule regime. It remains to be seen if these components can be used within optical circuits.

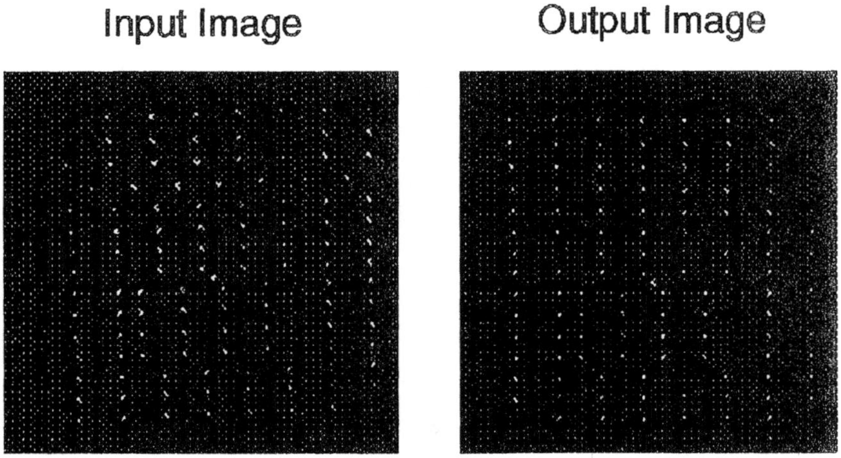

Figure 22. S-SEED 16 × 8-channel input and output- for scratch-removal.

A 16x8 version of O-CLIP was also constructed using s-SEED devices [35]. These can operate at pJ energies and, being differential devices, are more tolerant to fluctuations or spatial variations in the system parameters. The input and output, dual-rail images of a scratch removal algorithm are shown in figure 22. The cycle time potentially achievable was 1μs for an input power of 2μW per pixel; a processing rate of $100Mbs^{-1}$. With smaller devices the switch energies are reduced and Tbs^{-1} processing looks viable.

4.2.7. *EX-CLIP*. The CLIP architecture was based on that of existing electronic machines [36]. Without adaptation, it is unlikely to exceed their computational powers unless considerably lower switch energies are achieved. Two possible adaptations are the use of parallel input, real-time or by parallel data acquisition from a form of optical disc storage, and the use of nonlocal interconnects.

The EXtended-interconnect CLIP is designed to incorporate different optical connection options (via different fixed routes or using reconfigurability), some of which are nonlocal. This borrows from similar architectures proposed for specific tasks [37]. In order to quantify the potential performance of EX-CLIP with respect to a nearest-neighbour CLIP, simulations have been benchmarked for the number of cycles required to carry out certain algorithms. A perfect-shuffle based sort algorithm has been used as one example. For 128×128 bit-slice data, the speed-up of the routing steps achieved by exploiting a single-cycle perfect-shuffle is of order 600. It is very deceptive to use this number though because a considerable fraction of the sorting time is spent on control and processing of the data. If this is accounted for the net speed-up is reduced to a factor of only 4 [33]. The message is that in parallel processing it still may not be sensible to introduce communication speed-up unless the computational tasks are also accelerated. This would be achieved in dedicated modules by introducing task-oriented processing elements . At present the only way to do this is by means of smart pixels.

The *Smart-Pixel* concept is illustrated in figure 23. It combines an optical input and output with a measure of electronics. Given a pixel density of $10^4 cm^{-2}$, as described in section 2, and that the optical pixel size can be less than 10μm×10μm, then there is close to $(100\mu m)^2$ available area for electronics, per processing element. The current front runner in smart-pixel technology is the *FET-SEED* described by Miller herein. At present only limited functionality per pixel is achievable due to the large area of the GaAs FETs. Larger pixels are possible of course, but this raises the question of where the optimum is in the trade-off between pixel-complexity and pixel-parallelism.

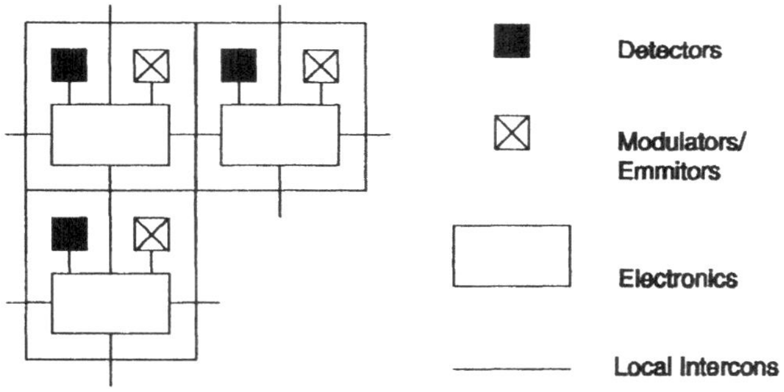

Figure 23. The smart-pixel concept. Detectors receive optical inputs via the third dimension and optical outputs are transferred to non-local sites through 3-D.

4.2.8. *SPE-4K*. The hybrid nature of smart pixel devices brings us back to the concepts of optically connected electronics discussed in section 2. The smart-pixel optoelectronic chip is envisaged above as a relatively fine grain system. In the course-grain limit one could, however, think of machines such as CosineIII as smart-pixel processors. SPE-4K is an intermediate-grain machine that has been constructed in Japan [38]. It has a 64x64 parallelism of processing elements that each contains three, 8-bit registers, a 1-bit arithmetic-logic unit, a 4-bit multiplier and multiplexing/demultiplexing. Each of these sensory processing elements (SPEs) receives input to phototransistors from LEDs ; the interconnects are nearest neighbour as in CLIP.

4.3 VECTOR-MATRIX ARCHITECTURES

In all of the architectures described so far the degree of signal fan-out can be kept to the minimum of two (nearest-neighbour schemes can be reduced to successive fan-out by two). This places low demand on contrast ratios and gain. There have been, however, a number of prototype digital optical systems that employ far greater fan-out (of the order of 100). These are based on the vector-matrix or generalised cross-bar architecture of figure 4b.

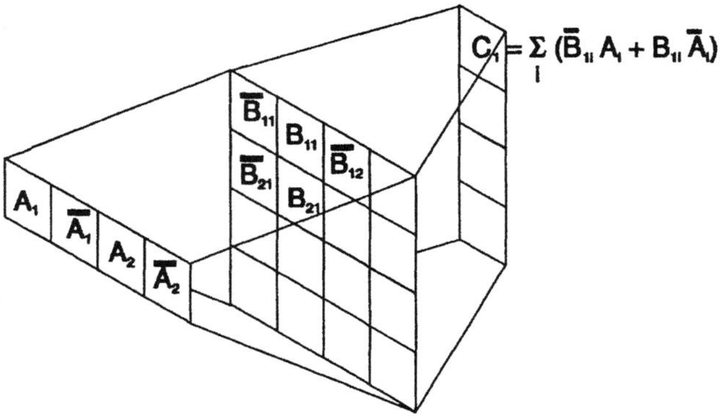

Figure 24. Principle of the Vector-Matrix page scanner.

4.3.1 *ODVM.* Figure 24 shows the Optical Digital Vector Matrix architecture as used to recognise data items stored in a list. The input is the binary, dual-rail version of the word/s to be recognised (input in complement form) and a spatial-light-modulator is set up with a spatial transmission that corresponds to the stored word/s. As shown the first output is :

$$C_1 = \sum \left(\overline{B}_{1i} A_i + B_{1i} \overline{A}_i + \right) \quad .$$

This output is zero if the items match but nonzero otherwise. This explains the use of the complement input. Therefore, provided that the noise level is low and the SLM contrast ratio is high, each null detection registers a recognition. Such a scheme has recently been implemented by using multichannel acousto-optic modulators for the SLM [39] and individually addressed diode laser input sources.

The *acousto-optic Bragg cell modulator* is a 1-D device in which a sequential RF signal is input via a piezoelectric transducer in order to produce propagating vibrations across the cell material (typically quartz) [40]. At any instant the vibrations form, through the photoelastic effect, a refractive index grating. Light incident on the cell will therefore diffract. The strength of the diffraction is determined by the grating strength and therefore by the original RF signal. The diffraction angle is determined by the grating spacing (the rf frequency). In the VM page-scanner the frequency is fixed. The cell length can be such as to accommodate a number of regions of differing grating strength, and so a word can in effect be laid out across the cell. The optics of the scanner is set up to detect only diffracted radiation. The ODVM scheme has 128 resolvable elements in the cell and 128 cells in parallel. Outputs are accepted when the database is in place; the dynamic nature of the Bragg cell mechanism means that the data flows across to be replaced with new data to be scanned. The system can make 10^{10} 16-bit comparisons per second.

4.3.2. *DOC II*. The same group has constructed a number of more general purpose computers based on acousto-optic devices. The Digital Optical Computer II, figure 25, uses point Bragg cells as input modulators, and multiple Bragg cells to produce masks that set up selected input combinations [41]. In this way a programmable logic array is constructed. The detectors receive signals of the form:

$$\sum\left(\xi_{ji}A_i+\overline{\xi}_{ji}\overline{A}_i\right) \quad ,$$

where the ξ-parameters take either zero or unity values. Thus a null signal at the j-th detector corresponds to a multi-input minterm. The system electronic output is set to be logic-1 if any one out of the set of detectors receives a zero input; this results in a sum-of-products relation. DOC II had a predicted 100MHz clock rate and 12.8 Gbs input into a 128 element linear array.

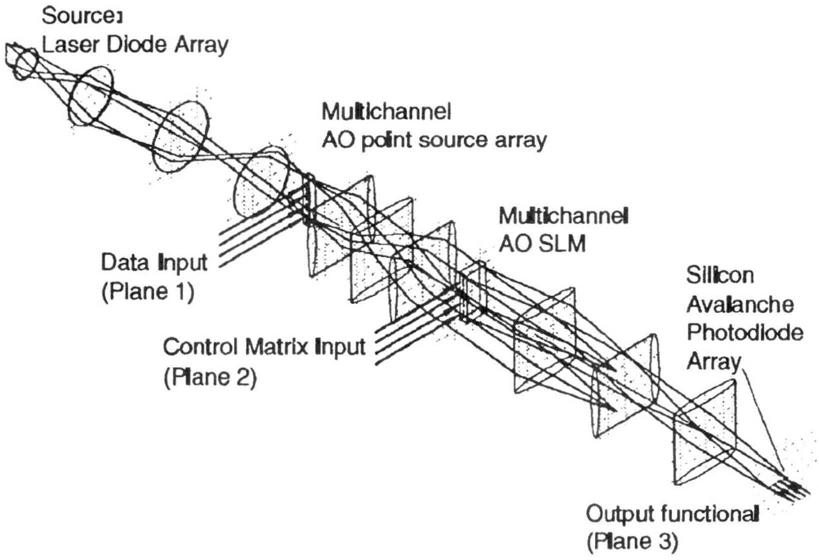

Figure 25. The DOC II machine, employing acousto-optic modulators.

4.3.3. *DOC III*. The next proposed machine in the DOC series takes the VM architecture one step further, figure 26a. Each input in a linear array (this time from individually addressed laser diodes), is fanned out via a dedicated hologram to chosen positions on the matrix plane [42]. The matrix consists of a 2-D array of detectors. Each detector therefore receives a null signal only if all those inputs that are directed to it are themselves zero. That means that the detector can recognise a specific minterm. The detected signal is amplified electronically, thresholded to select the zeros, inverted, and used to drive a surface-emitting laser output. Hence the SLM plane output is an array of minterms. Each column of the matrix is then focused onto a single detector. The linear array of detectors thereby provides not just one sum-of-products result, but a set of different combinations. The SLM plane elements are called *DANES* (detection, amplification, negation, emission).

The potential fan-out for DOC III is NxN/2, where the SLM size is NxN. A similarly high but still discrete fan-out (and fan-in) is proposed for a future DOC machine, figure 26b. The input is now 2-D parallel, and a planar array of sub-holograms replaces the linear array, so that on the first DANE plane minterms of up to NxN/2 variables could be formed. A second hologram array determines which of the minterms are to be ORed and a second detector array, of non-inverting DANES, produces a final optical output. The DANES each can provide gain so that this scheme could be placed in an iterative loop.

4.4 ANALOGUE PROCESSING

This topic does not form part of the present article, but for completeness it is noted that, as the fan-out becomes larger, one is approaching the regime of global fan-out as provided by bulk lenses in a Fourier- based processing system. It is well known that the effective bit rates of such processors are extremely high. Analogue correlators have been built as demonstrators but no commercial exploitation yet exists. Whilst these correlators require that the objects to be recognised are well defined and form a significant fraction (~1%).of the total image scanned, the necessary trade-off of fan-out versus dynamic range should not prevent use in such applications as label inspection or fault recognition on PC boards [43].

5. Summary

There is no doubt that optics will penetrate deeper in to the hierarchy of electronic computers. This will certainly include further development of sequential communication, exploiting the optical time-bandwidth. It also seems clear the parallel data acquisition from disc storage will make better use of the space-bandwidth. From this point one is left with a number of trade-off situations that are yet to be resolved, these include:

1. Parallelism-Smartness
   Cell parallelism (giving high nonlocality) versus cell dimension (giving high pixel smartness)

2. Parallelism - Speed
   Given similar energies per logic process, does one chose high parallelism with the interconnect bonus that implies , or small numbers of fast devices?

(a)

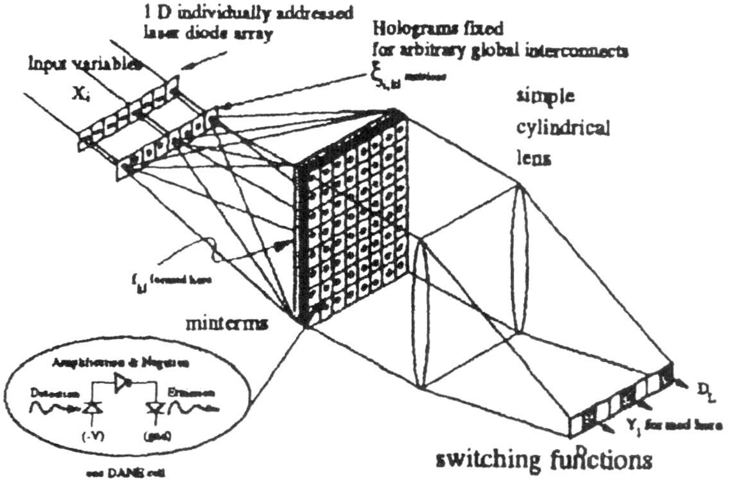

(b)

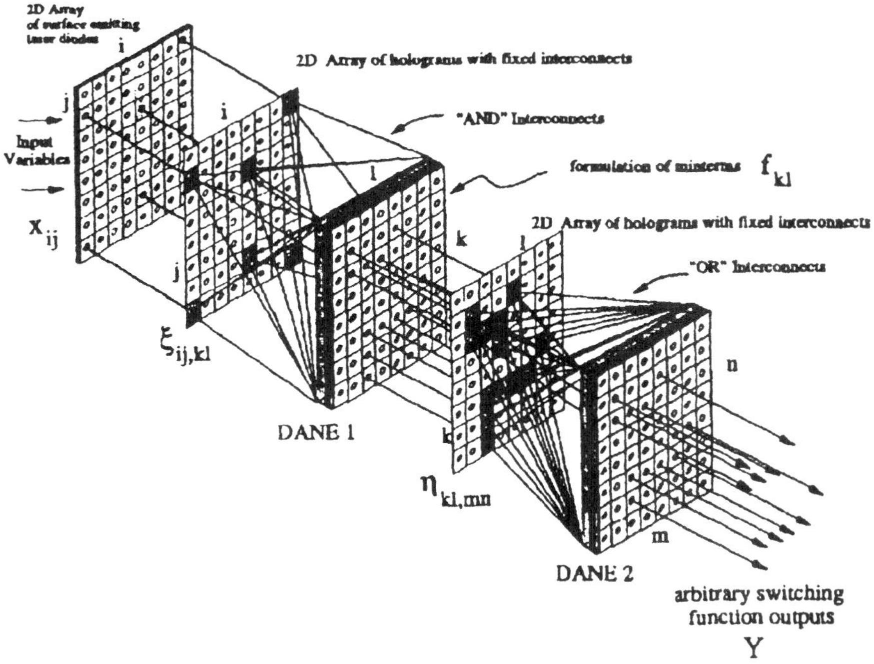

Figure 26. (a) The DANE-based DOCIII optical computer architecture and (b) a proposed extension of the DANE principle.

3. Fan-out - Accuracy.
   Is there a case for abandoning the high accuracy offered by digital optics and accepting the Fuzzy answers that perhaps machines such as neural network processors may be able to accommodate or may actually demand?

None of the architectures described here rely on the availability of all-optical switching or logic. The parallel schemes do, however, place a high demand on the optics and the optics-electronics interfacing. In the near future interfacing will be used in prototypes. It remains to be seen whether breakthroughs can be achieved for all-optical active components. These would certainly produce the more elegant systems but must operate with a performance (power-time product) comparable with electronics and with comparable tolerances and device yields.

This article has concentrated on the binary, digital processors. Most of those implementations described are prototypes that will be superseded rapidly by advances in optoelectronic technology. The active devices used in the demonstrators have been discussed briefly. The systems themselves demonstrate architectural principles and have been essential steps in increasing our understanding of the advantages, and limitations of optics. Finally the recognition that these architectures might become available opens up possibilities in computer science, for example in the design of new algorithms that exploit non local 2-D interconnects.

6. Bibliography

For general reference purposes a chronological listing of Optical Computing Conference Proceedings, Journal Special Issues, Edited and Authored texts is given.

1964 Proc. Symp. Optical & Electro-Optical Information Technology, Ed. Tippett, J., et al., MIT Press (1968), pp.777.

1975 Special Issue on Optical Computing; IEEE Trans. Comp., volume C-24, April, pp. 337-456.

1977 Special Issue on Optical Computing; Proc., IEEE, volume 65, January, pp. 1-157.

1980 Proc. Int. Conf. on Optical Computing, Washington, DC; SPIE, volume 231, pp. 316 and volume 232, pp. 231.

1983 Proc. 10th Int. Conf. on Optical Computing, Cambridge, Mass.; SPIE, volume 422, pp. 221.

1984 Special issue on Optical Computing; Optical Engineering, January, pp. 1-33.
Optical Computing, 5th SPIE Critical Reviews of Technology Series; SPIE, volume 456, pp. 159.
Special Issue on Optical Computing; Proc. IEEE, volume 72, July, pp. 758-975.
"Optoelectronic Switching Systems in Telecoms and Computers", Elion, H.A. & Morozov, V.N., Marcel Dekker Electro-Optic Series, volume 4, pp. 257

1985 Special Issue on Optical Computing; Opt. Eng., volume 24, January, pp. 1-90.
Technical Digest, Topical Meeting on Optical Computing, Lake Tahoe, Proc. OSA, March.

1986 Special Issue on Optical Computing; Applied Optics, volume 25, May, pp. 1541-1621.
Proc. Optical Computing Conference, Los Angeles; SPIE, volume 625, pp. 246.

Optical and Hybrid Computing; SPIE, volume 634, pp. 485.
Special Issue on Photonic Computing; Appl. Opt., volume 25, September, pp. 3014-3136.

1987 Special Issue on Optical Computing & Nonlinear Optical Signal Processing; Opt. Eng., volume 26, No. 1, January, pp. 1-59.
Special Issue on Optical Information Processing; Opt. Eng., volume 26, No. 5, May
Proc. OSA, volume 28, March, pp. 469-543.

1988 Proc. Int. Conf. on Optical Computing, Toulon; SPIE, volume 963, pp. 734.
"Optical Computing", Feitelson, D.G., MIT Press, Cambridge, Mass., pp. 393.

1989 Special Issue on Optical Computing; Opt. Eng., volume 28, No. 4, April, pp. 297-455.
Technical Digest, Topical Meeting on Optical Computing, OSA, Washington D.C., volume 9, pp. 420.
"Optical Computing", Eds. Wherrett, B.S. and Tooley, F.A.P., Edinburgh Univ. Press, pp. 317.
"Optical Computing: Digital and Symbolic", Ed. Arrathoon, R., Marcel Dekker Inc., New York, pp. 410.
"Optical Processing and Computing", Eds. Arsenault, H.H., Szoplik, T. and Macukow, B., Academic Press Inc., pp. 493.

1990 "Optical Computing in Japan", Ed. Ishihara, S., Nova Science Pub., New York, pp. 525.
"A Digital Design Methodology for Optical Computing", Murdocca, M., MIT Press, Cambridge, Mass., pp. 162.
"A Digital Optical Cellular Image Processor", Huang, K.-S., World Scientific, Singapore, pp. 258.

1991 Instigation of three journals dedicated to optical information processing:
Applied Optics: Information Processing
Int. J. Optical Computing, J. Wiley
Optical Computing & Processing, Taylor & Francis

An Introduction to Optics in Computers, Arsenault, H.H. and Sheng, Y., SPIE Tutorial Texts Series, TT8, pp. 126
Optical Computing Technical Digest, OSA, Washington D.C.; volume 6, pp. 317.
"Optical Computer Architectures", McAulay, A.D., J.Wiley & Sons Inc., New York, pp. 531.

1992 Proc. ICO Topical Meeting on Optical Computing, Minsk; SPIE, volume 1806, pp.
"Optical Computing: An Introduction", Karim, M.A. and Awwal, A.A.S., J. Wiley & Sons Inc., New York, pp. 360.
"Optical Signal Processing, Computing and Neural Networks", Yu,. F.T.S. and Jutamulia, S., J. Wiley & Sons Inc., New York, pp. 419.

1993 Optical Computing Technical Digest, OSA, Washington DC; volume 7, pp. 334.

7. References

1. Abbe, E. (1893) 'Beitrage zur theory des mikroskopes und der mikroskopischen wahrnehmung', Archiv. Mikroskopische Anat., 9, 413-468.
2. Cutrona, L.J., Leith, E.N., Palermo, C.J. and Porcello, L.J. (1960) 'Optical data processing and filtering systems', Trans. IRE, IT-6, 386-400.

3. Vander Lugt, A.B. (1964) 'Signal detection by complex spatial filtering', Trans. IEEE, IT-10, 139-145.
4. Rhodes, W.T. (1981) 'Acousto-optic signal processing: convolution and correlation', Proc. IEEE, 69, 65-79.
5. Abu-Mostafa, Y.S. and Psaltis, D. (1987) 'Optical neural computers', Scientific American, 256, 88-95.
6. Ed. Tippett, J. et al., (1964) Proc. Symp. Optical & ElectroOptical Information Processing, MIT Press, Boston, pp. 777.
7. Athale, R.A. and Collins, W.C. (1982) 'Optical matrix-matrix multiplier based on outer product decomposition', Appl. Opt., 21, 2089-2090.
8. Miller, D.A.B., Smith, S.D. and Johnston, A.M. (1979) Appl. Phys. Lett., 47, 197-199.
9. Lentine, A.L., Hinton, H.S., Miller, D.A.B., Henry, J.E., Cunningham, J.E. and Chirovsky, L.M.F., (1988) 'Symmetric self-electro-optic-effect-device - optical set-reset latch', Appl. Phys. Lett., 2, 1419-1421.
10. Okada, Y., Tajima, H., Hamazaki, Y. and Suzuki, M. (1990) 'A multiprocessor system with optical buses: Dialog. H' in 'Optical computing in Japan' Ed. Ishihara, S. Nova Sc. Pub. New York, 283-298.
11. Sakano, T., Noguchi, K. and Matsumoto, T. (1993) 'A multiprocessor system using interboard free-space optical interconnects: Cosine III' in Optical Computing Technical Digest, Optical Society of America, Washington, DC, 7, 276-279.
12. Parker, J.W. (1993) 'Progress in optical interconnection technologies and demonstrators under the ESPRIT II OLIVES programme' in OSA Proc. on Photonic Switching, Ed. Hinton, H.S. and Goodman, J.W. (OSA, Washington) 8, 186-189.
13. Charrier, M., Houssay, B., Lemoine, T. and Paineau, S. (1993) 'Massively parallel computers using holographic matrix as a backplane', in OSA Proc. on Photonic Switching, Ed. Hinton, H.S. and Goodman, J.W. (OSA, Washington) 8, 268-271.
14. (1989) 'IBM photonic chip link computers', Photonics Spectra, Oct., 36-38.
15. Bursky, D. (1992), 'Wafer-scale integration moves to three dimensions', Electronics Design, Dec. 17, 30.
16. M.A. Fischetti (1992) 'Integrated photonics comes out of the laboratory', Photonics Spectra, May, 98-108.
17. Monro, D., Dallas, J.A., Nicholls, J.A., Cripps, M.D. and Crossland, W.A. (1993) 'An optically connected parallel machine', in Optical Computing Technical Digest, Optical Society of America, Washington, DC, 7, 280-283.
18. Huang, K.-S. (1990) 'A digital optical cellular image processor', World Scientific, Singapore, pp. 258.
19. Wherrett, B.S. (1985) 'All-optical computing - a design for tackling a specific physical problem', Opt. Commun., 56, 87-92.
20. Hillis, W.D. (1985) 'The connection machine', MIT Press, Mass., pp. 190.
21. Main, P.T., Feuerstein, R.J., Heinz, V.P., Jordan, H.F., Fechrer, J.R. and Love, C.E. (1993) 'Implementation of a general purpose stored programme digital optical computer', in Optical Computing Technical Digest, Optical Society of America, Washington, DC, 7, 322-325.
22. Huang, K.-S., Sawchuk, A.A., Jenkins, B.K., Chavel, P., Wang, J.-M., Weber, A.G., Wang, C.-H. and Glaser, I. (1993), 'Digital optical cellular image processor (DOCIP): experimental implementation', Appl. Opt., 32, 166-172.

23. Fisher, A.D. (1990), 'Spatial light modulators: functional capabilities, applications and devices', Int. J. of Optoelectronics, 5, 125-167.
24. Murdocca, M.J. (1990) 'A digital design methodology for optical computing', MIT Press, Mass., pp. 161.
25. Prise, M.E., Craft, N.C., Downs, M.M., LaMarche, R.E., D'Asaro, L.A., Chirovsky, L.M.F. and Murdocca, M.J. (1991) 'Optical digital processor using arrays of symmetric self-electrooptic-effect devices', Appl. Opt., 30, 2287-2296.
26. Brenner, K.-H. (1988) 'Digital optical computing', Appl. Phys., B46, 111-120.
27. Murdocca, M.J. (1987) 'Digital optical computing with one-rule cellular automata', Appl. Opt., 26, 682-688.
28. Ichioka, Y. and Tanida, J. (1984) 'Optical parallel logic gates using a shadow-casting system for optical digital computing', Proc. IEEE, 72, 787-801.
29. Tanida, J. and Ichioka, Y. (1986) 'OPALS: optical parallel array logic system', Appl. Opt., 25, 1565-1570.
30. Yatagai, T. (1986) 'Optical cellular logic computers and space-variant logic gate arrays', SPIE, 634, 157-164.
31. Wherrett, B.S. (1986) 'Architectural aspects of optical computing', SPIE, 769, 7-26.
32. Craig, R.G.A., Wherrett, B.S., Walker, A.C., Tooley, F.A.P. and Smith, S.D. (1991) 'The optical cellular logic image processor: implementation and programming of a single-channel digital optical circuit', Appl. Opt., 30, 2297-2308.
33. Wherrett, B.S., Snowdon, J.F., Bowman, S. and Kashko, A. (1993) 'Digital optical circuits for 2-D data processing', Proc. Opt. Comp. '92, SPIE, 1806, 333-346.
34. Wherrett, B.S. and Hutchings, D.C. (1990) 'Optical bistability', Springer Series on Wave Phenomena, 9, 269-297.
35. Wakelin, S., Smith, G.R. and Tooley, F.A.P. (1992) 'Design and construction of a cellular logic image processor with symmetric-SEED arrays', SPIE, 1806, 575-583.
36. Duff, M. and Fountain, T. (1986) 'Cellular logic image processing', Academic, New York,
37. Murdocca, M.J., Huang, A., Johns, J. and Streible, N. (1988) 'Optical design of programmable logic arrays', Appl. Opt., 27, 1651-1660.
38. Ishikawa, M., Morita, A. and Takayanagai, N. (1993) 'Massively parallel processing system with an architecture for optoelectronic computing', in Optical Computing Technical Digest, Optical Society of America, Washington, DC, 7, 272-275.
39. Stone, R.V. and Hessenbruch, J. (1993) 'Data/knowledge base processing on an ODVM platform', in Optical Computing Technical Digest, Optical Society of America, Washington, DC, 7, 82-85.
40. Xu, J. and Stroud, R. (1992), 'Acousto-Optic Devices', J. Wiley, New York, pp. 652.
41. Guilfoyle, P.S., Rudokas, R.S., Stone, R.V. and Roos, E.V. (1991) 'Digital optical computer II (DOC-II): performance specifications', Optical Computing 1991, Tech. Digest Series, 6, OSA, Washington, DC, 203-206.
42. Guilfoyle, P.S., Zeise, F.F. and Hessenbruch, J. (1993) 'Smart optical interconnects for high speed photonic computing', in Optical Computing Technical Digest, Optical Society of America, Washington, DC, 7, 78-81.
43. Bagley, H. and Sloan, J. (1993) 'Optical processing: ready for machine vision?', Photonics Spectra, August, 101-106.

Subject Index